SUSTAINABLE AQUACULTURE TECHNIQUES

Edited by **Martha Patricia Hernández-Vergara and Carlos Ivan Pérez-Rostro**

Sustainable Aquaculture Techniques

http://dx.doi.org/10.5772/57089
Edited by Martha Patricia Hernández-Vergara and Carlos Ivan Pérez-Rostro

Contributors

Juan Fernando García-Trejo, Silvia Hurtado-Gonzalez, Genaro Martín Soto-Zarazúa, Gulnihal Ozbay, Grant Blank, Taworn Thunjai, Carlos IvÃ¡n Perez-Rostro, Jorge Alberto Perez-Fuentes, Martha Patricia Hernandez-Vergara, Yrielle Roets, Mulalo Edna Nemutanzhela, Rajesh Lalloo, Paola Navarrete, Dariel Tovar-Ramirez, Ghaneshree Moodley, Lethabo Mashigo, Kassim Zaleha, Héctor Cordero, Ăngeles Esteban, Alberto Cuesta

CBS, Edition **2017**

Published by InTech

Janeza Trdine 9, 51000 Rijeka, Croatia

Publishing Process Manager Iva Lipovic

Technical Editor InTech DTP team

Cover Designer InTech Design Team

Additional hard copies can be obtained from orders@intechopen.com

Sustainable Aquaculture Techniques, Edited by Martha Patricia Hernández-Vergara and Carlos Ivan Pérez-Rostro
p. cm.
ISBN 978-953-51-1224-2

INTECH

open science | open minds

Contents

Contents

Preface

As population continues to grow and consumers become increasingly convinced of the health benefits of eating seafood, market demands for seafood will expand in the years to come. Fisheries are stagnating and might even level off, so aquaculture will have to grow even faster. Fresh water will become increasingly scarce in the decades ahead and, considering available global resources for food extraction or production, it is clear that land for crops and pasture will come under serious pressure.

Because water is an ideal environment for microbial development, the role of bacteria—beneficial and harmful—in aquaculture systems requires much more research attention.

Beneficial bacteria, as probiotics and as part of biofloc systems in water quality control, will convert organic waste matter into nutritious biomass. This could greatly contribute to increased sustainability and more secure production systems, through reducing needs for water exchange and more efficient feed conversion.

That is why a better understanding of aquaculture ecosystem, better cultivation practices, and greater effort in the creation of prophylaxis and therapeutic measures should be priorities in aquaculture. The ultimate aim should be to develop new hatchery practices and apply innovative techniques for sustainable aquaculture.

Martha Patricia Hernández-Vergara [(1)] **and Carlos Ivan Pérez-Rostro** [(2)]

(1) Laboratory of Native Crustacean Aquaculture,
Technological Institute of Boca del Río, Boca del Río, Veracruz, Mexico

(2) Laboratory of Genetic Improvement and Aquaculture Production,
Technological Institute of Boca del Río, Veracruz, Mexico

Chapter 1

Impacts of Aquaculture on Habitats and Best Management Practices (BMPs)

Gulnihal Ozbay, Grant Blank and Taworn Thunjai

Additional information is available at the end of the chapter

http://dx.doi.org/10.5772/57471

1. Introduction

The demand for food must be met as the human population reaches an estimated nine billion people by the year 2050. This means we must increase overall food production by 70% and this increase must be sustainable and food price affordable (United Nations FAO 2009). Most of the population growth is expected to continue in underdeveloped countries with limited technologies and venues (United Nations FAO 2009). As a popular high protein food source, seafood contains omega-3 fatty acids that are required for healthy human development (UMD Medical Center 2013). Seafood is low in calories, total fat, and saturated fat, while high in vitamins and minerals including vitamins A and D, phosphorus, magnesium, selenium, and iodine (FAO FOCUS 2013). Fish have been shown to have numerous health benefits (Table 1). Seafood is a healthy, low-fat alternative to beef, poultry, and pork and significant omega-3 fatty acids much higher than vegetable-based diets (FAO FOCUS 2013). Specifically, omega-3 fatty acids contained within fish oil are critically important for infants and babies to develop a normal brain (FAO FOCUS 2013).

Population growth and economic development trends are the most important drivers for the demand for high quality and nutritional seafood products (Ewart 2013). With wild capture fisheries exceeding the maximum sustainable harvest capacity, aquaculture has become a bridge in closing the gap between rising demand and traditional seafood sources (Figure 1). Today, farmed seafood accounts for about 50% of overall production in the global marketplace (Bush et al. 2013). The United States aquaculture industry, valued at over $1.1 billion, produces a variety of fish and shellfish species for food, recreation, and industrial needs (Ewart 2013). However the United States is in a seafood deficit, importing more seafood to meet the demands for seafood consumption than it can produce (NOAA Office of Aquaculture 2013).

Seafood Nutrition Facts

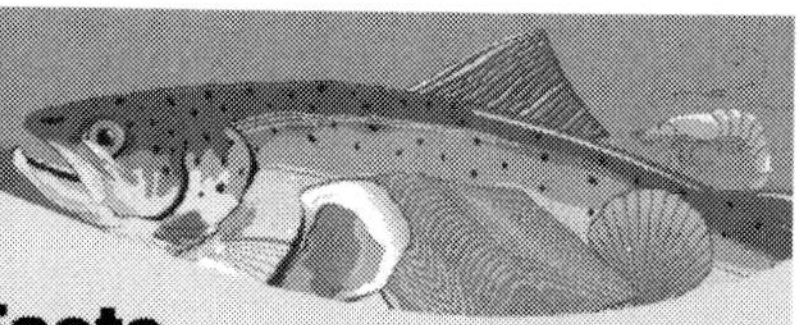

Cooked (by moist or dry heat with no added ingredients), edible weight portion. Percent Daily Values (%DV) are based on a 2,000 calorie diet.

Seafood Serving Size (84 g/3 oz)	Calories	Calories from Fat	Total Fat g	Total Fat %DV	Saturated Fat g	Saturated Fat %DV	Cholesterol mg	Cholesterol %DV	Sodium mg	Sodium %DV	Potassium mg	Potassium %DV	Total Carbohydrate g	Total Carbohydrate %DV	Protein g	Vitamin A %DV	Vitamin C %DV	Calcium %DV	Iron %DV
Blue Crab	100	10	1	2	0	0	95	32	330	14	300	9	0	0	20g	0%	4%	10%	4%
Catfish	130	60	6	9	2	10	50	17	40	2	230	7	0	0	17g	0%	0%	0%	0%
Clams, about 12 small	110	15	1.5	2	0	0	80	27	95	4	470	13	6	2	17g	10%	0%	8%	30%
Cod	90	5	1	2	0	0	50	17	65	3	460	13	0	0	20g	0%	2%	2%	2%
Flounder/Sole	100	15	1.5	2	0	0	55	18	100	4	390	11	0	0	19g	0%	0%	2%	0%
Haddock	100	10	1	2	0	0	70	23	85	4	340	10	0	0	21g	2%	0%	2%	6%
Halibut	120	15	2	3	0	0	40	13	60	3	500	14	0	0	23g	4%	0%	2%	6%
Lobster	80	0	0.5	1	0	0	60	20	320	13	300	9	1	0	17g	2%	0%	6%	2%
Ocean Perch	110	20	2	3	0.5	3	45	15	95	4	290	8	0	0	21g	0%	2%	10%	4%
Orange Roughy	80	5	1	2	0	0	20	7	70	3	340	10	0	0	16g	2%	0%	4%	2%
Oysters, about 12 medium	100	35	4	6	1	5	80	27	300	13	220	6	6	2	10g	0%	6%	6%	45%
Pollock	90	10	1	2	0	0	80	27	110	5	370	11	0	0	20g	2%	0%	0%	2%
Rainbow Trout	140	50	6	9	2	10	55	18	35	1	370	11	0	0	20g	4%	4%	8%	2%
Rockfish	110	15	2	3	0	0	40	13	70	3	440	13	0	0	21g	4%	0%	2%	2%
Salmon, Atlantic/Coho/Sockeye/Chinook	200	90	10	15	2	10	70	23	55	2	430	12	0	0	24g	4%	4%	2%	2%
Salmon, Chum/Pink	130	40	4	6	1	5	70	23	65	3	420	12	0	0	22g	2%	0%	2%	4%
Scallops, about 6 large or 14 small	140	10	1	2	0	0	65	22	310	13	430	12	5	2	27g	2%	0%	4%	14%
Shrimp	100	10	1.5	2	0	0	170	57	240	10	220	6	0	0	21g	4%	4%	6%	10%
Swordfish	120	50	6	9	1.5	8	40	13	100	4	310	9	0	0	16g	2%	2%	0%	6%
Tilapia	110	20	2.5	4	1	5	75	25	30	1	360	10	0	0	22g	0%	2%	0%	2%
Tuna	130	15	1.5	2	0	0	50	17	40	2	480	14	0	0	26g	2%	2%	2%	4%

Seafood provides negligible amounts of *trans* fat, dietary fiber, and sugars.

U.S. Food and Drug Administration
(January 1, 2008)

Table 1. Nutrition facts on various seafood species (The United States Food and Drug Administration 2008).

Commercial aquaculture is a young and rapidly expanding industry in the United States and the need for information on sustainable growth and development has increased dramatically during the past few decades (Wilson et al. 2002; FAO FOCUS 2013). Aquaculture in the simplest terms is the farming of aquatic plants and animals. Furthermore, the National Oceanic and Atmospheric Administration (NOAA) Office of Aquaculture (2013) describes aquaculture on a broader scale as the breeding, rearing, and harvesting of plants and animals in all types of water environments, including ponds, rivers, lakes, and the ocean. Similar to agriculture, aquaculture can take place in the natural environment or in a manmade environment where controlled cultivation and husbandry of aquatic plants and animals are achieved. Using aquaculture techniques and technologies, researchers, aquaculturists and the aquaculture industry are "growing," "producing," "culturing," and "farming" all types of freshwater and marine species (NOAA Office of Aquaculture 2013). According to Ewart (2013), aquaculture has a long history dating back a few thousands of years in China and Egypt. Aquaculture within the United States dates back to the late 1800s, when hatchery technologies were utilized to cultivate fish for restoration of depleted inland freshwater fishes (Ewart 2013). Ewart (2013) stated with a short commercial history (about 50 years), the United States aquaculture industry has a current annual farm gate value of $1.9 billion. Included in the domestic aquaculture production are variety of fish and shellfish species for food, recreation (stock enhancement, restoration, ornamental fish, aquatic plants, live bait), and industrial applications (food additives).

Aquaculture can benefit more than human economies and diets. Oyster shellfish aquaculture provides many of the same ecosystem services as natural oyster reefs (Dealteris et al. 2004; Erbland and Ozbay 2008). Unlike some finfish farming practices, rearing shellfish in high densities in shallow water with abundant phytoplankton concentrations can have positive effects on the environment and may promote biodiversity (Shumway et al. 2003; Dealteris et al. 2004; O'Beirn et al. 2004; Tallman and Forrester 2007; D'Amours et al. 2008; Erbland and Ozbay 2008; Taylor and Bushek 2008).

As stated by Emerson (1999), the process of aquaculture has been under increasing scrutiny as the world tries to supply food for a population which is currently over seven billion. This criticism is happening regardless of how aquaculture is perceived as an economic windfall for developing countries or potential food industries. Aquaculture is the fastest growing food production sector in the world but its sustainability is not fully satisfied (FAO 2013). This chapter will reassure the ultimate question we ask ourselves: is sustainable aquaculture our solution?

Emerson (1999) discussed how pollution, destruction of sensitive coastal habitats, threats to aquatic biodiversity and significant socio-economic costs must be balanced against the substantial benefits and how aquaculture has great potential for food production and the alleviation of poverty for people living in coastal areas where most of the poorest in the world live. He also stated a delicate balance between food security and the environmental costs of production must be achieved. This leads us to our second question: how do we make the world's fastest growing food sector environmentally and socially responsible?

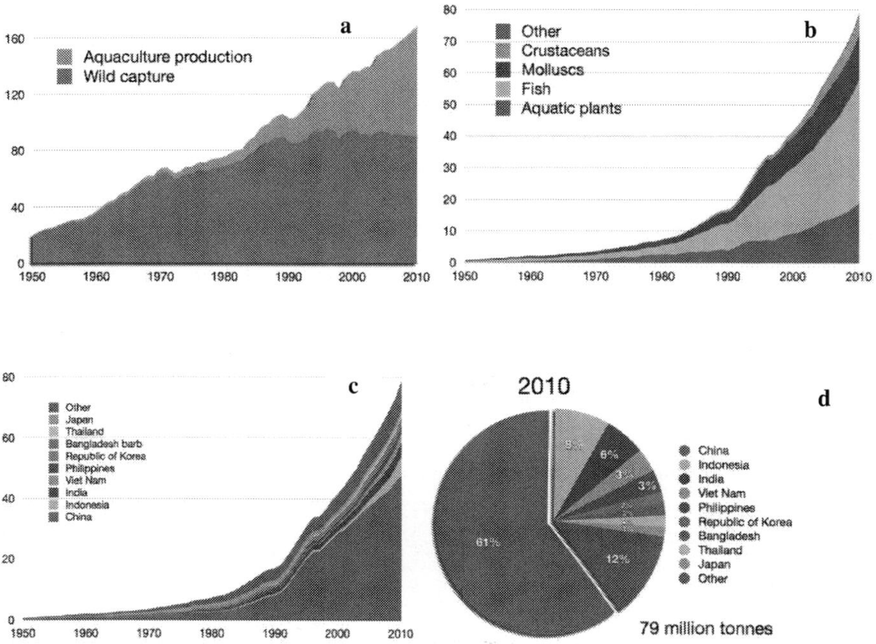

Figure 1. Global harvest of aquatic organisms in millions of tons between 1950-2010 (FAO 2011); b. Global aquaculture production in millions of tons between 1950-2010 (FAO 2011); c. Main aquaculture countries between 1950-2010 (FAO 2011); and d. Main aquaculture countries in 2010 (FAO 2011).

As we search for answers to these questions, the World Wildlife Federation (WWF 2013) gives additional reasons for why aquaculture must become more responsible. According to the WWF (2013), over 53% of the fisheries worldwide are exploited when over 32% are either depleted, overexploited or recovering including are top ten marine fisheries and as much as 30% of all capture fisheries production are either fully exploited or overexploited. Over 90% of large fish were overfished including are several important commercial fishes (i.e. tuna, skipjack, cod, sturgeon) to the point their survival is threated. Whether it is fully exploited or overexploited, by 2048 fish species harvested for food will be collapsed unless urgent management practices are taken to improve the current conditions (WWF 2013). Unwanted fish (by-catch), like many other animals, die due to inefficient, illegal, and destructive fishing practices every year. This destructive fishing practice along with overfishing largely results in poor fisheries management, pirate fishes, subsidies, and unfair fisheries partnerships (WWF 2013).

Over the past 50 years in the Unites States, the demand for seafood has increased as the population reached over 300 million people (NOAA 2011). Seafood import is over 86% of total seafood demand in the United States (NOAA 2010). Unfortunately, many economically and ecologically important fish species are disappearing from our oceans through over-harvest, loss of habitat, and pollution as we stated earlier. As our most important fisheries are collapsing, fishermen and seafood processors are losing their businesses. A solution to this issue lies in aquaculture, particularly marine aquaculture. Although technological advances enable safe, profitable, and environmentally sustainable culturing of aquatic organisms, sustainability seems to be the key to long lasting aquaculture practices that are profitable and environmentally sound.

Considering the majority of fish we consume are farm-raised fish with over 100 species cultured globally, various culturing technique have been used including traditional earthen ponds to high-tech tank systems, each culture method/technique yields its own environmental foot print (Monterey Bay Aquarium Foundation 2013). Although most aquaculture facilities manage with the best intent for stress reduction, beneficial health and fast growth, many larger intensive aquaculture systems are managed where the stock is raised under stressful environmental conditions where there is little ecological balance. In compensating for poor environmental and health condition, managers have often relied heavily upon chemical, antibiotic and water treatments to get their fish to harvest before the system becomes too stressful for optimum growth. Managers risk rising production costs, stock mortalities, and the degradation of habitats that receive liquid waste discharges (Briggs and Funge-Smith 1994). Dalsgaard et al. (1995) explained the concept of ecological sustainability as maximization of internal feedback within a culture system. They refer this maximization as the integrated resources management practice similar to the agro-ecological engineering approach to integrated agriculture-aquaculture farming used in China. Such a system would minimize inputs and wasted outflows of resources and maximize profits.

As the aquaculture industry grows, the use of treatments unapproved by the Food and Drug Administration (FDA) and/or the misuse of chemicals and treatment strategies administered to culture seafood also grows (FDA 2008). To protect consumers, it is important to ensure that both imported and domestic aquaculture seafood products are free from potentially harmful drug, microbial, and heavy metal residues. These residues in food can cause acute, chronic or microbial effects on people. An acute response from hypersensitivity or allergenicity may occur (FDA 2008). Chronic effects can be long term and they are difficult to detect because these events are typically underreported. Cancer is a potential chronic long term effect (Virtanen et al. 2008). Microbial effects caused by drug residues have an effect on human intestinal flora which limits the activity of intestinal bacteria (FDA 2008). Moreover, antibiotic drug residues can affect the development of resistant bacterial populations. FDA (2008) provides one such example "the unapproved use of fluoroquinolones, such as ciprofloxacin, poses the risk of increasing antibiotic resistant bacteria with the potential for serious human health consequences from untreatable infections. In addition, chronic dietary exposure to high concentrations of fluoroquinolone residues, particularly during early growth, may result in a number of toxicities including joint and testicular lesions." The use of unapproved compounds or misuse of FDA approved new animal drugs, will impact the safety of aquaculture products for the consumers in the United States.

In order to identify potential ways to decrease unnecessary outflows from aquaculture systems in the United States, the National Pollutant Discharge Elimination System (NPDES) permitting is utilized by the United States Environmental Protection Agency (EPA) on a case-by-case basis, typically in larger aquaculture operations. It is difficult to make correlations between aquaculture effluents and environmental impacts without accurate records from each facility. It is crucial to examine the aquaculture practices not only in the United States, but also practices world-wide that can minimize our impacts on aquatic ecosystems and while simultaneously increasing food production.

Some other areas of concern regarding aquaculture include, but are not limited to: eutrophication, benthic enrichment, habitat alteration, erosion, disease, water quality, and effective implementation of best management practices (Coastal Habitat Protection Plan 2005). With the addition of nutrients and phytoplankton, bacteria and viruses, become even more important in regards to aquaculture water quality concerns. There is a direct correlation between bacterial diversity and nutrient content. Naturally occurring bacteria from the environment and the guts of cultured fish stocks thrive in nutrient-rich waters and the surface layers of sediments (Garland Science 2011). There are various pathways humans can be infected by the zoonotic pathogens including food and contact with contaminated environments (Friend 2006). Viruses are a special concern in non-native stocks, where introduced species and hybrids may bring new viral strains into an area. Even if the potential for introduced viruses is reduced, periodic outbreaks of viruses are not uncommon (Yanong and Erlacher-Reid 2012)Y. *Vibrio* bacteria are major fish pathogens that are particularly problematic in aquaculture settings (Chatterjee and Haldar 2012). Uncontrolled proliferation within farm operations appears to have made a direct contribution in the dispersion of *Vibrio* pathogens in receiving water bodies (Yanong and Erlacher-Reid 2012).

Integrated aquaculture may provide solutions to many of aquaculture's problems. Since no organism lives naturally in a vacuum, stocking production facilities with complementary species is a logical way to integrate multiple species while simultaneously increasing production for a given area. For instance, to control algae and plant growth, grass carp or other herbivores may be raised along with primary stock. Suspension feeding bivalves are useful organisms in filtering phytoplankton. Mori (1979) found that phytoplankton concentrations decreased by 94% after water was passed through eleven oyster rafts. Not only are the secondary stocks beneficial in controlling water quality, they often are valuable food products as well. Integrated multi-trophic aquaculture systems yield not only greater profit and lower cost but also enhance economic stability and provide more acceptable management practices (Bastin 2013).

Various fish farming techniques have been used depending on the species and their growth stage. Some of these include but are not limited to: ponds, open net pens or cages, hatchery, bag and rack, raceways, recirculating systems, shellfish culture, submersible net pens, suspended culture, tuna ranching, and aquaponics. Although our discussion is limited primarily to inland aquaculture practices with particular emphasis on pond aquaculture in this chapter, recirculating aquaculture and aquaponics systems are also discussed as popular aquaculture practices that are frequently employed to eliminate potential nutrient loads to the surrounding environment. More specifically, our discussion on recirculating aquaculture and aquaponics systems is due to use of recirculating aquaculture systems for commercial

http://dx.doi.org/10.5772/57471

aquaculture species of high market value or application of aquaponics for their roles in minimizing nutrient loads from aquaculture water discharge and increase farm profits by growing alternate crops.

In order to address Best Management Practices (BMPs) in this chapter, we will explore studies from various countries including Thailand, South Africa, United States, Canada, and Australia. These studies address issues regarding how to best manage an aquaculture operation, while minimizing environmental effects and maintain profitable output. BMPs reflect the most technically practical and economically feasible methods which reduce environmental impacts and limit operation costs at aquaculture facilities. One primary goal is to discuss effluent treatment systems that reduce loads of organic matter, suspended solids, and nutrients to prevent polluting receiving waters. The best method to prevent soil and water quality problems includes selecting a site with appropriate soils and an adequate water supply, and maintaining moderate organism densities and feeding rates (Boyd 1989). Secondary management techniques to prevent soil and water quality imbalances include liming, fertilization, and aeration (Boyd 1989). Agricultural irrigation, created wetlands, settling basins, and biological filters also are practical methods for improving the quality of effluents from ponds that will be discussed within this chapter.

In the challenging area of integrating aquaculture Best Management Practices (BMPs), it is imperative that older, proven methods be incorporated with new and innovative ideas. Nearly 40 years ago at Woods Hole, MA and Fort Pierce, FL, Ryther et al. (1975) developed working integrated waste recycling systems utilizing commercially valuable mariculture stocks. Their systems proved so efficient that the final effluent of their system was incapable of supporting further growth or contributing to eutrophication. They suggested that similar systems can be developed for other aquaculture operations to desired needs and purposes. At the Eilat Laboratory in Israel, Neori et al. (1998) established a land-based integrated system that attempted to eliminate external food sources and water exchanges. Avnimelech (2012) provided detailed information on biofloc technology and how this technology can be used to increase farm profit and reduce the nutrient loads of the system. This manual discussed super-intensive biofloc shrimp farming and effects of biofloc technology on the sexual development of shrimp broodstock and other practices. The University of Virgin Islands Aquaculture Program (2013) established a biofloc system that produces tilapia every six months by using biologically active and suspended solids serving as the primary waste treatment in the farm. Additional management practices include good aeration, settleable solid removal, pH adjustments and anaerobic denitrification in this system. Even if this type of system may not be as profitable as growers would like, it is easy to see how the basic principles may be applied to a wide range of aquaculture systems. Unfortunately, there is little impetus to develop such systems unless discharge regulations are increased or the systems are shown to be profitable. Coupled with recirculating systems, aquaponics is described as a synergistic growing technique by the Aquaponics Association (2013) by growing fish and plants together in the same systems. The logic behind this system is that nitrate-nitrogen in fish waste serve as a fertilizer to grow the plants. Once the plants such as lettuce, basil, parsley remove nitrate-nitrogen, this water returns to the fish environment so no water has to be discharged to the environment (The Aquaponics Association 2013).

Integrated aquaculture is nothing new and has been used for thousands of years although their uses commercially are most recent (Bennett et al. 2012). With the demand for high protein food diet, limited resources and environmental concerns, integration provides a solution to maximize profits and reduce potential impacts on the surrounding habitat. By culturing multiple species, farmers can offset the negative impacts in the environment. In China, farmers have been using integrated farming practices for years, although not at the commercial scale at the present time, and have maximized the resource uses to feed their growing population (Bennett et al. 2012).

In this chapter, we will further discuss water quality, eutrophication and disease causing organisms concerns along with effective treatment methods for aquaculture effluents and best farm management strategies in the interest of giving aquaculture professionals, educational professionals, students, and decision-makers a better perspective on how to move forward in a rapidly-changing global market.

Aquaculture will play very important role in feeding about nine billion people by the year 2050 (Nutreco 2011). Meeting this demand can only be possible if seafood is farmed in a sustainable way, both environmentally and economically. As we work together we will find better ways to improve quantity, quality, and sustainability of food supply within the aquaculture sector.

2. Issues of special concern in aquaculture

2.1. Water quality and eutrophication

The highly variable nature of any aquatic environment is often held in a delicate balance by several mechanisms which are common in undisturbed habitats. When anthropogenic stressors (e.g. discharge from aquaculture, farming practices) are introduced into the environment this delicate balance can be disturbed. As a result of an increased aquaculture activity and related farming practices, the effects of seepage and discharge off farms can disturb the healthy conditions of aquatic ecosystems within entire watersheds As described by SAMS (2013), high concentrations of nutrients may lead to deleterious effects, especially in receiving water bodies with the limited water exchange such as lochs. The harmful effects that occur come about as a result of changes in microbial growth and community composition. These changes often result in toxic conditions arising from harmful algae blooms, de-oxygenation of water and sediments from excessive microbial growth, and the transfer/concentration of toxic compounds through the food web. Dissolved and particulate materials in estuaries and coastal environment increase from both natural and anthropogenic sources such as rivers, sewage outfalls, agriculture, and fish farms. These dissolved and particulate materials provide nutrients for phytoplankton and bacteria because they are sources rich in carbon, nitrogen, and phosphorus. Particulate and dissolved materials can also be carriers of heavy metals and drug residues harmful to aquatic life (SAMS 2013). Science Daily (2013) describes eutrophication as the enrichment of an ecosystem with nitrogen or phosphorus, or a mixture of both chemicals. Regarding eutrophication and healthy aquatic system, the water quality variables with the highest concentrations in pond effluents relative to the normal criteria allowed by National Pollution Discharge Elimination System (NPDES) permits are our major concern and discussed in detail throughout the chapter. This includes total dissolved solids, total phosphorus, and biochemical oxygen demand. Eutrophication is the leading problem associated with nutrient runoff of phosphorus (Boyd 2001). Resulting phytoplankton blooms often create an increase in organic matter by two to four times the original amount of metabolic wastes, multiplying the negative effects (Boyd and Queiroz 1997).

As we previously stated, total dissolved solids, total phosphorus, and biochemical oxygen demand are the water quality variables that have the highest concentrations in pond effluents relative to NPDES permits for standard water quality for effluents (Shireman and Cichra 1991; Schwartz and Boyd 1994a). These variables have especially high concentrations in the final 25% of effluent when ponds are completely drained (Boyd 1978; Schwartz and Boyd 1994b; Seok et al. 1995). According to Boyd et al. (2000), total suspended solids and total phosphorus are water quality variables consistently higher in concentration in aquaculture effluents than the typical concentration in effluents of other industries in the southern United States. In comparable studies of the effects of aquaculture effluents on water quality from catfish facilities between Alabama and Mississippi, Hariyadi et al. (1994) found greater concentrations of suspended clay, turbidity, dissolved inorganic phosphorus, total ammonia, and nitrite concentrations. Although effluents from aquaculture facilities with less commonly cultured species have not been studied as thoroughly as channel catfish pond effluents, it is reasonable to assume discharge off aquaculture ponds with other benthic species will have similar nutrient concentrations because of feeding and intensive culture methods. However, the methods of management will vary depending on the species cultured and life stage being cultured. Methods need to be developed for reducing effluent volume and improving the quality of aquaculture effluents in general. Developing specific procedures for removing or reducing suspended solids, total phosphorus and biochemical oxygen demand from pond effluent are especially important. The goal is to develop methods to treat aquaculture discharge so that the materials meet NPDES water quality criteria for effluents (Boyd et al. 1998).

Many techniques have been developed that can be effective in reducing the volume and enhancing the quality of aquaculture pond effluent. These methods include but are not limited to the use of proper site evaluation and design procedures, good construction practices, use of high quality feeds and good feed management, attention to erosion control, moderate stocking densities, reduction in water exchange, seine harvest, and the use of settling basins (Boyd and Tucker 1998). Suitable methods for removing aquaculture waste within effluents include sedimentation, filtration, and mechanical separation using screens, chemical and biological amendments, and using high quality fish-meal (Wheaton 1977, Boyd et al. 1998, Coloso et al. 2001). Boyd and Tucker (1998) summarized methods for using and improving effluents from ponds. These methods have advanced over the years and include hydroponics, irrigation, the development of culture medium for other aquatic organisms, constructed wetlands, settling basins, biological filters, nutrient removal by water hyacinths or other floating macrophytes, and fluidized-bed filters. Queiroz et al. (1998) tested the effectiveness of various bioorganic catalysts[1] on water quality, soil organic carbon, and channel catfish production and recorded higher concentrations of dissolved oxygen and a slight increase in phytoplankton productivity.

According Boyd and Tucker (1998), the most efficient procedures for treating effluents appear to be irrigation, settling basins, and wetlands. Filter-feeding fish, mollusks and certain plants have been successfully cultured in aquaculture pond effluents to reduce nutrient and organic matter loadings. Tucker et al. (1996) reported that harvesting fish without draining ponds between fish crops maintained water storage potential and reduced average annual nutrient and organic matter discharge by over 60% relative to annually drained ponds.

Unlike nitrogen or carbon, phosphorus can only enter the watershed via land-use runoff and coastal areas (Thompson and Polz 2006). Release of phosphorus into the aquatic environment is dependent on soil type, landscape slope, rainfall intensity, and the particle trapping capabilities of the watershed in question because phosphorus is considered a particle bound nutrient. Soluble Reactive Phosphorus (SRP) is a biologically available inorganic form of phosphorus often measured in estuarine systems to better assess the available phosphorus used by the aquatic organisms (Mitsch and Gosselink 2007). Through their bioactivity, oysters transport more phosphorus to sediments than they re-mineralize through metabolism (Dame et al. 1989). Mitsch and Gosselink (2007) stated that phosphorus removal within a system occurs through algal cell absorption and co-precipitation of phosphates in high pH waters. Therefore oysters and algae, both of which have been raised in an aquaculture setting, may provide an economical solution to improving the condition of certain effluents.

One of the most efficient methods for removing excess nutrients in water is seaweed culture. Seaweeds absorb the dissolved nutrients, nitrate and phosphate through their whole plant body. The nutrient absorption is very efficient as seaweeds are immersed and waste no energy for uptake and transport of either water or nutrients (SES 2013).

The most severe consequence of eutrophication on estuarine ecosystems is the depletion of dissolved oxygen (Becker et al. 2008) (Figure 2). Oxygen is consumed during the decomposition of organic matter, resulting in hypoxic and/or anoxic conditions unless dissolved oxygen is replaced. Excess organic matter increases microbial populations which utilize the available dissolved oxygen in order to break down the organic matter. Along with the increase in microbial populations, increases in nutrients from organic matter result in phytoplankton blooms. Phytoplankton cannot produce oxygen at night but instead uses up dissolved oxygen in the system that might otherwise be needed by various other resident organisms. If these conditions are sustained over time this can lead to low levels of dissolved oxygen referred to as a hypoxic condition (NOAA 1998). Low dissolved oxygen levels, specifically less than 5mg/L, can result in large fish kills in estuarine waters with limited tidal or water exchange activities and can have a detrimental impact on various commercially important species (Becker et al.

1 Bioorganic catalysts are catalytic compounds that enhance the biological conversion abilities that would otherwise occur naturally (ICAP Bio-Organic 2013).

2008). Boesch et al. (2001) stated "in addition to the obvious requirements for fish and shellfish growth, lack of oxygen also limits nitrification and subsequent denitrification, compounding the effects of eutrophication. Rivers, lakes, estuaries, and coastal areas receiving the nutrient rich water with low dissolved oxygen become impaired and ecosystem health is compromised more often." More often dissolved oxygen is the limiting condition in waters of intensive pond aquaculture facilities and this condition is mostly as a result of poor management and bad planning (Boyd 1998).

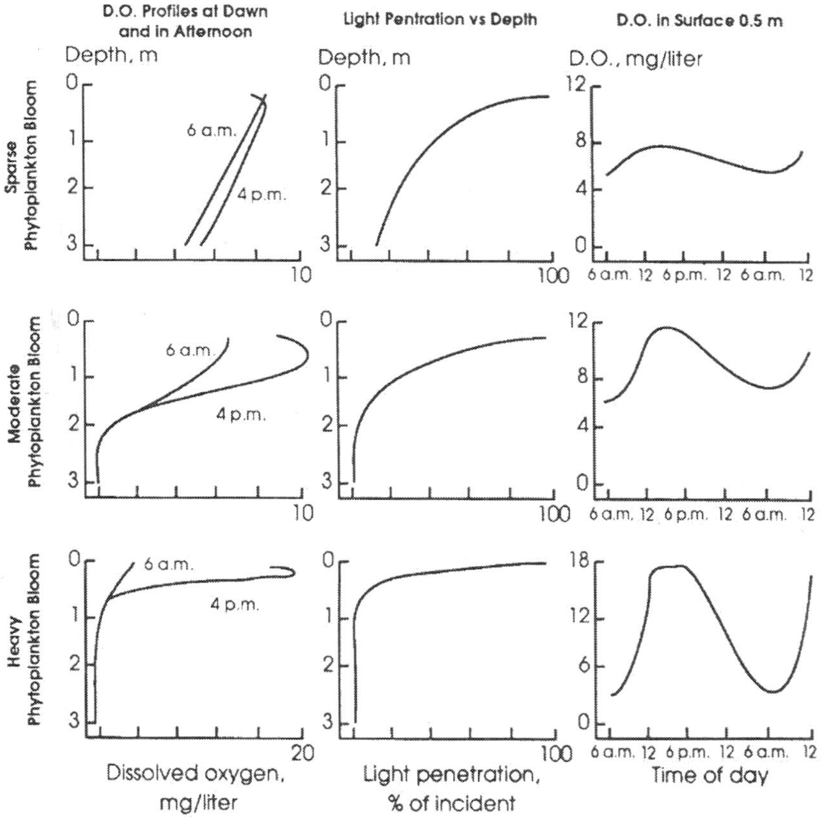

Figure 2. Relationships among phytoplankton density, dissolved oxygen, and light penetration in fish ponds (Boyd 1990).

Boyd and Musig (1992) summarized that the discharge of effluents below the permitted limits are very important. Effluent discharges by one farm may contaminate the water source of another farm downstream. Therefore, if intake water used for filling pond and for water exchanges are highly polluted, water quality problems can occur at even very low feeding rates. Poor water quality in incoming water may increase the risk of disease transfer and intensity of any potential diseases. Pollution load created by aquaculture should not exceed the assimilative capacity of the ponds and water supply of that area. Boyd and Queiroz (1997) stated that receiving stream waters assimilate pollutants through various physical, chemical, and biological processes. As long as the pollution load in the pond effluents does not exceed the assimilative capacity of a water body, adverse environmental changes should not occur.

Boyd (1995a) suggests that the best method to prevent soil and water quality problems is by selecting a site with good soils, an adequate supply of high quality water and to maintain moderate levels of fish densities and feeding rates. Secondary management techniques to prevent soil and water quality imbalances include liming, fertilization and aeration. Sedimentation basins may still needed to be considered to prevent ponds from discharging excess sediments.

Similar to nitrogen, phosphorus and dissolved oxygen, pH, alkalinity, hardness, salinity and ammonia are a few other water quality variables that require constant monitoring in modern aquaculture systems because these variables may become a threat to the habitat in receiving waters (Ozbay 2002). The Best Management Practices section at the end of this chapter describes in detail how the impacts of aquaculture farming are minimized and how striving for sustainability is the key for the long term profitable and environmental friendly farming practices.

2.2. Pathogens and disease risks

As we stated previously, aquaculture refers to culture of organisms (animals or plants) under controlled or semi-controlled conditions. In order to be commercially successful, aquaculture establishments generally have to operate at high density and under conditions which facilitate fast growth. Whatever the species or the type of aquaculture operation (i.e. pond, recirculation, aquaponics, and raceways) in question, maintaining good stock health is the key to successfully operating a profitable aquaculture facility (Bowser 2012). Even when present in low numbers, most disease-causing agents including bacteria, viruses, parasites, and fungi can cause problems and have significant impacts on the fish and associated habitat (Bowser 2012).

The presence of bacteria or viruses in the aquaculture system can be detrimental to the overall operation and surrounding environment. As Pietrak et al. (2010) stated, infection and disease can invade from multiple sources of water, wild fish or shellfish, newly-introduced farmed fish or shellfish, contaminated equipment, predators (i.e. birds, turtles), and human visitors. Newly introduced disease-causing pathogens can lead to production loss from mortality, lost marketability of products, and an inability to transport the product to other locations and farms (Pietrak et al. 2010).

Most diseases and related issues can be prevented by using proper management techniques. It is easier and more cost-effective to prevent disease-causing pathogens from entering the systems than it is treating the pathogens after they have already been introduced into the facility (Bowser 2012). As Bowser (2012) stated, maintaining optimum water quality conditions and keeping the facility clean and well organized are some of the key factors to reduce various stressors which fish are exposed to and will reduce the likelihood of a disease problem. Water quality problems listed as critical are: temperature, dissolved oxygen, pH, alkalinity, hardness, un-ionized ammonia-NH_3, nitrite, and potentially toxic substances including heavy metals, drug residues, pesticides, and CO_2 (Bowser 2012).

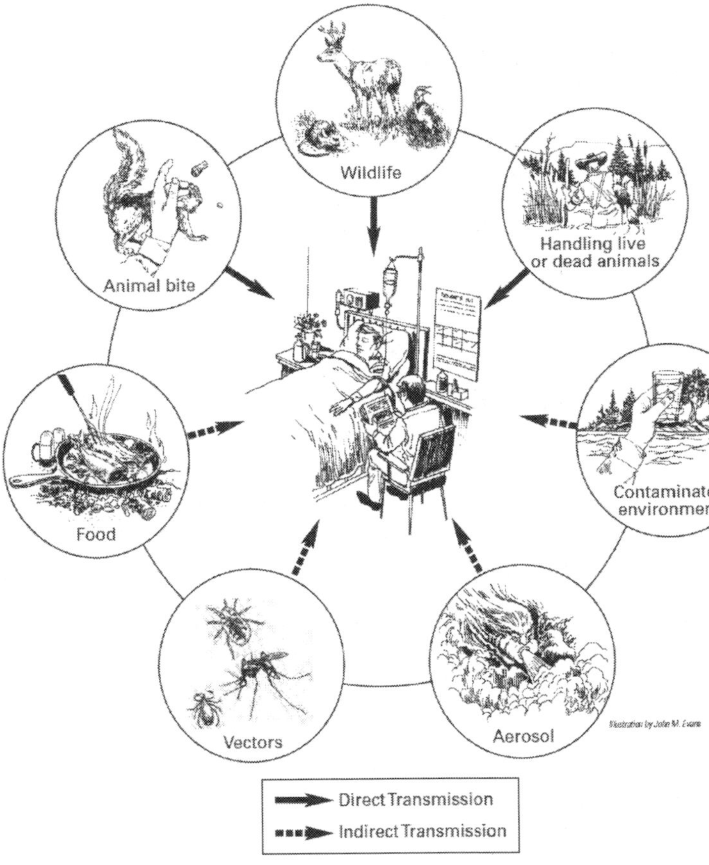

Figure 3. Common routes for potential transmission of infectious diseases and how they are transferred from animals to human and water to human and animal and others (Friend 2006).

Limited to the few intensively studied commercial aquaculture species, there is currently a large gap in our knowledge concerning diseases associated with other species with potential commercial and ecological importance. Included in this group are Enterobacteriaceae and fecal *Streptococci*, which threaten swimming beaches as well as wild fauna and can easily spread and persist in natural environment (Figueras et al. 2000). Viruses are a special concern in non-native stocks, where introduced species and hybrids may bring new viral strains into an area. Pathogenic bacteria such as ones belonging to the genus *Vibrio* have caused devastating disease outbreaks in shellfish larviculture (Thompson et al. 2004). These outbreaks resulted in substantial financial losses for commercial hatcheries and culture facilities (Austin 2010).

Numerous *Vibrio* species present in the aquatic environment are also common human pathogens including *V. cholerae, V.vulnificus,* and *V. parahaemolyticus* and can cause wound infections and gastro-intestinal disease (Austin 2010). Urakawa and Rivera (2006) and Austin (2006) reported other species such as *V.anguillarum, V. logei,* and *V. tapetis* as finfish and bivalve pathogens known to cause vibriosis, disease, and in some cases mortality in aquaculture facilities and hatcheries.

Similar to fish and shellfish pathogens, *V. shilonii* and *V. coralliilyticus* are a few *Vibrio* species linked to coral reef bleaching events, having detrimental impacts on the health and biodiversity of highly productive ecosystems (Thompson et al. 2004). Thompson and Polz (2006) reported that *Vibrio* play an important role in nutrient cycling in the aquatic environment by excreting different chitin-degrading enzymes when attached to zooplankton. This could also be devastating to commercial species such as crabs and lobster which rely on chitin exoskeletons. Thompson and Polz (2006) reported *V. cholerae* to occur as a free living form in the water column and attached to zooplankton. The direct relationship between nutrient enrichment via eutrophication and the occurrence of *V. cholerae* in estuarine and coastal environments calls for further investigations (Grimes 1991). Threats of shellfish-borne disease from *V. parahaemolyticus* (Vp) and *V. vulnificus* (Vv) are of significant public health concern in the United States (Baker-Austin et al. 2010). Increased *Vibrio*-related disease incidence and changes in *Vibrio* populations are a likely consequence of changing environmental conditions (Lipp et al. 2002). As Friend (2006) show in Figure 3, humans can be contaminated by eating seafood grown in infected waters. Escaped *Vibrio* from aquaculture can disrupt natural systems and can be a potential threat to wildlife or livestock. Infected wildlife or livestock entering into systems can be a threat the aquatic health as well (Friend 2006)Although our discussion in this chapter is limited to bacterial pathogens, other pathogens causing diseases in fish includes viral infections, fungal infections, water mold infections, such as *Saprolegnia* sp., metazoan parasites, such as copepods, unicellular parasites, such as *Ichthyophthirius multifiliis* (Moyle and Cech 2004).

Preventive measures are the most cost effective and practical ways to minimize disease outbreaks in these types of establishments. The common problems can be avoided by strictly adhering to the following practices: avoiding the movement of animals and farm traffic, having a good background check of the stocks brought into the farm, or certified stocks "pathogen free", utilizing good quality pasteurized feeds and tools to monitor water quality, and keeping good farm records (Pietrak et al. 2010).

Water quality has a direct and vital impact on the transmission of pathogens. Good water quality reduces the risks of transmission and mortality rates. Regardless of outbreak history at a farm, each farm should develop a biosecurity plan and the plan must be adapted to the specific farm and operation, location and culture method, consider existing threats in the area and avoid environmental contaminant risks (Pietrak et al. 2010).

Yanong and Erlacher-Reid (2012) stated biosecurity in aquaculture as the best practice to minimize the risk of introducing an infectious disease into a facility. Likewise biosecurity minimizes the risks where a diseased fish or infectious agents leaves the facility and is able to spread to other facilities and infect other susceptible species. The biosecurity goals they discussed include: animal management, pathogen management, and people management. — According to Yanong and Erlacher-Reid (2012), the main management practice is to obtain healthy animals (eggs, fry, juveniles, brood stocks) and optimize their health and immunity through good husbandry practices. Pathogen management primarily includes prevention, reduction and elimination of pathogens. While preventative practices can be cost-effective and easy to follow through, pathogen reduction and elimination can be very expensive and may cause further environmental and economic damages if the methods fail.

People management practices include educating everyone involved including visitors and suppliers. Well planned and coordinated facility work schedules and periodic worker trainings are the keys to ensure that people follow tight biosecurity plans and keep it in their minds as they complete daily tasks (Yanong and Erlacher-Reid 2012). There are various factors which play important roles for facilitating pathogen entries into a facility, spreading from unit to another, from one species to another in the facility and finally infect the whole facility. These factors depend on the species of concern, their immune status, life stages and susceptibility to pathogens, husbandry practices, and water quality conditions. In addition, understanding further characteristics of a particular pathogen (i.e. biology and life cycle of pathogen, reservoir potential), its survivorship in the facility, on the tools and equipment, application of the approved treatment options, understanding regulatory status and compliance with biosecurity protocols are additional biosecurity measures to minimize disease outbreak risks in a facility.

One application particularly useful for treating and disinfecting pond bottoms is to dry out ponds for one or two weeks, or longer if necessary (Boyd et al. 2012). As Boyd et al. (2012) states, parasites and disease organisms and their vectors survive in areas where puddles and wet areas remain or when the area has constant rain. When those areas cannot be dried they can be treated with burnt lime, calcium oxide, hydrated lime or calcium hydroxide. The purpose of these various chemical applications is to raise the pH above 10 to kill potentially harmful organisms. Boyd et al. (2012) suggested until natural food organisms have re-established, stocking shrimp or fingerling fish in the ponds should be avoided due to the toxicity risks of lime residues. Coagulation with alum, limestone or polyelectrolytes is effective in reducing virus counts (Boyd and Tucker 1998).

A well thought out biosecurity plan is necessary to minimize the potential for catastrophic losses from infectious disease in the facility. Knowledge is the key to understanding the risks associated with disease outbreaks. Knowing your animal, where your fish comes from, water source of the facility, how pathogens may potentially enter, live and persist in the facility, good husbandry practices, diagnostic tools and legal treatment options (Yanong and Erlacher-Reid 2012). Further practices include: having experts aid in the development of a biosecurity plan (production specialist, animal health professional, engineer, scientists…etc.), planning the facility sanitation, disinfection and system management schemes, good water quality monitoring, separating populations by their life stages, good planning of disposal and facility re-arrangement if necessary. Keeping good records of every operation in the farm (i.e. hazardous waste disposal, chemical use, water quality, fish growth and survivorship, feeding, vaccine application) is also critical in maintaining a productive and healthy facility (Yanong and Erlacher-Reid 2012).

Depending on the type of aquaculture operation (ponds, raceways, cages, or recirculating systems) specific biosecurity measures and management practices should be used. Although fundamental practices are generally the same for many biosecurity plans, practices may vary depending on the species cultured, life stage of the animals, pathogen, type of operation and many others listed earlier. The most important aspect of this plan is to prevent disease outbreaks so that economic and environmental risks are reduced. A biosecurity plan, along with the facility sanitation and disinfection practices, is part of the best management practices used in various successful modern aquaculture settings.

3. Methods of minimizing environmental impacts

3.1. Wetlands

There are various definitions on what wetlands are and what best describes wetlands. Kalff (2002) described wetlands as the transition zones between terrestrial and aquatic systems where the soils are waterlogged for at least part of the year or covered by shallow water, and which are typically occupied by rooted aquatic vegetation (*macrophytes*); not all wetlands are physically connected to lakes or lotic systems. Occupying three times the surface area of lakes, wetlands cover about 8.6 million km^2, or 6.4%, of Earth's land area (Shine and de Klemm 1999). There are tremendous benefits associated with the presence of wetlands (USEPA 2006).Figure 4a shows a healthy wetland and Figure 4b demonstrate the schematic representation of nutrient cycling in the soil-water column of a wetland. Many biogeochemical transformations occur in wetlands and mostly anaerobic conditions exist at the soil water interface. The plants also create an aerobic zone near the roots and different oxidation reduction mechanisms occur in the soil leading to nutrient cycling (USEPA 2008). Within the aerobic zone surrounding plant roots, ammonia is oxidized to nitrate by a process called nitrification; nitrate is then readily diffused into adjacent anaerobic soil. Nitrate is reduced to molecular nitrogen through denitrification, or may be reduced to ammonium under certain conditions through the dissimilatory nitrate reduction process (Ruckauf et al. 2004; Reddy and Delaune 2008). The nitrogen cycle is shown in Figure 4c.

Phosphorous enters wetlands in different forms (PO_4^-, PO_3^-...etc.); both biotic and abiotic mechanisms regulate accumulation and transformation of phosphorous compounds within the water column and soil. Biotic processes include assimilation by vegetation, plankton, periphyton, and microorganisms and abiotic processes include sedimentation, adsorption by soils, precipitation, and exchange processes between soil and the overlying water column (Reddy and Delaune 2008). Transformations of nitrogen, phosphorous, sulfur, iron, manganese, and carbon occur in the anaerobic environment and are mostly microbial mediated.

http://dx.doi.org/10.5772/57471

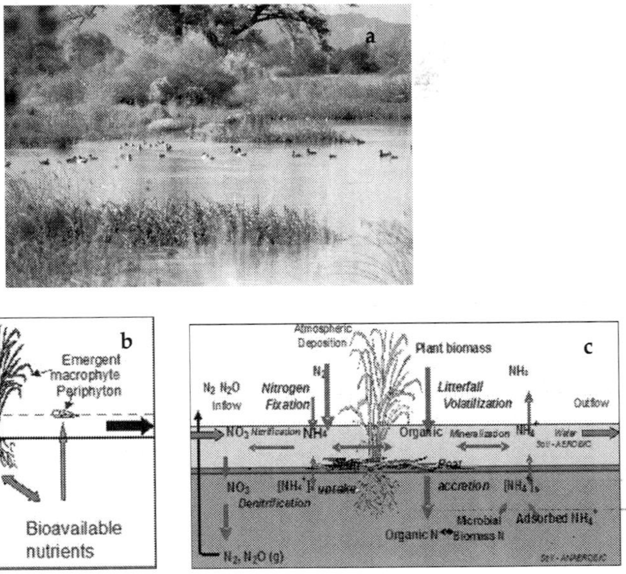

Figure 4. a. Healthy Wetland. Courtesy of http://www.newbedfordguide.com/wp-content/uploads/2011/02/wetlands1.jpg; b. Basics nutrient cycles in soil-water column of a wetland (USEPA 2008); and c. Nitrogen cycle in wetlands (USEPA 2008).

Transportation and translocation of transformed constituents occur in the oxidized layer, providing a barrier to translocate some reduced constituents (USEPA 2008). The value of wetlands for flood control, water storage, and water purification are estimated to be $15,000 ha/year (Kalff 2002). Their value as for fish and wildlife habitat, recreation, or maintaining biodiversity must also be considered (Mitch and Goselink 2007).

Unfortunately, half of these wetlands are disappearing and being converted for agricultural uses such as rice monoculture and aquaculture (Kalff 2002; Figure 5). Many nations restrict development in the wetlands because of the ecological value placed on the wetlands (Boyd and Tucker 1998). Considering their significant roles in removing excess nutrients, breaking down harmful metals and toxic substances via microorganisms living in soil, preventing soil erosion controls, capturing solids in flowing waters, providing habitat for many wildlife species, many countries insist restoration of degraded wetlands or the mitigation the creation of the new ones (Mitsch and Gosselink 1993). Mitsch and Gosselink (1993) stated that natural wetlands and constructed wetlands are very effective in reducing nutrient and organic matter concentrations in wastewater. Wetlands act as biological filters by removing suspended minerals and organic matters from water. Natural and constructed wetlands can be used for treating agricultural, municipal, and industrial wastewaters (Moshiri 1995). Boyd and Tucker (1998) describe the removal processes of suspended minerals and organic matter from water by a wetland as: sedimentation of suspended particles, filtration of suspended particles by plant materials, uptake of nutrients by plants and bacteria, decomposition of organic matter, denitrification, nitrification, and adsorption of ions by the soil. Macrophytes in wetland systems play a key role as substrate for periphyton and actively transport oxygen to the rhizosphere, which serves to facilitate chemical transformations in the sediment (Schwartz and Boyd 1995).

Figure 5. a. Degraded mangroves in Vietnam, courtesy © EJF/Thornton; and b. Shrimp Farm in South America courtesy WWF (http://www.worldwildlife.org/cci/aquaculture_photos.cfm)

Schwartz and Boyd (1995) passed pond effluent through constructed wetlands which drastically reduced concentrations of potential pollutants in channel catfish effluents. Concentrations of total settleable solids, total suspended solids, and total phosphorus were reduced 50% or more by the constructed wetlands except for total ammonia nitrogen due to low hydraulic residence time (HRT) of the wetlands in their study. The greatest removal of total phosphorus (TP), 84% and nitrate-nitrogen (NO_3-N), 75% were obtained in wetlands with a four-day HRT (Schwartz and Boyd 1995). Passing water through wetlands was more effective in removing pollutants than simply holding water in the ponds in their study.

There are two basic types of ponds used to raise channel catfish; levee ponds and watershed ponds (Boyd 1985). Levee ponds discharge little water following rains because of their limited watershed area. However, watershed ponds discharge larger water volume following heavy rains due to their larger watershed areas (Schwartz and Boyd 1994a). Most channel catfish farming is conducted in levee ponds where ponds consist of the inside slopes and tops of levees, resulting in high seepage rates especially from rain during the winter. Watershed ponds are usually located much farther apart than levee ponds, so it is typically not feasible to transfer water between ponds for reuse. Boyd and Tucker (1998) suggested that large wetlands could be used to treat effluent when ponds are drained. A smaller wetland could be used for treating the last 25% of highly concentrated effluent from watershed ponds. In a study by Shpigel et al. (2013), the authors demonstrated that nitrogen, phosphorous, and total suspended solids were efficiently removed using *Salicorna* as a biofilter within a constructed wetland. In another study by Lymbery et al. (2013), wetlands removed 60-90% of total nitrogen loads and at least 85% of total phosphorus, and orthophosphate loads from the aquaculture effluent.

Some advantages of utilizing wetlands in the wastewater treatment process include the elimination of chemical treatments, an inexpensive construction process, and wetlands contribute to wildlife habitat and plant communities and to local hydrologic processes. Therefore, using natural wetlands for aquaculture should be minimized to prevent them from disappearing (Schwartz and Boyd 1995; Kalff 2002). Because of the need for large areas, concern arises over the feasibility of using wetlands for treating aquaculture effluents (Schwartz and Boyd 1995; Boyd and Tucker 1998). Integration with pond effluent management procedures might reduce the area of wetland needed for treating fish farm effluents (Schwartz and Boyd 1995). One of best management options allows for the maximization of fish production while maintaining a good pond environment with minimal impacts on the adjacent coastal system including maintaining good stocking densities to improve food assimilation efficiency in a biogeochemical energy model (Serpa et al. 2013).

3.2. Settling basins and retention ponds

Settling basins can be built to remove turbidity and suspended solids from pond water supplies. Sediment ponds should be fairly deep to minimize land requirements and to provide hydraulic residence time. In general, a hydraulic residence time of at least 6-8 h is necessary but 2-3 days of retention is preferred (Boyd 1995b). Preliminary sedimentation studies on catfish pond effluents suggested that settleable solids and total phosphorus could be removed as effectively in settling basins as in wetlands (Seok et al. 1995; Boyd et al. 1998). Sedimentation can reduce biochemical oxygen demand by 40 to 50% (Boyd et al. 1998). Schwartz and Boyd (1994a) obtained information on the quality of effluent released from channel catfish ponds during pond draining and fish harvest in watershed ponds. The concentrations of total nitrogen, ammonia nitrogen, soluble reactive phosphorus, total phosphorus, and biochemical oxygen demand started increasing as early as the seining phase (Boyd et al. 2000). Schwartz and Boyd (1994a) suggested that the best way to minimize the pollution potential of aquaculture pond effluents is to harvest ponds as quickly as possible and not to discharge water during seining or to discharge fairly contaminated water into a settling basin or retention pond (Figure 6). Cathcart et al. (1999) suggested that harvesting catfish during late summer/ early fall can significantly decrease effluent discharge from the production ponds due to low water level. This may apply to other species cultured in the ponds such as shrimp and tilapia; however this practice may not be the right fit for other culture systems. Boyd and Musig (1992) found that settleable solids were seldom present in measurable quantities in effluents discharged at shrimp harvest, as seines are not used.

The maximum instantaneous settleable solids rate allowed by the EPA (1979) is 1 ml/L for 30-day average and 2ml/L for daily maximum (USEPA NPDES 2010). Boyd and Tucker (1998) found that the effluents from catfish ponds might contain settleable matter higher than the

Figure 6. a. Settling basin. Picture courtesy of Auburn University; b. Constructed wetlands in Mississippi (Tucker 2009) (BMP No. 6 2002).

allowable limit, and have a moderate oxygen demand at harvest. Boyd et al. (1998) demonstrated that a settling time of 8 hours sufficiently reduce total suspended solids and total phosphorus by 75%, and turbidity and 5-day Biochemical Oxygen Demand (BOD_5) by more than 40% in catfish pond effluents. Teichert-Coddington et al. (1999) studied the final effluent from draining shrimp ponds to settling ponds, and obtained near maximum sedimentation of most variables within 6 hours residence, with a removal of 88% of total suspended solids, 100% of settleable solids, 63% of BOD_5 and 55% of total phosphorus.

Using separate settling ponds to treat aquaculture effluent can be a problem on commercial fish farms because of the land requirements and construction costs. Production ponds can be utilized as settling ponds, but this would result in a loss of production capacity. Seok et al. (1995) suggested holding the last fraction of pond water for several days in production ponds before discharging it to the environment as a practical way to allay effluent impact.

The characteristics of solids in pond effluents have been studied to provide information for designing and operating pilot sedimentation basins to test their efficiency for treating pond effluents (Ozbay and Boyd 2004; Ozbay and Boyd 2003a; Ozbay and Boyd 2003b; Boyd 1999). Recommendation is made to lower water 25% of its full volume and settle pond effluents for minimum of 2 to 4 hours and more if necessary to remove over 90% of settleable matters, 75% of total suspended solids and over 50% phosphorus loads in the nearby pond used as a settling basin. Cathcart et al. (1999) studied the reduction of effluent discharge and groundwater use in catfish ponds in Mississippi. Deepening the settling ponds receiving overflows from adjacent production ponds reduced the effluent discharges of ponds by 40 – 90%. Hargreaves et al. (2003) summarized in a SRAC Report that over 50% of total suspended solids, total nitrogen, total phosphorus, and biochemical oxygen demand are related to particles less than 5 micrometers in diameter. Boyd (2000) suggested from the estimate of runoff from watershed studies that settling basins used to treat storm runoff from typical watershed type catfish ponds would need to have volumes of 30 to 40% of pond volume in order to provide a retention time of 8 hours. Thus, because of the large volume required, settling basins do not appear to be a feasible solution for treating storm runoff. Settling basins for treating intentional discharge for partial or complete draining would need to be around 10 to 20% of the volume of the largest pond on the farm (Boyd 2000).

Among the frequently applied practices for treating pond effluents such as coagulant application, water exchanges, and settling basins, many farmers advocate settling ponds for effluents. Even though settling has certain benefits in removing solids stirred into water during catfish harvesting, at other times, nutrients and organic matter in effluents are likely to be phytoplankton or dissolved substances, which do not settle easily (Boyd 2000). Schwartz and Boyd (1996) suggested that after seining, the last 25% of effluent water can be held in the pond for two to three days (depending on the farm operation and timing) to allow solids to settle before they are drained completely. This reduced the discharge of settleable solids by 96%, total nitrogen by 74%, and total phosphorus by 69% and organic matter by 59%. This level reduction is very effective but may not be feasible considering limited space availability in most aquaculture farms. Settling basins are not recommended to treat storm runoff of watershed type ponds because of the large volume of pond water required to reach desirable effluent qualification with a retention time of 8 hours (Hargreaves et al. 2003). Lutz et al. (2011) suggest additional buffer strips to allow plants to pick up excess nutrients and allow water to further slow down before it reaches any downstream creeks. Table 2 provides application discharge data on the wastewater treatment plant and recommended maximum daily loads on water quality parameters as main concerns to EPA (USEPA NPDES 2010). Depending on the type of operation and inflow or existing water conditions, outflow water quality parameters are recommended not to exceed the concentrations provided in the table 2 for NPDES Permit.

Parameter	Units	Discharge Data[(1),(2)]	
		Maximum Daily Discharge	Average Daily Discharge
Flow	MGD	--	0.06
pH	Standard Units	7.7-8.2 (min-max)	
Biochemical Oxygen Demand, 5-day (BOD_5)	mg/L	4.9	0.6
Total Suspended Solids (TSS)	mg/L	25	7
Ammonia (as N)	mg/L	6.76	4.26
Total Residual Chlorine	mg/L	0.03	0.01
Nitrate and Nitrite N	mg/L as N	0.68	0.35
Total Dissolved Solids (TDS)	mg/L	553.00	496.00

Table 2. Application discharge data (EPA NPDES 2010).

According to Tucker and Hargreaves (2008), uneaten feed and fecal wastes are the primary producer of solids that potentially degrade environmental conditions at a farm. Solid accumulation can deteriorate the conditions in a facility and create a threat to the aquatic species cultured. Solids can damage fish gills or block shellfish from filtering and increase dissolved oxygen demand due to increased microbial activity in the accumulated organic materials (Tucker and Hargreaves 2008). Excess phosphorus and nitrogen in the sediment with high solids accumulation has a drastic effect in receiving water bodies causing eutrophic conditions (Tucker and Hargreaves 2009). As the most frequently applied tool for solids removal of pond effluents, settling basins or retention ponds are used to mitigate aquaculture effluent or overflows (depending on size and availability) where the water from ponds or other types of aquaculture facilities are treated by the natural processes to minimize or eliminate pollutants (Setty 2013). Particles settle if given enough time by gravity and microbial community which break down excess nutrients and other pollutants into a less harmful or harmless form (Setty 2013). Coupled with other practices suggested for the best management practices, settling basins and retention ponds remove a significant portion of contaminants and excess nutrients as we discussed in this section and have been recommended for pond aquaculture facilities (Boyd and Tucker 1995).

3.3. Physical amendments

Sedimentation and filtration are two of the most commonly used particle removal techniques in aquaculture. The applications of this technology have become priceless because untreated effluents or discharges may pose a threat to the environment by carrying various materials in excess quantities including soils, nutrients, and minerals (Ozbay 2002). According to Ebeling and Vinci (2013), total suspended solids (TSS), settleable solids, 5-day biochemical oxygen demand (BOD_5), and total phosphorus (TP) are the four major pollutants found in aquaculture effluents/discharge. These pollutants are regulated to ensure that their concentrations can be minimized through the removal of solids containing feces and uneaten feed.

It is important that pollutant concentrations associated with specific particle size ranges in the effluents are considered during the physical removal stage and knowledge on the characteristics of these particles make the removal process more successful (Ozbay and Boyd 2003a). Water quality requirements are frequently discussed but physical characteristics and particle size distribution of the pollutants in the water are not known well. Analytical technology including size fractionation using sieves, laser diffraction, size fractionation using membranes, and characterization using the Coulter registered method have all improved and have been applied in different industries depending on the effluent and discharge characteristics of the particles in question (Cripps 1993). Boyd (2000) reported that about 40% of total suspended solids (TSS), total phosphorus (TP), total nitrogen (TN), and biochemical oxygen demand were associated with particles 51 μm or larger in sizes in catfish ponds. Table 3 shows differences in concentrations of water quality parameters of the pond effluents before and after filtration (Ozbay and Boyd 2003a).Cripps (1995) found that aquaculture wastewaters typically have low TSS concentrations, compared to various industrial and municipal wastewaters, and numerous small particles which clog the 45 μm membranes used to filter the solids out of suspension in aquaculture waters. Suitable treatment techniques should separate particles from the primary effluent flow. Cripps (1995) indicated that by using filters with pore sizes ranging between 200 – 5 μm, increased treatment effects were achieved through sequential decrease in pore size, which removed more nutrients. It is important for a treatment process to remove relatively larger particles, resulting in a reduction in nutrient loading of the effluent.

Cripps (1995) found lower concentration of TN than TP after filtration, and an increase in the filtration effort reduced both nutrients. Most of the plankton/ solids of eutrophic pond waters, with over 50% of the TSS, are found in particles smaller than 10 μm. Ozbay and Boyd (2003a) found that removing particles down to a very small size provided the required benefits targeted for the pond effluents. They recorded percentage removal of total phosphorus (TP), total nitrogen (TN), 5-day biochemical oxygen demand (BOD_5), and particulate organic matters (POM) associated with total suspended particles (TSS) removal using filters of different pore sizes (41, 30, 20, 10, 8 and 5μm). Most water quality parameters except total nitrogen were substantially reduced except nitrogen after the effluent water was passed through the filter with 41μm pore size. Further removal was achieved with the consecutive filtration using filters with smaller pore sizes. They only monitored a noticeable reduction of total nitrogen with successively finer filter sizes.

Time required to remove different size particles was also studied by Ozbay and Boyd (2003a) and they suggested a settling time of 24 hours to remove about 30% of TSS and TP and 35% of POM, 25% BOD_5 and 20% of TN. Considering the length of time, they do not recommend using settling basins for treating storm overflow and pond draining effluent. In their later study, Ozbay and Boyd (2003b) used turbidity and found it was strongly correlated with total suspended solids and inorganic suspended solids. The relationship was stronger between total suspended solids and inorganic suspended solids as compared to total suspended solids and particulate suspended solids due to the fact that fluctuations in phytoplankton concentration over time changes particulate organic matter concentration.

Filter Pore Size	Average Percentage Removal (cumulative)				
(μm)	TSS	POM	TP	TN	BOD_5
41	22.5	28.8	21.5	12.9	22.9
30	28.0	30.7	27.2	14.0	24.3
20	32.0	34.9	28.7	17.8	25.4
10	34.5	35.4	31.0	18.5	28.1
8	38.7	40.1	36.0	22.4	33.8
5	47.9	51.6	37.9	23.5	34.0

Table 3. Percentage of total suspended particles removed by filters of different pore sizes (41, 30, 20, 10, 8 and 5μm). Percentages of total phosphorus, total nitrogen, 5-day biochemical oxygen demand, and particulate organic matters are removed with the removal of total suspended solids (Ozbay and Boyd 2003a).

Ackefors and Enell (1994) found the majority of the phosphorus from fish farms is bound to the particles while nitrogen is not bound to the particles but more in a dissolved form in water. Similar to the phosphorus most of the biodegradable organic matter producing biochemical oxygen demand was in the particles in their study. Cripps (1992) studied the distribution of total phosphorus and total nitrogen in six serially filtered aquaculture effluent samples. Only the fraction containing particles smaller than 5 μm pore size added a disproportionately high nutrient load to the effluent in his study. Reduction rates of 60% for total phosphorus and 34% for total nitrogen were achieved using 5 μm pore size filter. Total nitrogen concentrations were greater in smaller particles than the large particles in his study. However, Cripps (1992) found 69% of the total phosphorus was associated with particles larger than 45 μm diameter. In the effluent the majority of suspended particles produced by the farms were within the size range of 4 – 120 μm diameters. Depending on the particle characteristics and farm effluent, excess nutrient removal can be achieved using the correct diameter in filtration. Cripps (1995) summarized the distribution of the particles for each successive size group and found phosphorus levels depend on the particle size distribution in the pond effluent. Removal of the relatively larger particles separated by the filters had little effect on the size distribution; hence changes in mean diameter were small but the effects were consequential. The phosphorus content in both suspended solids and particles increased significantly with decreased particle sizes (Cripps 1995). Bergheim et al. (1991) used screens with 200 μm or less pore sizes to remove particles. He found further reduction of phosphorus by using the screen with the filter pore size smaller than 5 μm produced negligible results, and in practice he found it difficult to implement. However, the phosphorus content of smaller particles (based on the total phosphorus concentration of water before the effluent was filtered) was significantly greater than larger particles. This difference may appear small but actually represents a large difference in filtration effort (5-200 μm pore size).

Particle size analysis, if used in conjunction with other techniques such as fractionation and nutrient analyses, can be used for the characterization of aquaculture wastes and for monitoring the improvement in wastewater treatment efficiency (Cripps 1994). Membranes can be used for fractionation; however these techniques on their own are limited in practical application. But when combined with other forms of analyses, such as nutrient concentration studies in a known volume of water, the determination of proper treatment techniques is simplified (Cripps 1996).

Although we provided a detailed overview on the feasibility of using filters with various pore sizes, specifically the effectiveness of filters with 5 μm or smaller pore sizes, for the purpose of removing phosphorus and nitrogen, sedimentation is probably the most practical application to remove the large particles in the effluent before further filtration is applied to remove the particles bound to smaller particles which cannot be effectively removed via sedimentation. Sedimentation is discussed in detail in the settling basin section of this chapter. Commonly, screens are placed in front of pond discharge areas to prevent fish, leaves, twigs, or other large debris from escaping in the pond effluent.

http://dx.doi.org/10.5772/57471

3.4. Chemical amendments

The smaller particles of colloidal clay settle slowly and they may impart unwanted turbidity to pond water (Boyd 1998). Some of the chemicals applied to aquaculture ponds to remove this undesired turbidity in pond waters include coagulants like alum (aluminum sulfate), ferric chloride, gypsum, lime and polymers. Although using organic matter to reduce turbidity has advantages from an environmental standpoint, this is difficult to obtain and apply to ponds (Boyd 1990). Coagulants are most often added to alter the physical state of dissolved and suspended solids, thereby facilitating their removal by filtration and sedimentation (Boyd and Tucker 1998; Pepper et al. 1996) (Figure 7). Coagulants destabilize colloids, thereby permitting suspended particles to form aggregates that can settle out of solution. Coagulation with alum, limestone or polyelectrolytes is very effective in removing suspended matter and phosphorus from water (Boyd and Tucker 1998).

Figure 7. Application of the chemical amendment, lime, to an aquaculture pond in Auburn, Alabama (Photo courtesy of Ozbay).

Boyd (1995) demonstrated that phosphorus precipitates from pond water as insoluble iron, aluminum, or calcium phosphates. Alum and ferric chloride are commercially available sources of aluminum and iron. Aluminum, calcium or iron based coagulants added to poultry litter reduced soluble phosphorus concentrations (Moore and Miller 1994). Gypsum (calcium sulfate) is a source of calcium, however, it is only suggested for use in low alkalinity waters (Boyd 1990). Boyd (1990) observed that alum treatment caused almost immediate flocculation of clay particles, and a great reduction in turbidity within 2 hours in all the treated ponds. However, application of alum produces a strong acidic reaction in water and its use should be limited. Boyd (1998 and 1995) suggested alum for pond treatment if alkalinity is 50 mg/L or above in the water. Alum can remove organic particles and clay colloids in association with phosphorus in water through coagulation and entrapment. Generally, aluminum precipitates with inorganic phosphorus as aluminum phosphate compounds (Gensemer and Playle 1999). Welch and Cooke (1999) applied alum to the surface of eutrophic lakes at rates ranging from 5.5 to 10.9 g Al/m^3. They found a 50% decrease in total phosphorus and chlorophyll a concentrations. Cooke et al. (1993) investigated phosphorus removal in order to control algae blooms by using salts of iron, aluminum or calcium (ferric chloride, aluminum sulfate or calcium hydroxide). They reported aluminum salts as being most frequently used in lake restoration. Welch and Cooke (1999) observed decreased in cyanobacteria bio-volume after treatment with alum in unstratified lakes. Jacoby et al. (1994) found the magnitude and blooms of cyanobacteria reduced after 2 consecutive years of alum and sodium aluminate treatments in a polymictic lake. Phosphorus, total phytoplankton, and chlorophyll a concentrations in hypereutrophic lakes were reduced in 3 years following liquid alum, 10 mg Al/L treatment. However, Rowan (2001) found that application of alum at the rate of 50 mg/L immediately after seining resulted in a somewhat greater removal rate of some pollutants during the first hours of settling, but did not result in significantly improved water quality. She suggested using alum after the first hour of settling from seining, and higher application rates of alum would have been necessary to precipitate significant amounts of phytoplankton. Masuda and Boyd (1994) used alum as low as 20 mg/L concentration for catfish pond water, and found significant removal of Soluble Reactive Phosphorus (SRP) in the pond water. They reported no residual effects of alum treatment if used at low concentration (20 mg/L). Boyd (1995a) noted that to increase the amounts of solids removed from water for shrimp ponds utilizing alum would necessitate alum treatment of water in a settling basin. In most situations, settling ponds may be adequate to remove suspended solids (Boyd 1995a).

In ponds if acidity results from increased carbon dioxide and exchangeable aluminum in soil after chemical treatment, total alkalinity and total hardness concentrations are buffered by the applications of agricultural limestone, burnt lime, and hydrated lime (Boyd 1995a). Liming is applied simultaneously to neutralize H^+ ions, and eliminate or reduce the risks associated with alum toxicity. Masuda and Boyd (1994) used agricultural limestone and alum in catfish pond water in order to reduce nutrient concentrations. Twenty mg/L alum treatment reduced soluble reactive phosphorus, 80%; total phosphorus, 50%; and turbidity, 45% in their study. Precipitation of phosphorus after calcium hydroxide was rapid and higher than agricultural limestone. Calcium carbonate or calcium hydroxide treatments were also applied to hard water lakes by Prepas et al. (1990) and they reported significant decreases in the concentrations of total phosphorus and chlorophyll-a, resulting increased calcite precipitation and higher phosphorus binding affinity to the sediments. On the other hand, Salonen and Varjo (2000) applied gypsum to a hypereutrophic lake and observed that the treatment reduced the chlorophyll-a concentration. Masuda and Boyd (1994) suggest agricultural limestone or burnt lime in removal of Soluble Reactive Phosphorus (SRP) in the ponds. Schwartz and Boyd (1996) suggested that application of hydrated lime or quick lime to pond bottoms enhances ammonia volatilization, and kills pathogens, and they should not be used very frequently because of their inhibition on microbial activity.

Gypsum has a neutral reaction in water but it has been the least effective of the three (alum, ferric chloride, and gypsum) coagulants used in removing clay turbidity (Boyd 2000; 1990). However, Boyd (1995) suggested that gypsum treatment is better for use in low alkalinity waters because gypsum is a good source of calcium, and is more soluble than liming materials (agricultural limestone, burnt lime, and hydrated lime). Masuda and Boyd (1994) reported drastic decrease in SRP concentration, and lower phytoplankton concentration when calcium concentration was elevated with gypsum application. The effects of gypsum treatment on water quality in sunfish ponds with high alkalinity and low hardness conditions were studied and the gypsum treatment reduced phosphorus concentrations and phytoplankton abundance (Wu and Boyd 1990).

The effectiveness of several different compounds to immobilizing soluble reactive phosphorus found in soil from constructed wetlands was studied by Ann et al. (2000). They found that ferric chloride had immobilized the highest percentage of phosphorus in comparison to other amendments; alum, $Ca(OH)_2$, calcite, and dolomite. These amendments were only effective if applied at higher rates in their study. Cooke et al. (1993) reported that phosphorus inactivation with iron salts has shown only short term effectiveness, and subsequent failure was attributed to sediment anoxia because phosphorus precipitation with iron salts is possible if the water - sediment column is aerobic. However, Boyd (1995) reported that alum generally is cheaper than ferric chloride for pond treatment, and commercially available.

Ferric chloride is not suggested for frequent use in lake restoration because of the potential effects to redox reactions and relevant changes of pH on the solubility of iron-phosphate compounds. Under anaerobic condition, phosphorus bonded to the hydroxyl complexes of ferric iron is solubilized and released to the solution. Under anaerobic conditions, phosphorus from the sediments will be released to the water column therefore ferric chloride treatment for phosphorus precipitation can only be possible in aerobic condition (Rowan 2001).

Gutcho (1977) summarized the uses of polyelectrolytes and concluded that anionic, cationic, and nonionic polyelectrolytes are practical flocculating and clarifying agents in the clarification of water and sewage treatment. They are used in the removal of solids from various industrial wastes (mining, papermaking industries). Gutcho (1977) stated that polyelectrolytes if applied with ferric chloride are more effective in removal of phosphate and organic solids from municipal and wastewaters. Non-ionic polymer is generally applied to remove algae, diatoms, and organic contaminants in lakes and pond waters. Ozbay (2005) studied the effectiveness of gypsum, alum with agricultural limestone, ferric chloride, and ferric chloride with non-ionic polymer (polyacrylamide) removing excess nutrients and solids in the pond waters. She found alum with agricultural limestone treatment removed turbidity, suspended solids, and phosphorus during the sedimentation of pond effluents used in a laboratory set-up. Her research outcome was confirmed during her field application of alum with agricultural limestone and 1 hour was sufficient to remove most of the pollutants in the ponds. Figure 8 below shows the significant reduction in turbidity, TSS and ISS after alum with agriculture limestone application.

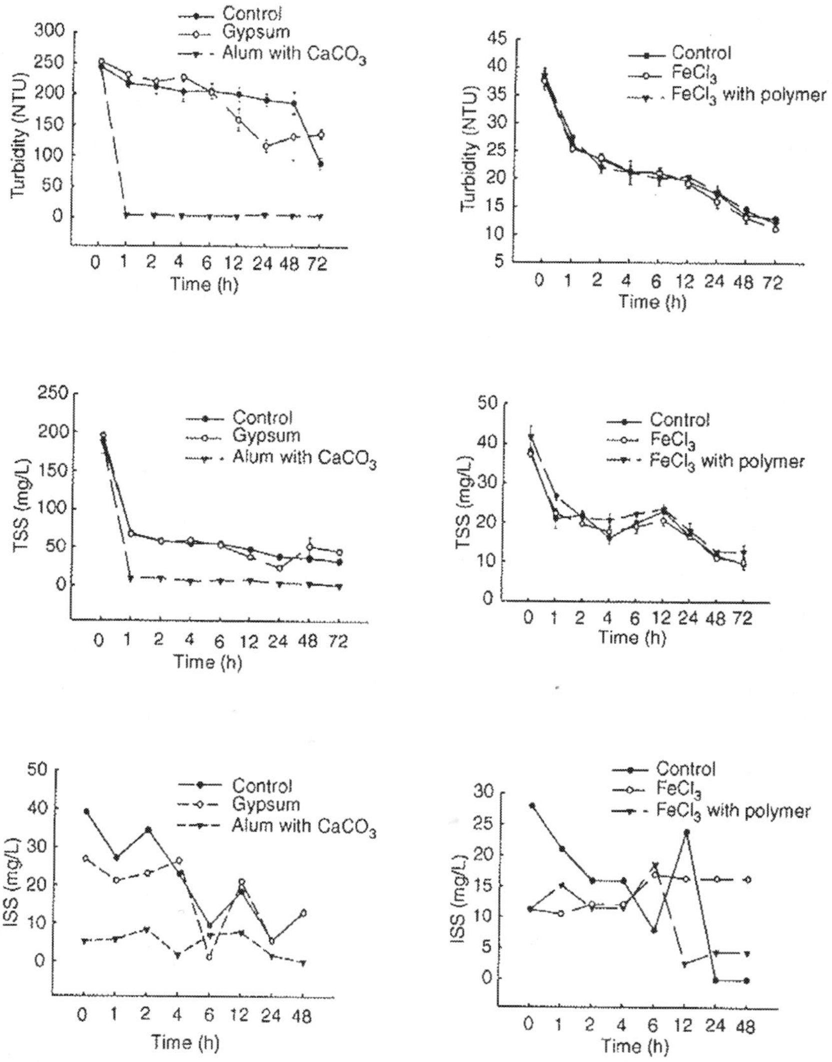

Figure 8. Means (±SE) of turbidity, total suspended solids (TSS), and inorganic suspended solids (ISS) concentrations in alum with agricultural limestone, gypsum, ferric chloride, and ferric chloride with polymer ((polyacrylamide) treated and control over a 72-72 hour sedimentation period, N=3 (Ozbay 2005).

Regardless of the chemicals selected for the aquaculture pond treatments, chemical treatment options should be minimized and carefully selected considering their effects on pond sediment pH and potential to increase solubility of various harmful metals.

http://dx.doi.org/10.5772/57471

3.5. Biological amendments

Increased awareness concerning environmental issues has been intensified by the constant pressure propagating from agricultural activities and more specifically from aquaculture farming. Many aquaculture farms have invested in alternative methods to minimize their environmental foot prints and keep their operations profitable. Phytoplankton is an important component of estuaries and coastal waters, reaching high population densities and accounting for a large fraction of the particulate matter in these systems (Wright et al. 1982). Considering high organic loading and detritus inputs from aquaculture effluents, further increase in phytoplankton abundance may become detrimental to estuarine health. Fish reared semi-intensively and intensively on formulated diets generate wastes containing organic particles and soluble nutrients. As a result, un-utilized feed and feces generate additional sources of nutrients, which result in higher abundances of phytoplankton (Lin et al. 1998). Phytoplankton blooms have a drastic effect on the water quality in receiving waters of estuaries and rivers.

During the last few decades, many studies have focused on reducing the secondary effects of poor water quality by means of introducing chemicals (copper sulfate, herbicides…etc.) or introducing herbivorous fish species (tilapia, carp, etc.), or introducing filter-feeding bivalves in order to eliminate the problems associated with heavy phytoplankton blooms (Ozbay 2002).

An alternative method which consisted of rearing manila clams to treat the marine fish pond effluents was attempted in Israel by Shpigel and Fridman (1990). The effluents from gilthead sea bream ponds which contained potential edible organic loading were then passed through manila clams in the effluent pond. They improved water quality by using manila clam as the filter-feeding bivalve in their study, and were able to simultaneously produce a high value product, the manila clam itself. Using manila clams to remove the organic loading, primarily phytoplankton, was an effective method to minimize the nutrient loads in pond effluents, and also produce an additional product for the market. Shpigel and Fridman (1990) also found manila clam to be very adaptable to changes in temperature, salinity, and high organic loading (i.e. phytoplankton) making them an ideal candidate for treating intensive aquaculture effluents. They suggested that this type of operation might have the potential for improving water quality depending on the pond conditions and species cultured.

Newell et al. (1999) found that the eastern oyster can exert a top-down control on phytoplankton stocks and also reduce turbidity, thereby increasing light available to benthic communities. Rehabilitation of natural oyster stocks has the beneficial effect of removing phytoplankton from the water column without stimulating further phytoplankton production. Rensel et al. (2011) investigated nutrient and phytoplankton removal and shellfish growth near the salmon pens. They monitored the highest oyster growth near the salmon pens due to food availability caused by the nutrients in fish feces. Although they did not find substantial differences in water quality parameters, phytoplankton was constantly available and removed by the Pacific oysters (*Crassostrea gigas*). Chrzanowski et al. (1986) investigated the ability of an oyster reef community to remove suspended microbial biomass and observed significant reduction in the suspended microbial biomass. Toro et al. (1999) found a significant negative relationship between oyster growth and amount of particulate inorganic and organic matter in water. Higher organic matters increase oyster growth via their filtration of phytoplankton in the organic matter. Ulanowicz and Tuttle (1992) observed that oyster abundance decreased phytoplankton productivity as well as stocks of pelagic microbes, ctenophores, medusae, and particulate organic carbon. Reduction in turbidity resulted from the removal of suspended solids including inorganic particles and phytoplankton by the oysters that oyster filtration plays an important role for increasing light penetration in the water column (Leffler 2001). Miura and Yamshiro (1990) recorded that phytoplanktivorous freshwater bivalves reduce phytoplankton blooms in the water outflow from fish tanks. Lowe et al. (1991) used mussels to increase water transparency in a lake and also observed a shift toward increased densities of benthic algae and recorded an increase in the visibility of water. Senichieva (1990) observed that actively filtrating mussels transform algae and microorganisms into feces and pseudo feces. Santelices and Martines (1986) found that the production of fecal material by filter feeders function as a fertilizer, and stimulated macroalgae growth that provides a venue to the farmers to integrate filter feeders and macroalgae.

Filter feeding bivalves provide a strong venue for the marine finfish farmers to cope with excess nutrient issues that are a result of un-eaten fish feeds and feces. Through their filtration activities, those filter feeders remove phytoplankton which results from additional nutrients introduced to the system. Various commercially and ecologically important species are dependent on oyster reefs for feeding, reproduction, and shelter from predators, including the naked goby (*Gobiosoma bosc*), skilletfish (*Gobiesox strumosus*), striped blenny, (*Chasmodes bosquianus*), and oyster toadfish (*Opsanus tau*) (Marenghi and Ozbay 2010a,b). There is a unique feeding cycle as these resident fishes feed primarily on benthic invertebrates and fish eggs and they also prey on other benthic fishes and will also eat each other while mud crabs (*Panopeus herbstii*) prey upon their eggs (Harding and Mann 2000). Although not oyster reef obligate, there are many other species that utilize oyster reefs including: black sea bass (*Centropristis striata*), northern pipefish (*Syngathus fuscus*), and Atlantic spadefish (*Chaetodipterus faber*) (Harding and Mann 2000). Oyster shells create habitat and serve as spawning substrate for the Florida blenny (*Chasmodes saburrae*), feather blenny (*Hypsoblennius hentz*) and the frillfin goby (*Bathygobius soporator*) (Tolley and Volety 2005). The larger, more transient, bottom-feeding species such as striped bass (*Morone saxatilis*), juvenile summer flounder (*Paralichthys dentatus*), juvenile winter flounder (*Pleuronectes americanus*), spot (*Leiostomus xanthurus*), black drum (*Pogonias cromis*), American eel (*Anguilla rostrata*), and Atlantic silverside (*Menidia menidia*) also utilize oyster reefs for feeding (Breitburg 1999).

Oyster reefs provide nursery habitat for many economically important juvenile species. Posey (1999) discussed why these reefs become important habitat for those species when natural seagrass beds are limited or absent because of environmental degradation. It is important to mention that 10 m^2 of restored oyster reef in the southeast United States is estimated to yield an additional 2.6 kg per year of production of fish and large mobile crustaceans (Peterson et al. 2003). Various ways reefs enhance fish production include increased recruitment, providing refuge from predation, and providing reef-associated prey (Peterson et al. 2003). Because an average size oyster filters 76 liters of water per day, they play a significant role in maintaining natural habitats (The Nature Conservancy 2013). Although aquacultured oysters provide limited but similar services as the natural oyster reefs, they can still effectively remove nutrients and control phytoplankton as they do in nature (Ozbay et al. 2013).

Cultured oysters can serve as broodstock, contributing to enhance and promote naturally occurring populations in the bays. Consecutive research by Marenghi and Ozbay (2010a) and Reckenbeil (2013) found newly settled juvenile oysters within floating oyster gear in man-made, residential canal systems, and on riprap shorelines for the first time around the Delaware Inland Bays (DIB). It appears that the small scale oyster aquaculture for restoration yields hopeful results within the impaired estuarine conditions as more signs of natural recruitment were observed at several locations within the DIB (Marenghi and Ozbay 2010b).

Rice (2008) discussed how biodeposition of filter feeders, such as bivalves, transfer organic nitrogen in phytoplankton and suspended particles in the water to the sediment. Filter feeding bivalves cycle nitrogen and phosphorus which play an important role in maintaining aquatic productivity (Rice 2008). Similar to Rice (2008), Lin et al. (1993) stated that shrimp-bivalve integrated culture systems in Thailand served as a biological control on phytoplankton populations, thus relieving the nighttime BOD stress. Wright et al. (1982) observed bivalve filtration of natural marine bacterioplankton and their reduction in the presence of bivalves.

Boyd and Queiroz (1997) investigated the feasibility of using salt-tolerant plants (halophytes) that were used as crop plants to remove nutrients from the effluent wastewater stream. The plant-soil system sequestered inorganic nitrite and phosphorus, and removed over 94% and 97% of the applied inorganic nitrogen and phosphorus. Wilson et al. (2002) summarized the use of plankton-feeding fish threadfin shad with channel catfish and monitored improved water quality conditions and enhanced catfish survival in the ponds with threadfin fish. Improvement in catfish production through the use of the Partitioned Aquaculture System (PAS) indicated that PAS offers the potential to eliminate blue-green algal dominance and associated off-flavor problems, while recovering wasted nitrogen and phosphorus discharges, which pose the threat of eutrophication to surface and groundwater supplies (Wilson et al. 2002).

In Yingbin Bay, China, the farmers set a large integrated aquaculture system that is capable of removing excess nutrients. By integrating seaweed and abalone into their main operation for shrimp culture, they were able to improve water quality. Pond bottoms are passed through seaweed and abalone to allow nutrients to be removed before using for shrimp grow-out ponds (Bennett et al. 2012). The authors found that farmers prefer seaweed farming because it reduces financial risks and leads to more consistent profits than shrimp farming however shrimp farming is more profitable for them.

Boyd and Tucker (1998) stated that the grass carp, the common carp, and certain species of tilapia have been evaluated for control of larger plant forms, including filamentous macroalgae. Plankton-feeding fish such as silver carp, bighead carp, tilapias, and gizzard shad are frequently employed in the ponds. Figure 9 shows the pictures of integrated aquaculture farm practices around the world and last picture is the illustration of the multi-trophic aquaculture system.

Figure 9. Pictures of various integrated farm practices around the world; a. tilapia culture with hydroponics herbs culture (http://land.allears.net/blogs/jackspence); b. TamilNadu Agricultural University horticulture fish farming integrated system (agritech.tnau.ac.in); c. Malawi fish farm and fruit trees along the edges (http://www.afap.org/africa_masasa); d. shellfish macroalgae culture (E & T Magazine, eandt.theiet.org); and e. Integrated Multi-trophic Aquaculture Schema (Government of Canada, dfo-mpo.gc.ca).

Tseng et al. (1991) reported that low concentrations of ammonia nitrogen and optimum algal density are better for controlling dissolved oxygen levels in tilapia ponds. Generally, microalgae stabilize pond water quality via either ammonia uptake or oxygen production. Burke and Bayne (1986) studied the effects of paddlefish on zooplankton, chlorophyll *a*, total ammonia nitrogen, and nitrite in a yearling paddlefish-catfish polyculture system. Higher seasonal mean chlorophyll *a* concentrations associated with lower zooplankton densities were measured in paddlefish treatment ponds. Smith (1985) found that filter feeders reduced algal biomass as much as 99%, increased phytoplankton diversity, and improved silver carp growth compared with other ponds without filter feeders, because filter feeders allowed high densities of zooplankters to remain and be available for fish. Fott et al. (1979) introduced carp in whitefish ponds and observed an increase in light penetration while primary production of phytoplankton and small zooplankton concentrations decrease substantially in the ponds.

During the last few decades integrated aquaculture practices have become a popular method to reduce the nutrient loads and pollutants entering natural waterways, and also increase profits by culturing more than one species of animal and/or plant. Canadian Aquaculture (2012) describes integrated multi-trophic aquaculture systems as the farming of various aquaculture species together where feces of one species serve as the feed to another, as demonstrated in fish /bivalve relationships. This system also increases profits for the farm and

http://dx.doi.org/10.5772/57471

decreases its negative environmental footprint. An aquaculture operation consisting of blue mussels and kelps located near pre-established Atlantic salmon aquaculture sites could substantially increase water quality and profits for the farmers in question. We provided a detailed overview on recirculating aquaculture systems and associated aquaponics systems, and their applications for integrated farming practices which ideally will result in economic and ecological benefits in our next sections.

3.6. Feeding and diet manipulation

Discharge of unused nutrients in effluents impacts eutrophication and different ecological measures. Impacts from aquaculture feed derived wastes have been observed on the natural environments (Gowen 1991). Boyd and Queiroz (1997) reported that in channel fish ponds in the United States, pond water quality was correlated to stocking and feeding rates. Water quality rapidly deteriorates at feeding rates above 100 - 120 kg/ha per day. In ponds utilized for fertilization and feeding, water quality is related primarily to nutrient input rates. Boyd and Queiroz (1997) stated that part of the nutrients in feeds and fertilizers is recovered in the harvested product, but the remaining nutrients enter the pond ecosystem as inorganic nitrogen, phosphorus, and carbon, dissolved and particulate organic matter. Therefore, relatively small percentages of nitrogen, phosphorus, and organic carbon are recovered in the harvested product. Consequently, the concentrations of nutrients and organic matter in the pond waters and the amount of organic matter settling onto bottoms increase as fertilization and feeding rates increase.

High quality feeds improve feed conversion ratios and reduce quantities of metabolic wastes and uneaten feed (Schwartz and Boyd 1996). Conservative feeding practices, and lower stocking rates also reduces feed inputs and improves feed conversion ratios (Boyd et al. 2000). Feeds are the ultimate source of organic matter pollution in catfish pond effluents (Boyd et al. 2000). The main types of wastes in aquaculture are residual feed particles, fecal matter, and metabolic by-products. Inefficient feed conversion results in poorer quality effluents and also decreases the concentrations of dissolved oxygen as shown in Figure 10. In a study by Filbrun and Culver (2013), dissolved oxygen levels in the ponds were increased by decreasing the feeding rates. The nitrogen in uneaten feed is transformed to ammonia by bacteria. Ammonia nitrogen tends to increase as feed application to a pond increases, and concentrations above 2 mg/L can be very harmful to aquaculture species at high pH (Gross et al. 1999). Ammonia is also added to ponds through fish excretions.

Cripps (1995) stated that it is likely that ponds containing specific sized particles would have elevated nutrient concentrations, resulting from their origin in the diet. Abou et al. (2012) demonstrated that using fern (*Azolla spp.*) as a fish meal substitute for Nile Tilapia had tremendously limited phosphorous loss in the effluent and is considered environment friendly. Coloso et al. (2001) found that soluble phosphorus discharge in effluent water can be reduced in fish fed diets that contained little or no fishmeal, or in diets that were supplemented with a low level of dietary phosphorus. In their rainbow trout study, the dietary combination of low phosphorus and high vitamin D_3 decreased soluble and fecal phosphorus levels in the effluents, indicating a strategy whereby effluent phosphorus concentrations can be reduced

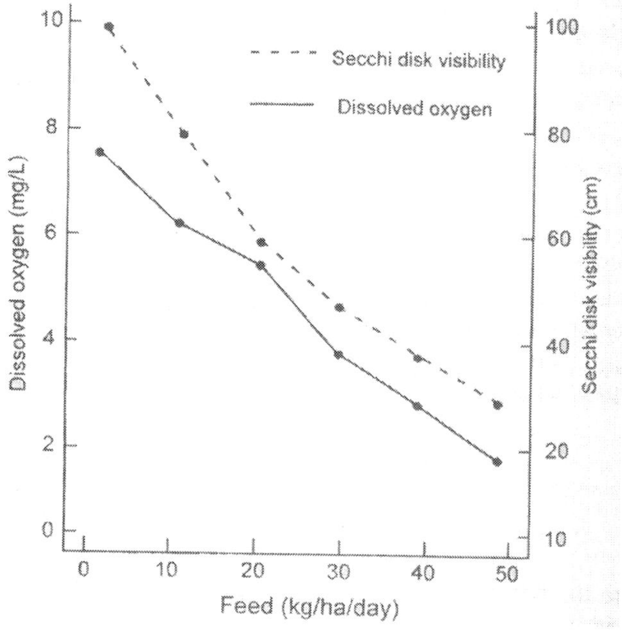

Figure 10. Effect of feeding rate on dissolved oxygen concentrations and Secchi disc visibilities at dawn (Boyd 2001).

by regulation of phosphorus metabolism. Increasing bioavailability of phosphorus will eliminate excess phosphorus in the effluent water.

According to Boyd and Queiroz (1997), increasing stocking density and feeding rates above assimilation capacity of pond water can harm the aquaculture pond in the long run. Heavy circulation, aeration, fish respiration and activities, plant abundance, feeding, fertilization, and stocking density in the ponds induced increases in concentrations of potential pollutants, which then require increased treatment efforts to reduce.

Nutrient manipulations were evaluated to promote more desirable phytoplankton communities by eliminating blue-green algae (Wilson et al. 2002). These methods include manipulating the ratio of nitrogen to phosphorus in the water, reducing the availability of phosphorus in bottom muds, enhancing the availability of inorganic carbon, increasing levels of salinity and potassium, and manipulating trace metal availability. Studies showed that various manipulations of waterborne plant nutrients have little promise for controlling phytoplankton community composition in catfish ponds with high feeding rates. Gross et al. (1999) found that the differences in phosphorus input among three feeds, containing 28, 32, and 36% crude protein, did not affect phosphorus concentrations in the effluents because most of the phosphorus from feed and fish excrement is absorbed by the soil. Gross et al. (1998) studied the phosphorus budgets for channel catfish ponds receiving one of five diets ranging from 0.60 to

http://dx.doi.org/10.5772/57471

1.03% phosphorus. They observed that low phosphorus diets did not decrease phytoplankton productivity or improve effluent quality. However, they suggested the use of low phosphorus diets in order to reduce the phosphorus load to bottom soils and to conserve the soils' ability to adsorb phosphorus.

Rangen Inc. (2013, Buhl, ID) provides important feeding tips of some of the commercially important aquaculture species. The tips that are species-specific would also minimize environmental impacts. Although recommended feeding practices differ by species, there are several common tips relevant to all species, including: 1) feeds should be broadcast well to allow all the fish to feed in the pond and minimize feed waste, 2) overfeeding should be avoided all costs to prevent from effluent pollution and gill damages, 3) feeding should be adjusted based on the percent body weight, 4) feed sizes should be adjusted as fish grows, 5) select the right feeding method for the species of interest, 6) feeding should be ceased before handling and shipping, 7) good record keeping is necessary to monitor fish growth and make the necessary adjustments, and 8) good storage and feeding management and feeding technique sanitation should be followed to avoid cross contamination and feed quality issues. In the last few decades, most fish farming has advanced from extensive rearing with few fish, to intensive rearing of high density populations in ponds and raceways. Cost effective good quality fish meals, proper feeding protocol, optimum growth and survival rates are the important goals of any fish farm operation.

3.7. Recirculating aquaculture and aquaponics

While many practices in aquaculture/mariculture (e.g., destruction of mangroves) have been criticized in years past for potential deleterious effects on the environment, the extent of any long term destruction due to aquaculture still remains debatable (Boyd and Schmittou 1990). Nevertheless in the United States and internationally, the most important environmental concern facing the aquaculture industry is the disposal of nutrient rich effluent water produced during the culture of aquatic animals (Goldburg and Triplett 1997; Frankic and Hershner 2003). Therefore, as aquaculture moves toward trying to feed the world, there is an inherent need to be stewards of the land, to protect, preserve and maintain conditions favorable to sustainability (Costa-Pierce 2002; United Nations FAO 2009). Recirculating aquaculture and its associated technology has largely developed out of concerns over water conservation and reducing environmental impacts (Martins et al. 2010). Besides growing fish, the purpose of a recirculating aquaculture system is to collect, concentrate, and process animal wastes rather than discharging wastes directly to the environment. Interestingly, an efficient recirculating aquaculture system is designed to reuse 90-99% of the water initially put into the system, while producing only a very small amount of waste or effluent (Chen et al. 2002, Hollingsworth et al. 2006; Badiola et al. 2012).

In their simplest form recirculating aquaculture systems (RAS) are similar to a home aquarium. Both a home aquarium and RAS have many of the same components including a tank or tanks, a pump to move water, some sort of filtration system, lighting, a heater or chiller, and fish. Also like a home aquarium, the RAS environment is very controlled to include lighting and room temperature good for the species of interest being cultured and other conditions in the facility. Fish are fed, water is added or taken out periodically, and water quality is monitored constantly and is often controlled through the addition of certain chemicals such as sodium bicarbonate. Unlike a home aquarium and relative to other types of aquaculture, RAS is very capital and energy intensive. It must rely on economic productivity for profitability, may require several additional components for processing water, and requires above average experienced personnel for successful management (Timmons et al. 2001; Timmons and Ebeling 2007; Ebeling and Timmons 2012).

With proper site selection, an advanced filtration capability, reduced water use and their small footprint, a recirculating aquaculture system lends itself to being a relatively environmentally friendly (Summerfelt and Vinci 2008; Ebeling and Timmons 2012, Losordo et al. 2013). Recirculating aquaculture systems do not rely on surface water for replenishment and with their ability to be located in close proximity to markets, they may be advantageous over other aquaculture systems (i.e., ponds, net pens, open ocean aquaculture) especially when comparing carbon footprint associated with food transport emissions (Martins et al. 2010). However, even with the positive attributes of a RAS, there is potential for it to be harmful to the environment and be considered unsustainable. Recirculating aquaculture systems are often described as the most effective way to grow large quantities of fish in a limited space. Furthermore, with their ability to control the environment they have the ability to grow fish year round (Hollingsworth et al. 2006; Ebeling and Timmons, 2012).

On a large production scale, recirculating aquaculture systems tend to be energy intensive and could be considered similar to other confined animal feeding operations, or CAFO's. In fact, under the 2004 United States federal aquaculture effluent limitation guidelines, recirculating aquaculture systems with an annual production exceeding 45,454 kg (100,000 lbs.) are classified as a concentrated aquatic animal production (CAAP) facility. Operations this large in scale are required to obtain a National Pollution Discharge Elimination System (NPDES) permit before discharging wastes into waters of the United States. Fortunately, the majority of recirculating aquaculture system operations in the United States choose alternatives to discharging effluent into natural waters, and instead either discharge into public municipal treatment works, collect the waste and apply it to nutrient deficient crops on land, or utilize treatment wetlands for processing effluent (Miller and Semmens 2002; Hollingsworth 2006; Summerfelt and Vinci 2008). A NPDES permit can be granted when the development of a facility specific Best Management Practices (BMP) plan specifies how discharges will be reduced of potential pollutants (Summerfelt and Vinci 2008).

The United States has a great deal of infrastructure that allows for regulation of discharge and more specifically, discharge into municipal treatment works, unfortunately the remainder of the world does not have this benefit. If recirculating aquaculture is to be adopted worldwide to raise fish in an environmentally and sustainable fashion, specific infrastructure is required. As previously mentioned, there is a plethora of literature available that describes RAS components and their efficiency at waste removal for large-scale fish culture. However, there is little information on dealing with the actual collected and concentrated solids that are generated from a large scale RAS. This is especially true when looking at recirculating aquaculture on an international capacity. Wetland ponds are often used in the United States and have been suggested on an international level, but wetland ponds have a limited lifetime and this is often a costly option. Another option for RAS effluent management that has been explored in other geographical areas is land application. Valencia et al. (2001) conducted a study to determine if effluent from a tilapia tank system could be used to replace nitrogen on guineagrass (*Panicum maximum*) managed as hay in a water limited area of the United States Virgin Islands. Interestingly, their results indicated that the tilapia tank system effluent could in fact serve as an adequate nitrogen and water replacement for guineagrass pasture, and hay production without changes in soil pH and phosphorous. Moreover, because this study used grass rather than row crops, it acts as a sink (similar to a wetland) with less risk of nutrient loss or leaching to the environment. The use of grass crops for assessing environmentally friendly ways to manage RAS effluent is but one step in the many ways research can explore repurposing and/or disposal of RAS effluent on and international level.

Overwhelmingly due to environmental concerns, but also to increase production efficiency, in recent years Best Management Practices have evolved across a number of industries from car manufacturing to food processing. Plain and simple, BMP's make sense and are a way of reducing multiple levels of risk. Within aquaculture, several entities including the Global Aquaculture Alliance (GAA) have created their own version of BMP's or Best Aquaculture Practices (BAP's). According to the GAA (2011), BAP's address environmental and social responsibility, animal welfare, food safety and traceability all on a national and international level. Through their BAP's the GAA further provides a certification program where they define elements of responsible aquaculture and evaluate adherence to these practices for each type of facility whether it be a hatchery, feed mill, farm or processing plant (www.gaalliance.org).

There are a number of BMP's that recirculating aquaculture system managers can use to make their facilities more environmentally friendly. Best management practices for recirculating aquaculture systems range in scope from choosing the right manager for the facility to using the most efficient types of filtration. Ensuring the use of high quality feeds with fewer fines will reduce nutrient levels and feed conversion ratios. Incorporating hybrid technology such as bioflocs which help to reduce feed costs and enhancing energy efficiency by using less and reusing energy where possible will all help the economic and environmental sustainability of recirculating aquaculture systems (Miller and Semmens 2002; Summerfelt and Vinci 2008; Hanna et al. 2010; Martins et al. 2010; Badiola et al. 2012). Summerfelt and Vinci (2008) have presented a thorough explanation of RAS BMP's beginning with site selection, working through solids storage, treatment and disposal, and complete facility operation and maintenance. Interestingly, Summerfelt and Vinci (2008) consider the point source waste stream to be the major facility level environmental issue (see also Hollingworth 2006).

Ultimately, for RAS to truly be environmentally friendly BMP's must be incorporated into their everyday function. Agriculture and its water counterpart aquaculture have been scrutinized due to various practices that have been employed over the many years of operation. In recent years, the colloquial buzzword has been "sustainable". You can't speak to anyone, go anywhere, or do anything anymore in any area of agriculture and natural resources without the mention of "sustainable". But with regard to agriculture what does sustainable mean? According to the United States Department of Agriculture, National Institute of Food and

Agriculture (USDA-NIFA 2013), sustainable agriculture is an integrated system of plant and animal production practices having site specific application that over the long term will be able to (1) satisfy human food and fiber needs, (2) enhance environmental quality and the natural resource base upon which the agriculture economy depends, (3) it should further make the most efficient use of nonrenewable resources and on-farm resources and integrate, where appropriate, natural biological cycles and controls, (4) sustain the economic viability of farm operations, and (5) enhance the quality of life for farmers and society as a whole.

Aquaponics developed from hydroponics, or the culture of plants with little to no soil. In hydroponics, plants are raised in some sort of structure where the roots are submerged in either water or some type of media base where they are fed via a solution containing all the nutrients (fertilizer, trace minerals, etc.) that they need (McMurtry et al. 1990). Aquaponics, however are virtually the same as any other RAS, except that they use the metabolic byproducts of one crop (i.e., finfish) to produce a secondary crop (plants), thereby adding value by producing two crops instead of one (Rakocy 2012). In turn, the plants in an aquaponics system filter potentially lethal nutrients (nitrite, nitrate) from the water and return it back to the fish culture tank (Losordo et al. 2013; Rakocy et al. 1992). With the mention of sustainability, increasing pressure from environmental groups, governmental regulations and the fact that aquaculture continues to play an ever increasing role in supplying the worlds' food supply it is not surprising that interest in aquaponics has begun to take a foothold with regard to RAS, especially in urban area of the US.

A typical aquaponics system is set up so that water flows from the fish culture tanks (Figure 12a) to a settling chamber, or clarifier where solids are removed from the waste stream, water then enters a biofilter where ammonia in the water from the fish and excess feed is converted to nitrite nitrogen and then nitrate nitrogen. Water then exits the biofilter and proceeds toward the plant component where there may or may not be several other components included (i.e., base addition, degasser). The plant component is either a raceway with floating rafts, or could be what is called an NFT (nutrient film technique). In general, this is where the plants feed off of the nitrate nitrogen before the water returns to the fish culture tank relatively free of nitrogen. In this system the plants receive trace minerals via the fish food; however, there is often the need to supplement with things like iron, calcium, and potassium. (For a complete description of an aquaponics system, refer to Rakocy 2012).

Aquaponics is beneficial for a number of other reasons including that the cycling of the fish water to the plant component in an aquaponics system reduces the amount of concentrated discharge coming from this system relative to other RAS. Also important is that while fertilizers, herbicides, and pesticides may be utilized in or around greenhouses housing hydroponics, these are highly discouraged around aquaponics systems because of the deleterious effects they would have on the fish. Similarly, because an aquaponics system is a form of RAS, the use of antibiotics is discouraged within the system so as not to kill beneficial bacteria that are involved in the natural nitrification process. For these and other reasons, aquaponics systems are considered to be broaching the realm of organic. Organic farming is often considered to be environmentally friendly and sustainable. Unfortunately, the United States Department of Agriculture (USDA) has yet to provide aquaculture with an organic

http://dx.doi.org/10.5772/57471

certification. While there are other private agencies worldwide that provide an organic certification for fish, the stringent guidelines provided by the USDA elevate this title to a higher level. Fortunately, in recent years there has been an ever increasing attempt at creating sustainable 'fishmeal reduced' and 'fishmeal free' diets for a growing number of fish species, especially with regard to highly prized carnivorous, saltwater species (see also Rhodes et al. 2013; Watson et al. 2013). The continued development of these diets may ultimately lead to a USDA organic certification for United States aquaculture.

While aquaponics systems are perhaps the most environmentally sustainable form of RAS to date, it does have drawbacks. Like any aquaculture venture, costs associated with initial investment, system components, their availability, construction and operation can have a significant impact on the economic sustainability of a system (Rackocy 2012). Hanna et al. (2010), for example have shown how different managers' management practices can affect the operation of identical RAS. It therefore becomes extremely important for aquaponics/RAS managers to follow Best management practices that will allow for a system to be profitable and sustainable. Best management practices for recirculating aquaculture and aquaponics systems have been described extensively in the literature (Chen et al. 2002; Summerfelt and Vinci 2008,). However, as we move toward trying to feed the world and keep RAS as environmentally friendly as possible, there are many important factors in operating a RAS. With regard to profitability and sustainability, perhaps offsetting initial investment, component and/or construction costs can be achieved by targeting highly sought after plant and fish species. Again, as fish meal free fish diets are developed for highly sought after marine species this reality becomes closer. One farmer at a recent national aquaponics conference suggested dealing directly with restaurants and "setting your price" rather than letting someone tell you how much something is worth (personal communication, National Aquaponics Conference, Tucson, AZ 2013). Similarly, it may also be advantageous for the owner/operator of a small scale RAS or aquaponics system to maximize profit and sustainability through raising high dollar plant and fish species as long as they have an established market (Frankic and Hershner 2003). With the recent surge in "farm to table" interest, it is very apparent that this concept can be profitable and environmentally friendly in the United States.

Ultimately, in trying to keep up with the worlds' population growth and food needs, RAS and aquaponics will continue to play a major role. Costa-Pierce (2002) and others suggest there has to be a behavioral shift in humans rather than technology in order for aquaculture to become truly sustainable. Many individuals only seem to see aquaculture in the sense that we need to produce as much fish as possible in as small an area as possible, however perhaps instead of trying to create RAS that are on the same level as a CAFO we instead look to systems that are sized according to the supporting the local community (Frankic and Hershner 2003). Again, one of the major advantages to a RAS is its' small footprint. By building a RAS with community size in mind it can be sized to feed the community on an ongoing basis. Having the RAS near or within a community would also reduce the carbon footprint by a reduction in fossil fuels needed for shipping etc. Figures 11 and 12 show a recirculating aquaculture system and an airlift aquaponics raft system in the Aquaculture Research and Demonstration Facility at Delaware State University, Dover, DE, USA.

Figure 11. a. An indoor multi-tank recirculating aquaculture system (RAS); b. A parabolic screen filter in a RAS (Photo courtesy of Blank).

Figure 12. a. Construction of an airlift aquaponics system; b. Aquaponics raft raceway showing aeration for plant roots; c. Planting herbs in a floating Styrofoam sheet (Photos courtesy of Blank).

4. Best Management Practices (BMPs) and sustainable aquaculture

Best Management Practices (BMPs) have been used as an important management tool for various aquaculture practices and management may vary based on the species cultured, type of aquaculture practice, location, surrounding habitat, economy and policy conditions of the area. Although there are differences in the application and level of engagement with BMPs, there are common issues through which the application of BMPs can be applied to all. While there are many types of different aquaculture operations, because worldwide ponds are the most prevalent the recommendation and management practices in this section will focus on the pond management.

BMPs reflect the most technically practical and economically feasible methods to reduce environmental impacts and limit costs at aquaculture facilities. One primary goal is to develop simple effluent treatment systems that reduce organic matter loads, suspended solids, and

http://dx.doi.org/10.5772/57471

nutrients in effluents to prevent polluting receiving waters. The best methods to prevent soil and water quality problems include selecting a site with appropriate soils and an adequate water supply, and maintaining moderate fish densities and feeding rates (Hargreaves et al. 2003). Secondary management techniques to prevent soil and water quality imbalances include liming, fertilization, and aeration (Boyd 1998). Agricultural irrigation, created wetlands, settling basins, and biological filters are also practical methods for reducing and improving the quality of effluent from ponds (Setty 2013; Tucker 2009; Ozbay and Jackson 2006).

Countries with regulations on farm practices and effluent standards follow guidelines and permit processes. However, countries without aquaculture regulations can also apply BMPs to minimize off-site water pollution and related environmental impacts. The code of conduct and codes of practice become useful tools for the farmers to adopt and serve as the guiding tool. As described by Boyd (2003) the code of conduct is a system of rules on how aquaculture should be conducted. The guiding principles for responsible aquaculture by the Global Aquaculture Alliance (GAA) serve as the code of conduct. In order to avoid social and environmental problems, codes of practice are used to solve the problems through management activities. Overall the goal of these codes is to minimize, or remove the negative impacts of an aquaculture operation (Boyd 2003). Boyd (2003) describes BMPs as management practices that are the most effective at reducing pollution levels and other environmental impacts which meet water quality or resource management goals.

Although not comprehensive, Boyd (2003) highlights some BMPs for pond aquaculture: a. use stocking and feeding rates that do not exceed the assimilative capacity of ponds; b. avoid overfeeding and apply a strict feeding plan; c. do not use fertilizers unless it is absolutely necessary to promote healthy phytoplankton growth; d. reduce water exchange; e. reuse water as much as possible; f. use a settling basin if available to treat pond effluents before final discharge. Application of these recommendations is based on farm operation and design, species cultured, and culture methods used (Boyd 2003). As Boyd states, selection of BMPs is case specific. One such example includes 10 codes of practice established for responsible shrimp farming by the GAA including mangroves, site evaluation, design and construction, feeds and feed use, shrimp health management, effluents, solids waste and few others. Although one BMP may be sufficient for one small farm, multiple BMPs may be necessary for others. ALEARN (www.alearn.info) listed over 20 best management practices for ponds, raceways, cages, effluents, safety, and others.

Shrimp aquaculture is a rapidly expanding field and is being closely scrutinized by environmentalists and government agencies. Due to the need for saltwater, discharge from shrimp facilities often flows into fragile coastal ecosystems. Problems associated with discharge include eutrophication due to nitrogen loading and detritus, low dissolved oxygen levels, sedimentation, along with other problems (Villalon 1991). These problems however are not without solutions; treatment of effluent should be regarded as an opportunity rather than just an obligation. Our goal is to provide broader perspectives on how basic principles and natural solutions can make shrimp aquaculture longer lived and be more sustainable. We discuss some of the management strategies current shrimp aquaculture operations along with best management practices for reducing potential impacts of shrimp aquaculture.

Although this may not be the ultimate solution, one particular recommendation discussed in this section would be to improve shrimp management practices. The mariculture of shrimp may provide one of the best opportunities for polyculture and integrated systems. Shrimp require higher water quality standards than many other cultured species and, thus, would benefit from a more stable ecosystem. The ability of shrimp to utilize a broad spectrum of the food web would allow them to be cultured with a number of other species. Feed and fertilization management can be geared toward supporting the food web to produce food items which shrimp prefer, rather than relying on the direct consumption of pelleted feed. In addition, Hopkins et al. (1991) discussed how dissolved oxygen levels were higher in polyculture ponds presumably due to a healthier phytoplankton community. Thus, polyculture may actually reduce aeration costs.

Hollingsworth et al. (2006) suggests growers may develop a farm-specific Standard Operating Procedure (SOP) manual and apply the code of conduct for significant farm practices in their SOP manuals. Although not required for all farms, the development of farm specific procedures will promote efficient management decisions including trouble shooting problems, training employees, planning future expansions or developing biosecurity and emergency procedure plans.

Many states in the United States have adopted BMPs over the years and some states and countries have implemented further policies and regulations based on scientific knowledge to sustain the environment and aquaculture industry. One specific example is that of Louisiana's aquaculture producers (Lutz et al. 2011). By implementation and application of best management practices, producers minimize potential pollutants (i.e., mainly excess nutrients) to the state's water resources and by doing so they reduce the cost that would be incurred to treat water quality problems, potential disease outbreaks and wild fish stock mortality related costs. Lutz et al. (2011) suggested that sediment runoff reduction should be one of the most important practices a pond aquaculture farmer must adopt to save money and reduce the environmental foot print of their operation. As an example, in Thailand shrimp aquaculture, scientists and policy-makers have developed new ways to improve the quality of the culture system, ecosystem, as well as the efficiency of regulations. It is critical that advances such as this and many other practices are discussed and maintained with integrity and strong regulations to improve the quality of our shared water resources for future generations. The key is to make aquaculture an asset to the environment while continuing to food production simultaneously.

Initial efforts and guidance on BMPs have been developed by Hargreaves et al. (2003). He has provided guide sheets on various topics and issues of concerns including reducing storm runoff into ponds, managing ponds to reduce effluent volume, erosion control on watersheds and embankments, pond management to minimize erosion, control of erosion by effluents, settling basins and wetlands, feed management, fertilization of catfish ponds, water quality protection to improve effluents, water quality enhancers, therapeutic agents, fish carcasses, general operations and worker safety, emergency response and management, and a few others added as the technology advanced in recirculating, bioflocs and aquaponics systems and other aquaculture operations.

The Best Aquaculture Practices (BAP) standards developed by the Global Aquaculture Alliance (GAA 2011) "address environmental and social responsibility, animal welfare, food safety and traceability in a certification program voluntary for aquaculture facilities." Certification for BAP ensures aquaculture operation is responsible and operates by the quantitative guidelines by which the farm operation is evaluated based on those practices. There are various standards developed in aquaculture sector including fish farm, hatchery, feed mill, and processing plant. The standard for the multi-species farming opens whole new area of attention with the new aquatic species used in integrated culture condition. Species BAP Standards used include channel catfish, shrimp, tilapia, and *Pangasius* initially and seabass, sea bream, cobia, seriola, trout, grouper, barramundi, perch, carp, flounder, turbot, striped bass, crabs, freshwater prawns, mussels, crawfish recent. According to GAA (2011), the new multi-species farm standards apply to all types of culture systems for finfish and crustaceans but not including cage-raised salmonids since this operation requires separate BAP standards. Seven of the most recent BAP standards listed in the GAA website (http://www.gaalliance.org/bap/standards.php) include Seafood processing/repacking plant, seafood processing plant, finfish and crustacean farm, salmon farm, mussel farm, shrimp hatchery, and feed mill.

The United States Environmental Protection Agency (USEPA 2004) initiated a new rule called the "effluent limitations guidelines (ELGs)' for concentrated aquatic animal production facilities including aquaculture facilities. This rule is applied to all commercial aquaculture facilities, with the below mentioned specifications, that discharge their wastewater from their farms directly into waters of the United States. According to the final rule, aquaculture facilities that "produce at least 45,360kg a year in flow-through and recirculating systems that discharge wastewater at least 30 days a year (used primarily to raise trout, salmon, hybrid striped bass and tilapia); at least 45,360 kg a year in net pens or submerged cage systems (used primarily to raise salmon)." The whole expectation with implementation of this rule is that the ELGs will help reduce discharges of conventional pollutants, primarily total suspended solids. As the solids are removed, it is expected that non-conventional pollutants such as nutrients will also be reduced. Other contaminants not discussed in this chapter include heavy metals, drug residues and other hormonal chemicals used in facilities to manage fish health and chemicals and better growth and this regulation is expected to be effective for reducing those contaminants in discharges of the facilities. With the implementation of this rule through National Pollution Discharge Elimination System (NPDES) permits the discharge of total suspended solids are expected to be reduced more than 226,796 kg per year, and the biochemical oxygen demand and nutrients in discharge is to be reduced by about 136,078 kg per year. With the application of this rule it is expected that water quality conditions will be improved and provide increased opportunity for other users, swimmers, fisherman and environmentalists concerned about keeping biodiversity in the streams, rivers and estuaries.

There are many definitions for sustainability and sustainability with regard to a catfish farm may not be sustainable for a shrimp farm. For aquaculture, sustainable aquaculture is an ultimate goal with the application of all the best aquaculture standards and management practices. Sustainability is described by the Northwest Earth Institute (2012) simply as meeting the needs of the present without compromising the ability of future generations to meet their own needs" (taken from UN World Commission on Environment and Development, Our Common Future). According to the Monterey Bay Aquarium (2013), environmental impact of fish farming varies depending on the species cultured, location of the farm, life stage of the organism, methods of culture and culture technique. Creating a sustainable farm should ensure species cultured will last long and habitat damage be minimal. The key factor with aquaculture sustainability is to operate with sound environmental management practices in place (FAO FOCUS 2013). There are tremendous efforts being made to use integrated aquaculture-agriculture farming systems to sustain both aquaculture operations and maintain the healthy environmental conditions for aquatic life in rivers, streams and estuaries. Environmentally friendly methods are also beneficial to the species cultured and farm operation (FAO FOCUS 2013). Sustainable aquaculture should utilize the most readily available technology to produce high protein food diets while applying the same exact principle to reduce its environmental impact using similar technology. Sustainability is not a practice, it is a life style and condition we must grasp.

Although it is beyond the scope of this chapter, the application of innovative technology in sustainable aquaculture such as Geographic Information Systems (GIS) would be an effective tool for selecting sites for bivalve culture and farm management. Coupled with ecosystem models, this technology can assist in predicting the carrying capacity of estuaries (Newell et al. 2013). Similar to shellfish site selection and farm management, Clearwater Seafoods has utilized GIS to take an informed approach to harvesting which minimizes the impact of fishing activities and promotes sustainability both at sea and on land. By investing in GIS, this company saved and minimizes their impacts on ocean ecosystems and promotes a sustainable approach to fishing (ESRI News 2013).

5. Case Study: Sustainable aquaculture culture in Thailand

Fisheries have long been integral to the Thai way of life. Management of fisheries in Thailand began in 1901 with the establishment of the Thailand Department of Fisheries (DOF). In 1901, the ministry of interior issued 3 guidelines to manage fisheries resources: 1) produce fisheries production for population in country, 2) produce fisheries production for country income, and 3) taxation for capture fisheries. However, only taxation was implemented because there was no fisheries biologist at that moment. In 1923, Dr. Hugh M. Smith, MD., LL.D (Commissioner of Fisheries USA) was invited as an advisor in fisheries to His Siamese Majesty's Government. After finishing his survey research, he published a book called "A Review of the Aquatic Resources and Fisheries of Siam, with Plans and Recommendation for the Administration, Conservation and Development". In 1926, the Department of Fisheries was established and Dr. Smith was appointed to be head official. Under his guidance, management systems were implemented and continue to be conducted. Fisheries production in the beginning relied mainly on wild capture since Thailand has many natural freshwater resources scattered all over the country, such as rivers, swamps, and reservoirs. Thailand also has a 3,500 kilometer-long coast line including both the Gulf of Thailand and Andaman Sea, including more than 900 islands. However, drastic changes to these habitats and overfishing have negatively impacted the wild capture fisheries. Therefore, production from aquaculture has gradually begun to play a more important role in maintaining total fisheries production (Figure 13).

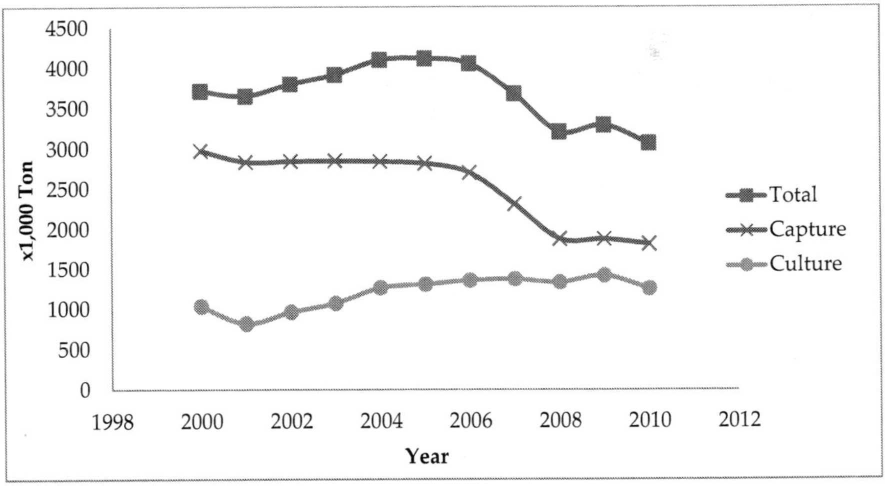

Figure 13. Thailand fisheries production from 2000-2010 (Thailand DOF Information System Center 2011).

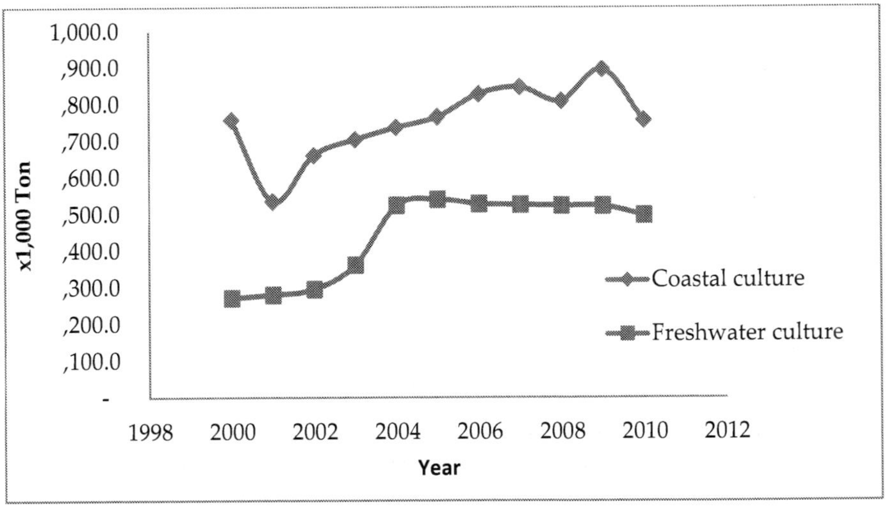

Figure 14. Freshwater and coastal aquaculture production from 2000-2010 (Thailand DOF 2011).

Aquaculture in Thailand evolved from traditional practice to modern science-based practices as aquaculture technology and innovations developed. Therefore, aquaculture in coastal areas, which contribute more national economics and provide more benefit to farmers, tended to increase and contribute more when compared to freshwater species, especially brackish water-cultured shrimp and prawn. However, aquaculture in freshwater areas has also increased due to population growth and market demands in the country (Figure 14) (Thailand DOF Information System Center 2011).

Of the three main groups of brackish water aquaculture – fish, shrimp, and shellfish – shrimp culture has increased dramatically while fish culture has decreased and become steady from 2000 – 2010, while shellfish culture production dropped because of shellfish diseases and natural changes (Thailand DOF Information System Center 2011) (Figure 15). However, shrimp culture is the most cultured species within brackish water aquaculture, globally.

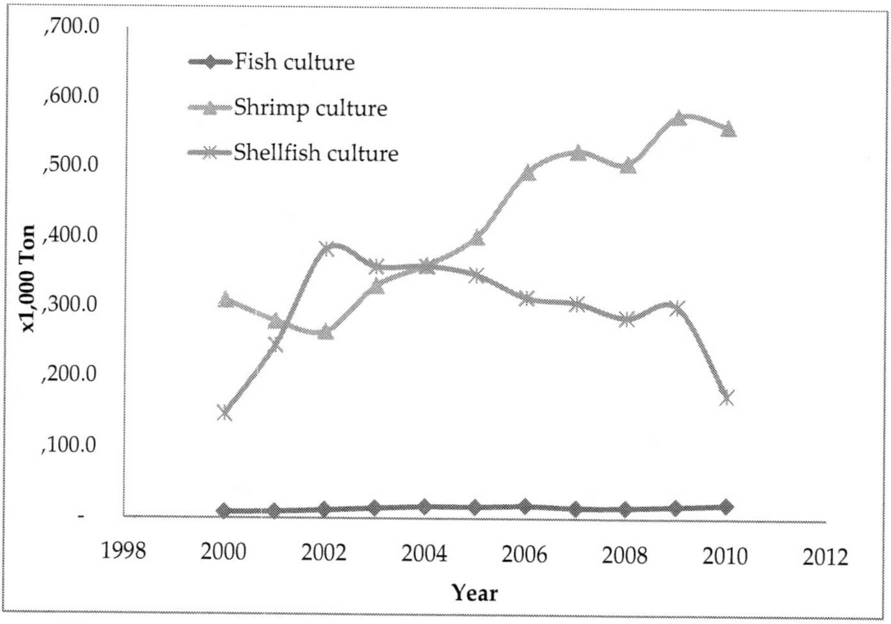

Figure 15. Brackish water aquaculture production by group from 2000-2010 (Thailand DOF 2011).

Shrimp culture in Thailand started more than 50 years ago. Production has been greatly increased within the last three decades as intensive farming techniques were developed and applied. In the beginning of brackish water shrimp culture the major shrimp species used was tiger prawn (*Peneaeus monodon*), which was substituted by white shrimp (*Peneaeus vannamei*) by more than 95%, by the year 2000 (Bureau of Agricultural Economic Research, 2011). Thailand now exports shrimp products at about 500,000 tons annually in the

http://dx.doi.org/10.5772/57471

year 2010 (Center for Agricultural Information 2011), making Thailand one of the largest shrimp exporters in the world.

Cultured shrimp, in particular, has come under threat in key export markets due to adverse publicity concerning the environmental and social impact of some aquaculture activities (Kongkeo, 2001). Some of this publicity has served to highlight some of the negative environmental and social impacts that have occurred in the development of aquaculture. These include the destruction of wetlands and mangrove forests, water pollution, reduction of biodiversity, waste of natural resources, and loss of access to fishing grounds by artisanal fisherman (Boyd and Tucker, 1998). Therefore, management inputs that remedy these problems that occur between culturing period and after harvesting such as chemicals used, water management, water discharge, etc. must be considered in order to encourage sustainable growth and practices within the shrimp aquaculture sector.

Shrimp culture in Thailand is performed under intensive culture conditions which consumes heavy feed, water supply, and aeration; therefore, management must play a key role in helping to reduce problems and the impacts of effluent from culturing systems. Sustainable shrimp culture, economically speaking, is less about increasing production but more about the ability to maintain steady production, customer satisfaction and reliability, and mitigate social and environmental impact concerns.

Long before Good Aquaculture Practices (GAPs) and Food and Agriculture Organization (FAO) Code of Conduct for Responsible Fisheries (CoC) were integrated in Thailand, best management practices were implemented by the Department of Fisheries to increase farm productivity while minimizing the environmental impacts. The environmental issues caused by the effluent discharges of the shrimp culture facilities have been one major concern of the aquaculture operation. Some of these management practices facilitated by the Department of Fisheries at the National and local level have focused on increasing feed conversion ratios, better water exchange, aeration, and pond management, and, if available, applying integrated multi-trophic aquaculture technology to reduce nutrient loads from farm effluents and increase profits. Specifically, aquaponics using commercial crop systems become popular and practical to the farmers while the operation is sustained and become environmental friendly.

The Thailand DOF integrated GAP and CoC in 1998 with support from the World Bank. Under GAP standard, the requirement are farm registration;, farm management; use of veterinary drugs chemical, hazardous substances, and probiotics used in aquaculture; farm sanitation; harvest and post-harvest prior to distribution; effluent and sediment management; energy source and fuel use; social and environmental responsibility; and record keeping. Code of Conduct for shrimp culture has two components: operational guidelines for hatcheries and farms, and guidelines for harvesting and transport. These practices - GAP and CoC - are standard for shrimp culture to ensure that shrimp culture has minimal to no chemical residues which protects consumers and applies environmental friendly practices. Three year GAP or CoC certificates are issued by the Thailand DOF to shrimp farms after they meet qualifications and comply with annual surveillance.

Shrimp culture systems in Thailand also have a traceability system from farm to product which initially started as a form of hatchery management. Some hatcheries complied with bio-secure systems to ensure that larvae produced are healthy and viable before selling to grow-out farmers. Hatcheries must provide Fry Movement Documents (FMD) to their customers to indicate the number of fry that a farmer purchases in addition to other hatchery information since this document is checked if any problems occur during grow-out. Shrimp farmers must also provide Movement Documents (MD) which indicate the weight of shrimp in the shipment in addition to other farm information to processing plants or their customers. A DOF officer checks MD, Hazard Analysis and Critical Control Point (HACCP), GMP, and product quality at the processing plant before issuing product health certificates.

Moreover, many policies and projects are established for sustainable aquaculture support. The agencies involved at the national level include, the Department of Fisheries, Pollution Control Department and Department of Marine and Costal Resources, while there are Provincial government and Local Administrative Organization involve at local level. Water quality testing program pond and discharge water responsible by DOF and water quality in natural water responsible by Pollution Control Department, etc. Mangrove rehabitation projects are established by the national and local government sectors, private sector, and Non-Governmental Organizations (NGOs) to increase mangrove forest area along the costal zones. Thailand also supports Non Illegal Unreported Unregulated Fishing (Non-IUU) and issued several programs for fish resource conservation, which include combatting IUU fishing, prohibiting certain fishing gears within spawning season, and expanding fish conservation areas. Although wild fisheries are not a part of the discussion in this chapter, it does play a part serving as a source of fish meal industry which is used in aquaculture feed. Therefore, control of IUU and certain fishing gear, will support sustainable aquaculture.

In conclusion, fisheries production in Thailand has decreased while aquaculture production has increased and plays a vital role for providing a high protein food source for economic development in the future. Aquaculture in Thailand evolved from traditional practice to science-based due to a number of policies and regulations put in the place to sustain both the aquaculture industry and the environment. Shrimp aquaculture in Thailand is an excellent example of why sustainable aquaculture practices are necessary and how they have become implemented. With the establishment of GAP and CoC, Thailand has ensured that shrimp farming results in production of a high quality product, safe from chemical residues, that is environmental friendly. However, truly sustainable aquaculture will only be attainable when the balance between food security, economic benefit, social benefit, and a reduction of environmental impact is achieved. Figure 16 below shows the steps involved in best management practices in shrimp aquaculture farm.

http://dx.doi.org/10.5772/57471

Figure 16. a. Clean and dried pond before start; b. Water quality monitoring; c. Water drained into treatment ponds; d. Aeration in culturing period using paddle wheel aerators; e. Shrimp health sampling with a cast net; f. Shrimp monitoring with a lift net; and g. Collected specimens are assessed for growth and survival rate.

6. Final remarks

In recent years aquaculture has gone through the "blue revolution" in which there has been rapid growth worldwide in aquaculture production of both fresh and saltwater fish and shellfish species. In part this is due to the fact that the natural fisheries are close to their maximum sustainable yield. However, this rapid growth in aquaculture may also be attributed to the ever increasing world population and an increase in demand for high protein sources of seafood. In the past, aquaculture has been demonized for destruction of mangroves worldwide for shrimp production as well as causing potential eutrophication through unwelcomed discharges of nutrient rich effluents.

With increased environmental awareness and the general populations increased concern over its food sourcing, aquaculture has stepped up to the proverbial plate to try and fulfill the worlds seafood demands through increased production while trying to maintain more environmentally friendly practices of culturing fish through many technological advancements. Unfortunately, aquaculture has not yet truly reached its sustainability goal. However, in addition to much technological advancement, aquaculture has begun to incorporate best management practices to create a more environmentally friendly way of producing fish. In this chapter we have gone over several areas associated with BMPs and described them with regard to how their incorporation can impact or reduce the impacts of aquaculture on the environment. Many of the BMPs discussed are simple and rely on common sense approaches to nutrient problems. Others are more technologically advanced and require additional components and or descriptions that are beyond the scope of this chapter.

In the end we are all trying to get to a point at which aquaculture can be considered a sustainable farming entity so that its impacts are minimal at best to the surrounding environment, there is a continuous supply of food, and it is profitable for all of those who are involved. As we consider moving ahead we must continue to remember that the world's resources are there for all and we want to maintain them for future generations to come. Again, we quote Costa-Pierce (2002) in suggesting there has to be a behavioral shift in humans rather than technology in order for aquaculture to become truly sustainable.

7. Technical summary

Tremendous efforts have been made to improve aquaculture farm practices through disease prevention and treatment, planning and management of facilities, feeding, and advances in aquaculture technology and sustainable practices. However, the industry is not without its issues and faults. Although significant accomplishments have been made in minimizing the negative impacts of aquaculture operations on the environment, it is not reached sustainability worldwide. Further efforts ranging from husbandry practices to policy and regulations are essential to ensure the sustainability of aquaculture on a global scale. As aquaculture moves from feeding millions to billions of people in the last century, intensive culture practices have become common and require better management and monitoring efforts. Intensive production of fish farming requires significant inputs of nutrients in the form of inorganic fertilizers or feeds. Of these inputs, typically only 25% of the chemical constituents of the feed are assimilated into fish biomass while the rest is released into the water as metabolic wastes. In pond culture, fish are usually harvested after draining the pond partially or fully. The waste water expelled from these ponds into watershed, laden with organic matter and nutrients, concerns regulatory agencies as a point-source of pollution. In addition, in most countries, including the United States, statistics are lacking on the amounts of chemicals used, and as a result, regulations cannot truly be effective. Furthermore, it is difficult to make correlations of aquaculture effluents to environmental impacts without accurate records. The main goals of effluent management or, more often referred to as best management practices, are to minimize impacts to the environment while maintaining productivity. Fortunately, most of these strategies are as beneficial to the aquaculturists/ farmers as they are to the environment. Both production costs and effluent can be reduced by using stock-specific feeds applied in smaller quantities several times a day, good aeration, improved husbandry practices, and paying good attention not to exceed to the carrying capacity of the system. By lowering concentrations of phytoplankton, savings on herbicides and aeration are inevitable. Limited water exchange, integrated aquaculture, and good monitoring are further best management practices measures.

Acknowledgements

The authors would like to thank Dr. Karuna Chintapenta, Brian Reckenbeil, Kris Roeske, and Cory Janiak for their editorial assistance and review of the chapter. The authors would also like to thank Dr. Claude Boyd for the resources and guidance he provided both authors for their research and professional development. Some of the projects discussed in this chapter were funded by SRAC, USDA-NIFA, USDA Evans-Allen, NSF-EPSCOR, AU, DSU, and Government of Thailand.

Author details

Gulnihal Ozbay[1], Grant Blank[1] and Taworn Thunjai[2]

1 Department of Agriculture and Natural Resources, Delaware State University, Dover, USA

2 Department of Fisheries, Kasetsart University, Ladyao, Jatujak, Bangkok, Thailand

References

[1] United Nations Food and Agriculture Organization. UNFAO: 2009. How to Feed the World 2050. High-Level Expert Forum. Global agriculture towards 2050. http://www.fao.org/fileadmin/templates/wsfs/docs/Issues (accessed 10 September 2013).

[2] University of Maryland Medical Center. UMD: 2013. Omega-3 fatty acids. Overview. http://umm.edu/health/medical/altmed/supplement/omega3-fatty-acids (accessed 3 October 2013).

[3] United Nations Food and Agriculture Organization. UNFAO: 2013. FOCUS. Fisheries and food security. Sustainable aquaculture development. http://www.fao.org/focus/e/fisheries/sustaq.htm (accessed 5 September 2013).

[4] Ewart JW. 2013. Shellfish Aquaculture in Delaware's Inland Bays: Status, Opportunities, and Constraints.In Delaware Sea Grant Program; 2013. p 43.Available from http://darc.cms.udel.edu/ibsa/Inland (accessed 7 November 2013).

[5] Bush SR, Belton B, Hall D, Vandergeest P, Murray FJ, Ponte S, Oosterveer P, Islam MS, Mol APJ, Hatanaka M, Kruijssen F, Ha TTT, Little DC, and Kusumawati R. Global Food Supply. Certify Sustainable Aquaculture? Science 2013;341(6150) 1067-1068.

[6] National Oceanic and Atmospheric Administration. NOAA: 2013. 10 Myths about Marine Aquaculture. http://www.nmfs.noaa.gov/aquaculture (accessed 7 November 2013).

[7] United States Food and Drug Administration. USFDA: 2008. Taken from Seafood Health Facts: Making Smart Choices. Balancing the Benefits and Risks of Seafood Consumption. http://seafoodhealthfacts.org/seafood_nutrition/practitioners/composition_chart.php (accessed 3 September 2013).

[8] Wilson JL, Stone NM, Bayne D, Popma TJ, Boyd C, Brune DA, Collier JA, Schwedler TE, Burtle GJ, Lewis GW, Styer EL, Romaire RR, Hargeaves JA, Kingsbury SK, Nagle DG, Daniels HV, Hodson RG, Shrader KK, Zimba PV, Grimm CC. Control of Blue-green Algae in Aquaculture Ponds. SRAC Fifteenth Annual Progress Report 2002;30.

[9] United Nations Food and Agriculture Organization. FAO: 2011. FAO STAT. http://faostat.fao.org/site/629/default.aspx (accessed 11 December 2013).

[10] Dealteris JT, Kilpatrick BD, Rheault RB. A comparative evaluation of the habitat value of shellfish aquaculture gear, submerged aquatic vegetation and a non-vegetated seabed. Journal of Shellfish Research 2004;23(3) 867-874.

[11] Erbland PJ, Ozbay G. A Comparison of the macrofaunal communities inhabiting a Crassostrea virginica oyster reef and oyster aquaculture gear in the Indian River Bay, Delaware. Journal of Shellfish Research 2008;27 757-768.

[12] Shumway SE, Davis CR, Downey R, Karney J, Kraeuter J, Parsons R, Rheault, Wikfors G. Shellfish aquaculture: in praise of sustainable economies and environments. World Aquaculture 2003;34 15-18.

[13] O'Beirn F X, Ross PG, Luckenbach MW. Organisms associated with oysters cultured in floating systems in Virginia, USA. Journal of Shellfish Research 2004;23(3) 825-829.

[14] Tallman JC, Forrester GE. Oyster grow-out cages function as artificial reefs for temperate fishes. Transactions of the American Fisheries Society 2007;136(3) 790-799.

[15] D'Amours O, Archambault P, McKindsey CW, Johnson LE. Local enhancement of epibenthic macrofauna by aquaculture activities. Marine Ecological Progress Series 2008; 371 73-84.

[16] Taylor J, Bushek D. An Assessment of Habitat Value of Constructed Intertidal Oyster Reefs and Oyster Aquaculture Systems in Delaware Bay, USA. Journal of Shellfish Research 2008;27(4) 1056-1056.

[17] Emerson C. Aquaculture impacts on the environment. Cambridge Scientific Abstracts 1999.

[18] World Wildlife Foundation. WWF: Unsustainable fishing. Our oceans are being plundered. http://wwf.panda.org/about_our_earth/blue_planet/problems/problems_fishing/ (accessed 2 November 2013).

[19] National Oceanic and Atmospheric Administration. NOAA: Fish watch U.S. Seafood Facts: Outside the U.S. 2011 http://www.fishwatch.gov/farmed_seafood/outside_the_us.htm (accessed 6 October 2013).

[20] National Oceanic and Atmospheric Administration. NOAA: 2010. U.S. Seafood Consumption Declines Slightly in 2009. http://www.noaanews.noaa.gov/stories2010/20100909_consumption.html (accessed 8 December 2013).

[21] Monterey Bay Aquarium Foundation. MBAF: Fishing and Farming Methods. http://www.montereybayaquarium.org/cr/cr_seafoodwatch/sfw_gear.aspx (accessed 2 December 2013).

[22] Briggs MRP, Funge-Smith S. Unsustainable shrimp culture- causes and potential solutions from experience in Thailand. In: Development of Strategies Shrimp Farming. In report to the Overseas Development Administration, Research Project R4751;1994. p31.

[23] Dalsgaard JPT, Lightfoot C, Christensen V. Towards quantification of ecological sustainability in farming systems analysis. Ecological Engineering 1995;4 181-189.

[24] Food and Drug Administration. FDA: Food and Drug Administration Compliance Program Guidance Manual, 2008. http://www.fda.gov/downloads/Food/ComplianceEnforcement/ucm073192.pdf (accesses 10 October 2013).

[25] Virtanen JK, Mozaffarian D, Chiuve SE, Rimm EB. Fish consumption and risk of major chronic disease in men. *American Journal of Clinical Nutrition* 2008;88(6) 1618–1625.

[26] Soft Bottom. In Coastal Habitat Protection Plan; 2005. p365-448. Available from http://ncfisheries.net/habitat/chppdocs/G_Soft%20Bottom.pdf (accessed 3 August 2013).

[27] Garland Science. In: Microbes in the Marine Environment; 2011. p23. Available from http://www.fisheries.go.th/library/app/ (accessed 3 November 2013).

[28] Friend M. Disease Emergence and Resurgence: The Wildlife-Human Connection: Reston, Va., U.S. Geological Survey. 2006;400.

[29] Yanong RPE, Erlacher-Reid C. Biosecurity in Aquaculture, Part 1: An Overview. Southern Regional Aquaculture Center 2012;(4707) 16.

[30] Chatterjee S, Haldar S. Vibrio Related Diseases in Aquaculture and Development of Rapid Accurate Identification Methods. Journal of Marine Science 2012; 7.

[31] Mori K. Effects of artificial eutrophication on the metabolism of the Japanese oyster Crassostrea gigas. Marine Biology 1979;53(4) 361-369.

[32] Bastin L. International Aquafeed. Incorporating Fish Farming Technology 2013;16(1) 26.

[33] Boyd CE. Water quality management and aeration in shrimp farming 1989: Auburn University, Department of Fisheries and Allied Aquacultures, Auburn, AL, Series No. 2. p183.

[34] Ryther JH, Goldman JC, Gifford JE, Huguenin AS, Wing JP, Clarner LD, Williams, Lapontie BE. Physical models of integrated waste recycling of "Marine Polyculture Systems." Aquaculture 1975;5 163-177.

[35] Neori A, Ragg NLC, Shpigel M. The integrated culture of seaweed, abalone, fish and clams in modular intensive land-based systems: II. Performance and nitrogen partitioning within an abalone (Haliotis tuberculata) and macroalgae culture system. Aquacultural Engineering 1998;17(4) 215-239.

[36] Avnimelech Y., editor. Biofloc Technology - A Practical Guide Book. The World Aquaculture Society, Baton Rouge, Louisiana; 2012. p272.

[37] The University of Virgin Islands Aquaculture Program. 2013. Biofloc Systems. http://www.uvi.edu/research/agricultural-experiment-station/aquaculture-home/biofloc-systems/default.aspx.

[38] The Aquaponics Association. AA: What is aquaponics? http://www.aquaponicsassociation.org/ (accessed 8 November 2013).

[39] Bennett L, Diana J, Qiuming L. Learning From Tradition. In: Integrated Aquaculture Practices in Yingbin Bay;2012. http://pdf.gaalliance.org/pdf/GAA-Bennett-Nov12.pdf (accessed 10 October 2013).

[40] Nutreco. Feeding the Future? How can aquaculture contribute to feeding 9 billion people in 2050 in a Sustainable way? In Nutreco Prins Frederiklaan; 2011. http://www.nutreco.com. (accessed 8 November 2013).

[41] Scottish Association for Marine Science. SAMS: Aquaculture Impacts and Coastal Eutrophication. http://www.sams.ac.uk/microbial/aquaculture-impacts-and-coastal-eutrophication#sthash.pZzJ4ask.dpf (accessed 9 December 2013)

[42] Science Daily. In Science Reference; 2013. Eutrophication. http://www.sciencedaily.com/articles/e/eutrophication.htm (accessed 8 November 2013).

[43] Boyd CE. Soil and Water Quality Considerations in Shrimp Farming. In: Haws MC. and Boyd CE. (ed.) Methods for Improving Shrimp Farming in Central America. In Central American University Press; 2001. p17-20.

[44] Boyd CE, Queiroz J. Aquaculture pond effluent management. Aquaculture Asia 1997;2 (2) 43-46.

[45] Shireman JV, Cichra CE. Water quality from Florida catfish and fingerling culture ponds. In: Joseph St. (eds) 1996: national livestock, poultry, and aquaculture waste management, Proceedings of the National Workshop, American Society of Agricultural Engineers, St. Joseph, MI. p206-210.

[46] Schwartz MF, Boyd CE.Channel catfish pond effluents. The Progressive Fish-Culturist 1994a;56(4) 273-281.

[47] Boyd CE. Effluents from catfish ponds during fish harvest. Journal of Environmental Quality 1978; 7 59-63.

[48] Schwartz MF, Boyd CE. Effluent quality during harvest of channel catfish from watershed ponds. The Progressive Fish-Culturist 1994b;56(1) 25-32.

[49] Seok K, Leonard S, Boyd CE, Schwartz ME. Communications: Water Quality in Annually Drained and Undrained Channel Catfish Ponds over a Three-Year Period. The Progressive Fish-Culturist 1995;57(1) 52-58.

[50] Boyd CE, Queiroz J, Lee J, Rowan M, Whitis GN, Gross A. Environmental assessment of channel catfish, Ictalurus punctatus, farming in Alabama. Journal of the World Aquaculture Society 2000;31(4) 511-544.

[51] Hariyadi S, Tucker C, Steeby J, Van der Ploeg M, Boyd CE. Environmental conditions and channel catfish (Ictalurus punctatus) production under similar pond management regimes in Alabama and Mississippi. Journal of the World Aquaculture Society 1994;25 236-249.

[52] Boyd CE, Gross A, Rowan M. Laboratory study of sedimentation for improving quality of pond effluents. Journal of Applied Aquaculture 1998;839-48.

[53] Boyd CE, Tucker CS. Pond Aquaculture Water Quality Management. In Kluwer Academic Publishers; 1998. p541-575.

[54] Wheaton FW. Aquaculture Engineering. In Robert E. Kreiger; 1977. Malabar, Florida, USA.1977. p414-462.

[55] Coloso RM, Basantes SP, King K, Hendrix MA, Fletcher JW, Weis P, and Ferraris RP. Effect of dietary phosphorus and vitamin D_3 on phosphorus levels in effluent from the experimental culture of rainbow trout (*Oncorhynchus mykiss*). Aquaculture 2001;202(1) 145-161.

[56] Queiroz JF, Boyd CE, Gross A. Evaluation of a bio-organic catalyst in channel catfish, Ictalurus punctatus, ponds. Journal of Applied Aquaculture 1998;8(2) 49-61.

[57] Tucker CS, Kingsbury SK, Pole JW, Wax CL. Effects of water management practices on discharge of nutrients and organic matter from channel catfish (Ictalurus punctatus) ponds. Aquaculture 1996;147(1) 57-69.

[58] Thompson JR, Polz MF. Dynamics of Vibrio populations and their role in environmental nutrient cycling. The biology of Vibrios. In ASM; 2006. p90-203.

[59] ICAP Bio-Organic. 2013. Technology. http://www.icapbio-organic.com/technology.html (accessed 8 November 2013).

[60] Dame RF, Spurrier JD, Wolaver TG. Carbon, nitrogen and phosphorus processing by an oyster reef. Marine ecology progress series 1989;54(3) 249-256.

[61] Seaweed Energy Solution. SES: Seaweed is simple. http://www.seaweedenergysolutions.com/seaweed/ (accessed 8 November 2013).

[62] Becker V, Huszar VLM, Crossetti LO. Responses of phytoplankton functional groups to the mixing regime in a deep subtropical reservoir. *Hydrobiologia*; 2008; 628(1) 137-151.

[63] National Oceanic and Atmospheric Administration. NOAA: State of the Coastal Environment. Oxygen Depletion in Coastal Waters. http://oceanservice.noaa.gov/websites/retiredsites/sotc_pdf/HYP.PDF (accessed 8 December 2013).

[64] Boesch D, Brinsfield R, Magnien R. Chesapeake Bay eutrophication: Scientific understanding, ecosystem restoration, and challenges for agriculture. Journal of Environmental Quality 2001;30 303-320.

[65] Boyd CE. Water quality for pond aquaculture. Alabama Research and Development Series 1998; 43 37.

[66] Boyd CE. Water Quality in Ponds for Aquaculture: Auburn University; 1990; p482.

[67] Boyd CE, Musig Y. Shrimp pond effluents: observations of the nature of the problem on commercial farms. In: Proceedings of the Special Session on Shrimp Farming. World Aquaculture Society 1992, Baton Rouge, Louisiana 2006; 195-197.

[68] Boyd CE., editor. Bottom Soils, Sediment, and Pond Aquaculture. Chapman and Hall; 1995a. p348.

[69] Ozbay G. Studies of Channel Catfish *Ictalurus punctatus* Pond Effluent Characteristics and Treatment Options. PhD thesis. Auburn University Auburn; 2002; p109.

[70] Ebeling JM, Vinci B. Solids Capture. Recirculating Aquaculture Systems Short Course; 2013. http://ag.arizona.edu/azaqua/ista/ISTA7/RecircWorkshop/Workshop%20PP%20%20&%20Misc%20Papers%20Adobe%202006/5%20Solids%20Capture/Solids%20Control.pdf (accessed 15 November 2013).

[71] Bowser PR. General Fish Health Management. Northeastern Regional Aquaculture Center 2012;111 p16.

[72] Pietrak M, Leavitt D, Walsh M. Biosecurity on the Farm-Guidelines & Resources for Developing a Biosecurity Plan. Northeastern Regional Aquaculture Center 2010;208 p6.

[73] Figueras MJ, Borrego JJ, Pike EB, Robertson W, Ashbolt N. Sanitary Inspection and Microbiological Water Quality. In: Bartram J. and Rees G. (eds.) Monitoring Bathing Waters-A Practical Guide to the Design and Implementation of Assessments and Monitoring Programmes. In E and FN Spon; 2000. p311.

[74] Thompson FL, Iida T, Swings J. Biodiversity of Vibrios. Microbiology and Molecular Biology Review 2004;68(3) 403-431.

[75] Austin. Vibrios as causal agents of zoonoses. Veterinary Microbiology 2010;140(3-4) 310-317.

[76] Urakawa H., Rivera ING. Aquatic Environment. In: Thompson FL, Austin B, Swings J. (eds.) *The Biology of Vibrios*, D.C. ASM Press; 2006. p175–189.

[77] Austin, B. Miscellaneous animal pathogens. In: Thompson, F.B., Austin, B. Swings, J. (eds.) The Biology of Vibrios In ASM Press; 2006. p423.

[78] Grimes DJ. Ecology of estuarine bacteria capable of causing human disease: a review. Estuaries 1991;14(4) 345-360.

[79] Baker-Austin CL, Stockely R, Rangdale, Martinez-Urtaza J. Environmental occurrence and clinical impact of *Vibrio vulnificus* and *Vibrio parahaemolyticus*: A European perspective. Environmental Microbiology Reports 2010;2 7-18.

[80] Lipp EK, Huq A, Colwell RC. Effects of Global Climate on Infectious Disease: the Cholera Model. Clinical Microbiology Reviews 2002;15(4) 757-770.

[81] Moyle PB, Cech JJ. 2004. Fishes: an introduction to ichthyology. Prentice Hall; 2004. p 465, 615.

[82] Boyd CE, Boyd C, Chainark S. From The Shrimp Book: Shrimp Pond Soil and Water Quality Management. Global Aquaculture Advocate 2012;15(1):28-31.

[83] Kalff, J. Limnology: inland water ecosystems. Prentice Hall; 2002. p592.

[84] Shine C, de Klemm C. 1999. Wetlands, Water and the Law. Using law to advance wetland conservation and wise use. IUCN Environmental Policy and Law Paper 1999; p330.

[85] United States Environmental Protection Agency. USEPA: Economic Benefits of wetlands. http://www.epa.gov/owow/wetlands (accessed 8 October 2013).

[86] United States Environmental Protection Agency. USEPA: Nutrient Criteria Technical Guidance Manual- Wetlands. http://www2.epa.gov/sites/production/files/documents/wetlands_full.pdf (accessed 6 November 2013).

[87] Rückauf U, Augustin J, Russow R, Merbach, W. Nitrate removal from drained and reflooded fen soils affected by soil N transformation processes and plant uptake. Soil Biology and Biochemistry 2004;36(1) 77-90.

[88] Reddy KR, Delaune RD. Biogeochemistry of Wetlands: Science and Applications. CRC Press; 2008. http://www.crcpress.com. (accessed 15 October 2013).

[89] Mitsch WJ, Gosselink JG. Editors. Wetlands. In John Wiley and Sons, Inc., New York, NY. USA. 2007. p600.

[90] Moshiri S. Identification of resistance source in barley to certain fungal pathogen in Khorasan 1995: conference proceedings, September 2-7, 1995, Karadj.

[91] Schwartz MF, Boyd CE. Constructed wetlands for treatment of channel catfish pond effluents. The progressive fish-culturist 1995;57(4) 255-266.

[92] Boyd CE. Hydrology and pond construction, In: Tucker CS. (ed.) Channel Catfish Culture. Elsevier; 1985. p107-134.

[93] Shpigel M, Ben-Ezra D, Shauli L, Sagi M, Ventura Y, Samocha T, Lee JJ. Constructed wetland with Salicornia as a biofilter for mariculture effluents. Aquaculture 2013;412 52-63.

[94] Lymbery AJ, Kay GD, Doupé RG, Partridge GJ, Norman HC. The potential of a salt-tolerant plant (Distichlis spicata) to treat effluent from inland saline aquaculture and provide livestock feed on salt-affected farmland. Science of The Total Environment 2013;445 192-201.

[95] Serpa D, Pousão-Ferreira P, Caetano M, Cancela da Fonseca L, Dinis MT, Duarte P. A coupled biogeochemical-Dynamic Energy Budget model as a tool for managing fish production ponds. Science of the Total Environment 2013;463 861-874.

[96] Boyd CE. Potential of sodium nitrate to improve environmental conditions in aquaculture ponds. Journal of the World Aquaculture Society 1995b;26 38-40.

[97] Cathcart TP, Pote JW, Rutherford DW. Reduction of effluent discharge and groundwater use in catfish ponds. Aquacultural Engineering 1999;20(3) 163-174.

[98] USDA/Natural Resources Conservation Service. [DVD]: Alabama Aquaculture Best Management Practices (BMP). Settling Basins and Wetlands 2002;BMP No. 6.

[99] Tucker CS. Aquaculture Effluents and the Environment. Aquaculture Meeting; 2009 NC USA p61.

[100] United States Environmental Protection Agency. USEPA: NPDES. Final Permit Fact Sheet. http://www.epa.gov/region9/water/npdes/pdf/ca/GoletaSdNpdesPermitFinal-Order.pdf (accessed 11 October 2013).

[101] Teichert-Coddington DR, Rouse DB, Potts A and Boyd CE. Treatment of harvest discharge from intensive shrimp ponds by settling. Aquacultural Engineering 1999;19 147-161.

[102] Ozbay G, Boyd CE. Treatment of Channel Catfish Pond Effluents in Sedimentation Basins. The World Aquaculture Society Magazine 2004;35(3)10-13.

[103] Ozbay G, Boyd CE. Particle Size Fractions in Pond Effluents. The World Aquaculture Society Magazine 2003a;34(4) 56-59.

[104] Ozbay G, Boyd CE. U.S. Study Finds Turbidity Useful in TSS, ISS Estimates. Global Aquaculture Advocate 2003b;6(4) 58-60.

[105] Boyd CE. Aquaculture sustainability and environmental issues. World Aquaculture 1999;30(June) 10-15.

[106] Hargreaves J, Cathcart T, Boyd C, Engle C, Romaire R, McClain R, Boardman G, Browdy C. Management of Aquacultural Effluents from Ponds. SRAC Sixteenth Annual Progress Report 2003; p18.

[107] Boyd CE. Farm effluent during draining for harvest. Global Aquaculture Advocate 2000;3(4) 26-27.

[108] Schwartz MF, Boyd CE. Suggested management to improve quality and reduce quantity of channel catfish pond effluents. Alabama Agriculture Experiment Station; 1996. Auburn University, Alabama, USA. Leaflet 108.

[109] Lutz G, LeBlanc BD, Sheffield R, Nix KE. Aquaculture Environmental Best Management Practices. BMPs. LSU Ag Center Communications; 2011. Louisiana State University, LA, USA. p28.

[110] Tucker CS, Hargreaves JA. (eds.). Environmental Best Management practices for aquaculture. John Wiley and Sons: Blackwell Publishing; 2008. p582.

[111] Setty K. Design Manual: Retention Basin. Bren School of Environmental Science and Management; 2013. University of California, Santa Barbara, CA USA. p7.

[112] Boyd CE, Tucker CS. Sustainability of channel catfish farming. World Aquaculture 1995;2645-53.

[113] Cripps SJ. The application of suspended particle characterization techniques to aquaculture systems. In: Wang, J. (ed.) Techniques for Modern Aquaculture; 1993. p26–34.

[114] Cripps SJ. Serial particle size fractionation and characterization of an aquacultural effluent. Aquaculture 1995;133 323–339.

[115] Ackefors H, Enell M. The release of nutrients and organic matter from aquaculture systems in Nordic countries. Journal Applied Ichthyology 1994;10(4) 225-241.

[116] Cripps SJ. 1992. Characterization of an aquaculture effluent based on water quality and particle size distribution data 1992: conference proceedings, May 21-25, 1992. Orlando, FL, USA.

[117] Bergheim A, Tyvold T, Seymour EA. 1991. Effluent loadings and sludge removal from land based salmon farming tanks. In: Bredene EAS (ed.) Europe'91: Aquaculture and the Environment International Aquaculture Conference: proceedings of Aquaculture, Dublin, Ireland, 10–12 June, 1991. Belgium: EAS Special Publication.

[118] Cripps SJ. Minimizing outputs: treatment. Journal of Applied Ichthyology 1994;10 284-294.

[119] Cripps SJ., Kelly L. Reduction in wastes from aquaculture. In: Baird DG, Beveridge M, Kelly GA, Muir BF. (eds.) Aquaculture and Water Resource Management. Blackwell Scientific; 1996. p219.

[120] Pepper VA, Sutterlin AM, Nicholls T, Collier C. Newfoundland experience with development of all-female and non-reproductive Atlantic salmon for marine aquaculture. Bulletin-Aquaculture Association of Canada 1996;14-23.

[121] Moore PA, Miller DM. Decreasing phosphorus solubility in poultry litter with aluminum, calcium, and iron amendments. Journal of Environmental Quality 1994;23(2) 325-330.

[122] Gensemer RW, Playle RC. The bioavailability and toxicity of aluminum in aquatic environments. Critical reviews in environmental science and technology 1999;29(4) 315-450.

[123] Welch EB, Cooke GD. Effectiveness and longevity of phosphorus inactivation with alum. Lake and Reservoir Management 1999;15(1) 5-27.

[124] Cooke GD., Welch EB., Peterson SA., Newroth PR. editors. Restoration and Management of Lakes and reservoirs. Lewis Publishers: CRC Press; 1993. p548.

[125] Jacoby JM, Gibbons HL, Hanowell R and Bouchard DD. Wintertime blue-green algal toxicity in a mesotrophic lake. Journal of Freshwater Ecology 1994;9(3) 241-251.

[126] Rowan M. Chemical phosphorus removal from aquaculture pond water and effluent. PhD thesis. Auburn University Auburn; 2001. p180.

[127] Masuda K, Boyd CE. Effects of aeration, alum treatment, liming, and organic matter application on phosphorus exchange between soil and water in aquaculture ponds at Auburn, Alabama. Journal of the World Aquaculture Society 1994;25(3) 405-416.

[128] Prepas EE, Murphy TP, Crosby JM, Walty DT, Lim JT, Babin J, Chambers PA. Reduction of phosphorus and chlorophyll a concentrations following calcium carbonate

and calcium hydroxide additions to hypereutrophic Figure Eight Lake, Alberta. Environmental Science and Technology 1990;24(8) 1252-1258.

[129] Salonen VP, Varjo E. Gypsum treatment as a restoration method for sediments of eutrophied lakes–experiments from southern Finland. Environmental Geology 2000;39(3-4) 353-359.

[130] Wu R, Boyd CE. Evaluation of calcium sulfate for use in aquaculture ponds. The Progressive Fish-Culturist 1990;52(1) 26-31.

[131] Ann Y, Reddy R, Delfi JJ. Influence of chemical amendments on phosphorus immobilization in soils from a constructed wetland. Ecological Engineering 2000;14 157–167.

[132] Gutcho SJ. Waste Treatment with Polyelectrolytes and Other Flocculants No. 31. Noyes Data Corp 1977; Technology & Engineering. University of Minnesota, p24.

[133] Ozbay G. Effects of Coagulant Treatments on Aquaculture Effluent Quality. Journal of Applied Aquaculture 2005;17(4) 1-23.

[134] Wright RT, Coffin RB, Ersing CP, Pearson D. Field and laboratory measurements of bivalve filtration of natural marine bacterioplankton. Limnology and Oceanography 1982;27(1) 91-98.

[135] Lin CK., Shrestha MK., Thakur DP., Diana JS. Management to minimize the environmental impacts of pond draining. In: Burke D., Baker J., Goetze B., Clair D. and Egna H. (eds.) Fifteenth Annual Technical Report; 1998. p132-139.

[136] Shpigel M, Fridman R. Propigation of the Manila clam, Tapes semidecussatus, in the effluent of fish aquaculture ponds in Eliat, Israel. Aquaculture 1990;90 113-122.

[137] Newell RIE, Cornwell JC, Owens M, Tuttle J. Role of oysters in maintaining estuarine water quality. National Shellfisheries Association. Journal of Shellfish Research 1999;18(1) 300-301.

[138] Rensel J, Bright K, Siegrist Z. Integrated Fish-Shellfish Mariculture in Puget Sound. Final Report; 2011. p82.

[139] Chrzanowski TH, Spurrier JD, Dame RF, Zingmark RG. Processing of microbial biomass by an intertidal reef community. Marine Ecology-Progress Series 1986;30 181-189.

[140] Toro JE, Paredes PI, Villiagra JD, Senn CM. Seasonal variation in the planktonic community, seston and environmental variables during a 2-year period and oyster growth at two mariculture sites, southern Chile. Marine Ecology 1999;20(1) 63-89.

[141] Ulanowicz RE, Tuttle JH. The trophic consequences of oyster stock rehabilitation in Chesapeake Bay. Estuaries 1992;15(3) 298-306.

[142] Leffler M. Improving Bay Water Quality: The role of oyster reefs. Maryland Sea Grant. Maryland Aquafarmer 2001. http://www.mdsg.umd.edu/Extension/Aquafarmer/ Winter01.html (accessed 10 October 2013).

[143] Miura T, Yamashiro T. Size selective feeding of Anodonta calipygos, a phytoplanktivorous freshwater bivalve, and visibility of egested algae. Japanese Journal of Limnology 1990;51(2) 73-78.

[144] Lowe R, Reich LA, Sierra J. Algal herbivory by the zebra mussel: Fate of algae in feces and pseudofeces. Journal of Shellfish Research 1991;10(1) 250-251.

[145] Senichieva MI. Dynamics of phytoplankton community in the region with mussel mari-farming. Hydroores 1990;7(8) 86-89.

[146] Santelices B, Martinez. The effect of filter feeders on macroalgae spores. 2. Indirect effects and landscape determination. Biota 1986;1:102.

[147] Marenghi FP, Ozbay G. Floating Oyster, *Crassostrea virginica* Gmelin 1791, Aquaculture as habitat for fishes and macroinvertebrates in Delaware Inland Bays: The comparative value of oyster clusters and loose shell. Journal of Shellfish Research 2010a; 29(4) 889-904.

[148] Marenghi FP, Ozbay G. Preliminary habitat assessment of floating oyster (Crassostrea virginica) gardens (Delaware). Ecological Restoration 2010b;28(3) 254-257.

[149] Harding JM, Mann R. Estimates of naked goby (*Gobiosoma bosc*), striped blenny (*Chasmodes bosquianus*) and eastern oyster (*Crassostrea virginica*) larval production around a restored Chesapeake Bay oyster reef. Bulletin of Marine Science 2000;66(1) 29-45.

[150] Tolley SG, Volety AK. The role of oysters in habitat use of oyster reefs by resident fishes and decapod crustaceans. Journal of Shellfish Research 2005;24(4) 1007-1012.

[151] Breitburg DL. Are three-dimensional structure and healthy oyster populations the keys to an ecologically interesting and important fish community. Oyster reef habitat restoration: a synopsis and synthesis of approaches; 1999. Virginia Institute of Marine Science Press, Gloucester Point, Virginia. p239-250.

[152] Posey DA. Developing sui generis options for the protection of living aquatic resources of indigenous and local communities. In: ICLARM Conference Proceedings, 1999.http://www.fao.org/ag/againfo/programmes/en/genetics/documents/ITWG-AnGR4/MontpellierAnGR-Papers.pdf (accessed 8 November 2013).

[153] Peterson CH, Grabowski JH, Powers SP. Estimated enhancement of fish production resulting from restoring oyster reef habitat: quantitative valuation. Marine Ecology-Progress Series 2003;264 249–264.

[154] The Nature Conservancy. 2013. Oyster Restoration Reaches New Depth. New Hampshire. Do you know how many gallons of water an oyster can filter in a day? http://

www.nature.org/ourinitiatives/regions/northamerica/unitedstates/newhampshire/explore/oyster-restoration-reaches-new-depths.xml.

[155] Ozbay G, Reckenbeil B, Marenghi F, Erbland P. Can't Go Wrong With Eastern Oysters (*Crassostrea virginica*): Enhancing Biodiversity of the Delaware Inland Bays. Journal of Biodiversity and Endangered Species 2013;1 2.

[156] Reckenbeil BA. Assessment of oyster restoration along human altered shorelines in the Delaware Inland Bays: An examination of riprap stocked with the eastern oyster (Crassostrea virginica). Master's thesis Delaware State University; 2013. p.174.

[157] Rice MA. Environmental effects of shellfish aquaculture in the Northeast. NRAC Publication Fact Sheet 105; 2008.

[158] Lin CK, Ruamthaveesub P, Wanuchsoontorn P. Integrated culture of the green mussel in wastewater from an intensive shrimp pond: concept and practice. World Aquaculture 1993;24 68-72.

[159] Tseng KF, Huang JS, Liao I. Species control of microalgae in an aquaculture pond. Water Research 1991;25(11) 1431-1437.

[160] Burke JS, Bayne DR. Impact of paddlefish on plankton and water quality of catfish ponds. The Progressive Fish Culturist 1986;48 177-183.

[161] Smith NP. The suitability of routine weather data for estimating local estuarine heat energy fluxes. Estuaries 1985;8 270-278.

[162] Fott J., L. Pechar and M. Prazakova. Fish as a factor controlling water quality in ponds. In: Barica J., Mur LR. (eds.) SIL Workshop on Hypertrophic Ecosystems, Vaxjo, Sweden; 1979. Sept. 10-14.

[163] Canadian Aquaculture. Integrated Multi-trophic Aquaculture. What is it? Canadian Aquaculture Industry Alliance; 2012. http://www.aquaculture.ca/files/species-multi-trophic.php. (accessed 11 October 2013).

[164] Gowen RJ. Aquaculture and the environment. In: De Pauw N., Joyce J. (eds.) Aquaculture and the Environment. European Aquaculture Society Special Publication 1991;16 23 – 48.

[165] Filbrun JE, Culver DA. Can Reduced Provision of Manufactured Feed Improve Fish Production Efficiency in Ponds? North American Journal of Aquaculture 2013; 1 64-76.

[166] Gross A, Boyd CE, Wood CW. Ammonia volatilization from freshwater fish ponds. Journal of Environmental Quality 1999;28(3) 793-797.

[167] Abou Y, Saidou A, Mama D, Fiogbé ED, Micha JC. Evaluation of Nitrogen and Phosphorus Wastes Produced by Nile Tilapia (Oreochromis niloticus L.) Fed Azolla-Diets in Earthen Ponds. Journal of Environmental Protection 2012;3(6) 502-507.

[168] Gross A, Boyd CE, Lovell RT, Eya JC. Phosphorus budgets for channel catfish ponds receiving diets with different phosphorus concentrations. Journal of the World Aquaculture Society 1998;29(1) 31-39.

[169] Rangen. Salmon Feeds/ Trout Feeds/ Tilapia Feeds/ Quality Feed for Catfish. Rangen Inc; 2013 Buhl, ID, USA. http://www.rangen.com.

[170] Boyd CE, Schmittou HR. Achievement of sustainable aquaculture through environmental management. Aquaculture Economics and Management 1999;3 59-69.

[171] Goldburg R. Triplett T. Murky water: environmental effects of aquaculture in the United States. Environmental Defense Fund, New York, USA; 1999.

[172] Frankic A, Hershner C. Sustainable aquaculture: developing the promise of aquaculture. Aquaculture International 2003;11 517-530.

[173] Costa-Pierce, B. A. Ecology as the paradigm for future of aquaculture. In: Barry AC., (ed.) Ecological Aquaculture: The evolution of the blue revolution. Blackwell; 2002 Science, MA USA. p. 339-372.

[174] Martins CIM, Eding EH, Verdegem MCJ, Heinsbroek LTN, Schneider O, Blancheton JP, Roque D'Orbcastel E, Verreth JAJ. New developments in recirculating aquaculture systems in Europe: a perspective on environmental sustainability. Aquaculture Engineering 2010;43 83-93.

[175] Chen S., Summerfelt S., Losordo T., Malone R. Recirculating systems, effluents, and treatments. In: Aquaculture and the environment in the United States; 2002. p29.

[176] Hollingsworth CS, Baldwin R, Wilda K, Ellis R, Soares S. Best management practices for finfish aquaculture in Massachusetts. Publication AG-BPFA; 2006. Western Massachusetts Center for Sustainable Aquaculture. University of Massachusetts Extension.

[177] Badiola M, Mendiola D, Bostock J. 2012. Recirculating aquaculture systems (RAS) analysis: main issues on management and future challenges. Aquaculture Engineering 2012;51 26-35.

[178] Timmons MB, Ebeling JM, Wheaton FW, Summerfelt ST, Vinci BJ. Recirculating Aquaculture Systems. Northeastern Regional Aquaculture Center. In Cayuga Aqua Ventures; 2001. Ithaca, NY 14850.

[179] Timmons MB, Ebeling JM. Recirculating Aquaculture Systems. In Northeastern Regional Aquaculture Center; 2007. 2nd Ed. Cayuga, AquaVentures LLC. Ithaca, NY 14850.

[180] Ebeling JM, Timmons MB. Recirculating Aquaculture Systems. In: James HT. (ed.) Aquaculture Production Systems. John Wiley and Sons; 2012. p245-277.

[181] Summerfelt ST., Vinci. Better Management Practices for Recirculating Aquaculture Systems. In: Craig ST., John AH. (eds.) In. Environmental Best Management Practices for Aquaculture. Blackwell Publishing; 2008. p389-426.

[182] Losordo T, Haider D, Tran H. Recirculation meets aquaponics: Engineering flexibility into fish/plant production systems. The Global Aquaculture Advocate 2013;16 88-90.

[183] Miller D, Semmens K. Waste management in aquaculture. Publication # AQ02-1. Aquaculture information series: 2002. Agriculture and Resource Economics Program, Division of Resource Management. College of Agriculture, Forestry, and Consumer Sciences. West Virginia University, Morgantown, WV.

[184] Valencia E, Adjei M, Martin J. Aquaculture effluent as a water and nutrient source for hay production in the seasonally dry tropics. Community Soil Science Plant Annals 2001;32 1293-1301.

[185] Global Aquaculture Alliance. GAA: BAP Standards. www.galliance.org. (accessed 5 October 2013).

[186] Hanna K, Wheaton F, Lazur A, Vankeuren S. Water quality in identical recirculating systems managed by different aquaculturists. International Journal of Recirculating Aquaculture 2012;11 55-74.

[187] United States Department of Agriculture, National Institute of Food and Agriculture. USDA-NIFA: Legal definition of sustainable agriculture (US code Title 7, Section 3103). www.csrees.usda.gov/nea/ag_systems/in_focus/sustain_ag_if_legal.html (accessed 8 November 2013).

[188] McMurtry MR, Nelson PV, Sanders DC, Hodges L. Sand culture of vegetables using recirculated aquaculture effluents. Applied Agriculture Research 1990;5 280-284.

[189] Rakocy JE., Aquaponics-Integrating Fish and Plant Culture. In: James HT. (ed.) Aquaculture Production Systems. John Wiley and Sons; 2012. p343-386.

[190] Rakocy JE, Losordo TM and Masser MP. Recirculating aquaculture tank production systems: integrating fish and plant culture. SRAC Publication 1992: Southern Region Aquaculture Center, Mississippi State University, Stoneville, MS. p1-11.

[191] Rhodes MA, Hanson T, Alavi S, Davis DA. High-soy, fishmeal-free diets support Florida pompano growth. The Global Aquaculture Advocate 2013;16 100-103.

[192] Watson AM, Barrows FT, Price AR. Taurine supplementation of plant derived protein and n-3 fatty acids are critical for optimal growth and development of cobia, *Rachycentron canadum*. Lipids 2013;48 819-913.

[193] Ozbay G, Jackson AR. Aquaculture Effluents: Perspectives on Best Management Practices. Global Aquaculture Advocate 2006;9(5) 68-70.

[194] Boyd CE. Guidelines for aquaculture effluent management at the farm-level. Aquaculture 2003;226(1) 101-112.

[195] Villalón JR. Practical Manual for Semi-intensive Commercial Production of Marine Shrimp. Texas A&M University Sea Grant College Program; 1991. TAMU-SG-91-501. College Station, Texas, USA. p1-104.

[196] Hopkins JS, Stokes AD, Browdy CL, Sandifer PA. The relationship between feeding rate, paddlewheel aeration rate and expected dawn dissolved oxygen in intensive shrimp ponds. Aquacultural Engineering 1991;10(4) 281-290.

[197] United States Environmental Protection Agency. USEPA: Effluent Guidelines: Aquatic Animal Production Industry. http://water.epa.gov/scitech/wastetech/guide/aquaculture/index.cfm (accessed 8 November 2013).

[198] Northwest Earth Institute. 2012. Discussion Course on Choices for Sustainable Living. Northwest Earth Institute, Portland, OR USA. p138.

[199] Newell CR, Hawkins AJS, Morris K, Richardson J, Davis C, Getchis T. ShellGIS: A Dynamic Tool for Shellfish Farm Site Selection. World Aquaculture 2013;44(3) 50-53.

[200] ESRI News. 2013. Esri News for Agriculture. Clearwater Seafoods Achieves Sustainable Operations through GIS. Courtesy of Esri Canada Limited. Spring 2013.

[201] Thailand DOF Information System Center. Fisheries Statistics of Thailand 2010; 2011. Department of Fisheries. Bangkok, Thailand. http://www.fisheries.go.th/dof/en/ (accessed 10 October 2013).

[202] Bureau of Agricultural Economic Research.Situation of Agricultural products and Trend in the Year 2011. Office of Agricultural Economics, Bangkok, Thailand.

[203] Center for Agricultural Information. Thailand Foreign Agricultural Trade Statistics 2011; 2011. Office of Agricultural Economics, Bangkok, Thailand.

[204] Kongkeo H. Current Status and Development Trends of Aquaculture in the Asian Region. In: Subasinghe RP, Bueno P, Phillips MJ, Hough C, McGladdery SE, Arthur JR. (eds.) 2001: Aquaculture in the Third Millennium. Technical Proceedings of the Conference on Aquaculture in the Millennium, 20–25 February 2000, NACA, Bangkok and FAO, Rome, Italy.

Chapter 2

Development of Freshwater Native Species with Aquacultural Potential

Fernando Garcia-Trejo, Silvia Hurtado-Gonzalez,
Genaro M. Soto-Zarazua and P. J Gutierrez-Yurrita

Additional information is available at the end of the chapter

http://dx.doi.org/10.5772/57215

1. Introduction

Aquaculture is an activity practiced by Chinese people since 2800 B.C. The first culture species were the carp (*Cyprinus carpio*) and mullet (*Mugil cephalus*), the first documented knowledge about fish culture in captivity belong to the carp. This activity has been supporting human demands for fish products for centuries and now is an important worldwide industry. Over the years this practice has become more technical with the objective to make the work easily but mainly in order to increase production. Nowadays this activity has grown to an entire industry that handles both supplements as the product itself. Global production from aquaculture now supplies one third of seafood consumed worldwide. With this massive increase in world production the current aquaculture industry is one of the fastest growing sectors in food production [1]. However, this rapid growth in the aquaculture industry has generated ecological damage due to a huge extractive use of water, land, and feeds. Besides, is important to consider that these requirements are associated to another impacts such as; polluting, salinization of soils, nutrient-loading, clearing of natural habitat, overexploit of ground water reserves, introduction and transmission of diseases [2]. At this respect Bailey [3] establish a new term "blue revolution", which describes the expansion of fish-farming in tropical regions, according to this idea aquaculture must provide huge quantities of fish and help to solving problems of world food security and alleviating poverty. However, production increase in aquaculture demands feeds, energy for the cultured species and almost always is obtained through catch, so the fastest growing sectors in food production threatens its ability to continue to provide increasing yields in a sustainable manner, and concerns with the resulting from fish-farming have led to calls for the "greening of the blue revolution" [4].

The situation in México is not an exception; the expansion of aquaculture has been accompanied by degradation of the natural environment, especially on marine aquaculture [5]. As it happens in the world directly impacts of fisheries and aquaculture are:

- Introduction of nonnative fish species:
- Introduction of exogenous parasites:
- Nutrient pollution:
- Habitat modification:
- Overcollection of wild seed stock:
- Changes of food webs
- Increase of interspecific competition:

It is clear that current food productions techniques in aquaculture are good just under financial point of view but always leave aside the environment aspects. The relationship between aquaculture and environment is complex specifically the biodiversity topic. Many examples of positive and negative impacts have been documented, however until now there is no solution which allows the development of a relationship between food production and the environment. This solution must be adopting a new paradigm based in ecological concepts of extreme resource efficiency and the closing of nutrient and waste cycles, resource-use optimization [6]. As can be seen, this is not an easy task since it requires the creation of multidisciplinary teams which can see the problem holistically and try to give a solution that benefits all parties involved in the process. The efforts and the perception of the environment are different between countries, but ultimately the problem to be addressed holistically. However, studies on fish typically focus on species that currently have commercial value, causing species that lack such market value to be ignored. This is the case of several freshwater native species, which can be founded in central and South America. Some attempts to cultivate native species have occurred mainly in areas or rural communities, where in addition to enhance the conservation of species protein contributes to the diet of the community [7]. One of the most interesting case studies in Mexico is growing "white fish" (*Chirostoma estor*) with the aim repopulate some areas where the introduction of their populations has been declined [8]. Most documented is that of the native Central American cichlid (*Cichlasoma urophthalmus*), of which there have been many studies to support its culture [9]. In southern region some attempts to grow some silversides in Argentina (*Odontestes Basilichthys*), some Characidae family members have been grown in Brazil and more recently three species of carnivorous cichlids aquarium purposes. In Peru it has favored the cultivation of called piracucu (*Arapaima gigas*), one of the largest fish of fresh water. Possibly the Cichlidae family members are those that show the greatest potential for cultivation [10, 11].

The main purpose of this chapter is to show the experience of three studies with native species; one refers to a small native species located in the state of Querétaro and with a great ecological importance, *Girardinichthys multiradiatus* [12]. The study of this fish focused on the description of its habitat throughout a hydrologic cycle in which ecophysiological responses were determined in order to stablish guidelines for its management and to preserve its population. In this work population structure and dynamics were getting and trophic and ecophysiological responses to fluctuations in environmental factors were also identified in order to have the possibility of laboratory reproduction and growth. On the other hand, native mojarra *Herinchthys cyanogutattus* founded on the basin Pánuco river. In this case the purpose was to evaluate its useful in fishery and later in the aquaculture. The work consisted of two stages: First, the characterization of their environment in order to locate stable populations of the mojarra and to characterize ecologically its habitat. Second, the mojarra was moved to the laboratory to try different forms of acclimatization for its future use giving them tried food. Once acclimated, the stock was use to carried out density studies of individuals for culture (capacity of load), as well as of ideal thermal for its production. Finally, *Procambarus digueti* which faces severe ecological problems (over fishing, no control of heights neither of sexes, there is not articulated extraction methods neither fishing seasons and restrains), since they are captured as food and as curative remedies from pre-Hispanic eras. To this situation the strong environmental pressure is added by the disturbance of their habitat, what has carried a decrease in numerical abundance. The objective of this work was determined the optimum cultivation conditions with respect to the load capacity and diet in the growth of *P. digueti* in intensive production systems. The controlled production of this specie will reduce the fishing pressure and it will be able to serve to repopulate the sites where may have been decline the natural population.

To successfully achieve the cultivation of a native species, compared to the technological advantages offered by more exotic trading requires knowledge of the biology, ecology and aquaculture potential (ability to live at high densities, accept food encapsulation, and withstand high environmental variations) of each species. If aquaculture potential studies are performed with ecophysiological and bioenergetics approach may be developed predictive models of how to develop a population under different environmental factors, and even develop experimentally testable hypotheses [13].

2. The experience with Girardinichthys multiradiatus

Studies on the biological aspects of fish typically focus on species that currently have commercial value, causing species that lack such market value to be ignored. This is the case of several freshwater fish, specifically of several members of the Goodeidae family. This is a diversified and small family of cyprinodontoid fish, confined to the central plateau of Mexico where its dispersion center lies in the well-isolated Lerma basin. Four species of the Godeidae family have been reported in Querétaro: three species are distributed widely in the Lerma basin (*Goodea atripinis, Xenotoca variata, Goodea gracilis*), and one species (*Girardinichthys multiradiatus*) can only be found in one body of water in the municipality of Amealco [14]. Scientific knowledge about it focuses on sexual dimorphism, peculiar courtship rituals, and viviparity [15], taxonomic aspects [16, 17], ethology [18], biology [19], and trophic ecology [20].

2.1. Environmental conditions

G. multiradiatus was founded on San Martín Dam, located at 60 km south of Queretaro city, near to Amealco municipality (100° 09′ 43″ W; 20° 15′ 02″ N), at 2600 meters above sea level. The climate is subhumid with summer rains (Cw1) with an average temperature of 15.1°C, the months of May and June are those with the highest temperatures. The average annual rainfall is 659.5 mm, occurring mainly during the summer [21]. The main contribution to the dam water is from rain. Sampling was made over a full hydrological cycle (one year, beginning in February) in which the *G. multiradiatus* population was monitored once every two months, at the same time physical factors were measure (T°, pH, dissolved O_2, turbidity, depth). The physicochemical parameters of water showed stable behavior during the studied hydrological cycle (pH = 7 to 9; dissolved oxygen = 6.5 to 7 ppm). On the other hand, the temperature showed significant variation, with the highest temperatures (20-25°C) recorded between April and August, with the lowest (10-18°C) recorded between October to February. Also, due to seasonal differences in water usage, the water level of the dam was low from April to August and high from October to March (Fig. 1).

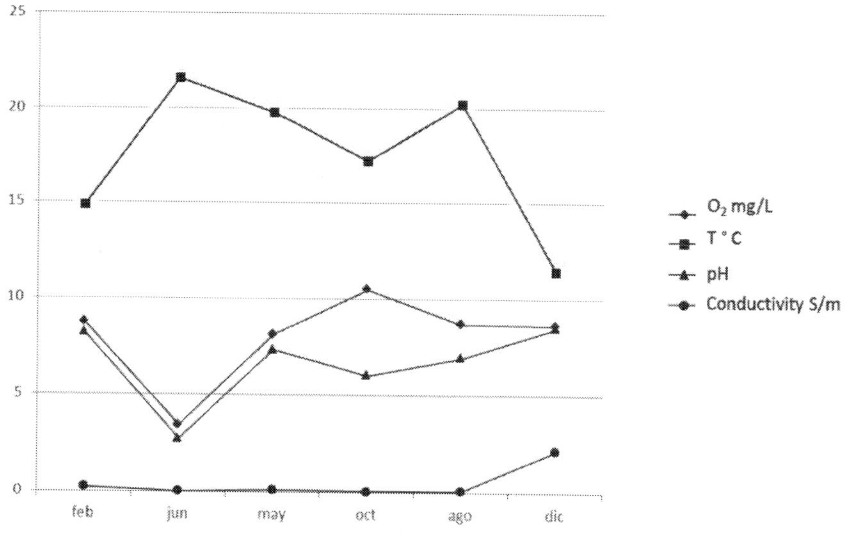

Figure 1. Environmental parameters during a hydrological cycle in San Martín Dam.

2.2. Population ecology

According to the Cassie method, the population of *G. multiradiatus* consists of 12 classes, ranging from 8 to 48 mm standard length. Figure 2 shows the general structure of the mexcalpique population of San Martín. The numbers in parentheses indicate the percentage of each size class of the total population obtained through a year. Two of these size classes were found only in laboratory studies due to their small sizes; these sizes were smaller than what the nets in situ could catch.

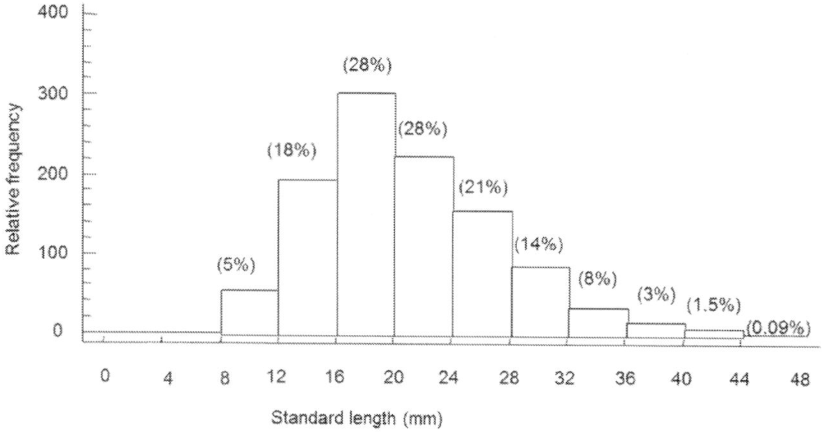

Figure 2. General structure of the population of San Martín mexcalpique. The numbers in parentheses indicate the percentage of each size class of the total population obtained through a year.

Figure 3 shows the bimonthly structure of the population of *G. multiradiatus* in San Martín and the seasonal pattern of population growth, demonstrated in Von Bertalanffy equation with $L\infty = 47$; $K= 0.8870$ and, $t0 = -0.2103$. The variations throughout the year are present in both, in the structure of the population and its growth rate, with the shorter pattern cohorts having a higher growth rate and no increase in length during the period of February to June.

2.3. Feeding habits

To assess the daily feeding activity of the *G. multiraditus* in San Martín Dam samples were collected with spoon nets every four hours during a period of 24 hours (10:00, 14:00, 18:00, 22:00, 02:00 and, 06:00 hours). These catches allowed determining the feeding ecology of the species (feeding time, type of diet at different times of day, food components). The relative density in activity was measured using the catch per unit effort method (CPUE) based on the number of individuals caught by dragging. The fish were fixed in 70% alcohol and then transported to the laboratory (Nielsen et al., 1983). From 1022 stomachs analyzed, (1022), 18 food components were identified. The most abundant component found were insects (47%), especially the Diptera order, followed by detritus (24.0%) and Cladocera (17.5%), with the remaining components accounting for 10.7%. Unusual food components (less than 10% of the total), were only found at a specific times of the hydrological cycle, Table 1. The benthic review showed 20 components and trophic index was calculated indicating that *G. multiradiatus* is a

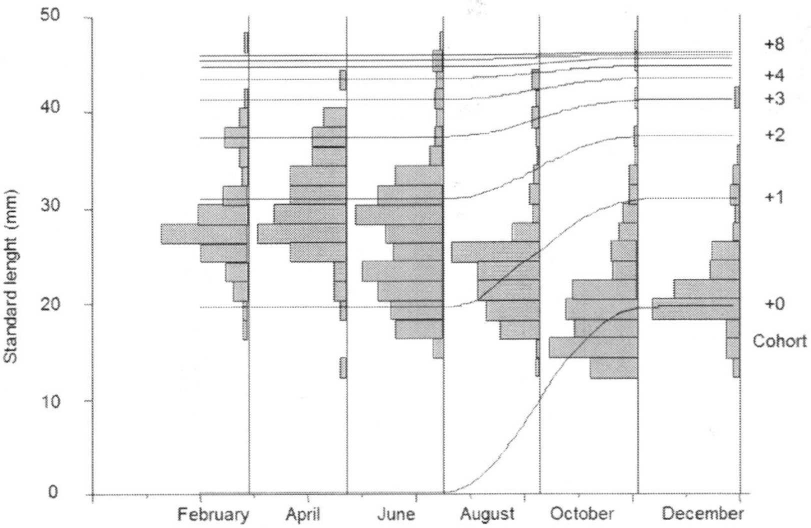

Figure 3. Bimonthly structure of the population of G. multiradiatus in San Martin and the curves of growth with the von Bertalanffy model for highly seasonal cycles. Upon reaching the asymptotic curve determines the final class of each age cohort.

polytrophic specie ($H \neq 0$). The maximum numbers of categories found in their stomachs were eight while minimum was two. Circadian sampling showed that this specie is polytrophic with two daily feeding periods (5:00 to 8:00 and 14:00 to 18:00 hours).

2.4. Bioenergetic

To quantify the aerobic metabolism and nitrogen excretion, animals were placed in a semi closed system with (0.5L) respirometric chambers, using a method which assumes that the reduction of oxygen and the increase of nitrogen in the chamber depend on the weight of the animal, the volume of water, the period of time in which no water circulated in the chamber and the ambient temperature [22]. After each cycle, sample was oven-dried in an electric oven between 70-80°C until the samples had constant weight. From each composite sample 2 g were measured and taken as analytical sample; the samples were digested with concentrated nitric acid. The determination of the percentage proximate composition was chemically analyzed according to the method of analysis described by the Association of Official Analytical Chemist [23]. While other 2 g sample tissue were combusted in a Parr bomb calorimeter to obtain oxycalorific measurements. The flow of energy that was used to determine the energy efficiency and assimilation is show in the next equation:

$$C = P + G + ER + EU + F$$

Where; C is the total energy content of food consumed, P and G are the energy equivalents of somatic and gonadal growth respectively, ER is the energy utilized in respiration, EU is the

Category	Youth N = 654		Males N = 195		Females N = 173	
	n	%	n	%	n	%
Detritus	230	16.08	141	37.03	127	38.49
Diptera	517	55.31	146	34.32	142	34.41
Cladocera	231	21.5	51	13.73	38	13.73
Copepoda	53	3.78	25	2.26	31	2.26
Animal remains	41	1.27	29	2.56	47	2.56
Amphipoda	20	0.67	27	3.1	34	3.01
Hemiptera	16	0.52	27	2.96	38	2.96
Odonata	13	0.38	21	3.21	28	3.21
Himenoptera	1	0.06	4	0.11	12	0.66
Coleoptera	0	0	2	0.04	6	0.6
Crustacea	1	0.004	4	0.41	4	0.14
Ephemeroptera	3	0.06	0	0	5	0.38
Gasteropoda	3	0.05	1	0.02	6	0.17
Plecoptera	3	0.03	2	0.11	0	0
Thrichoptera	0	0	1	0.1	0	0
Homoptera	0	0	1	0.05	0	0

Table 1. Occurrence of food components (N = number sampled, n = number of organisms that have the category, % occurrence rate) by sex.

energy lost as nitrogenous and other waste compounds excreted in the urine, and F is the unabsorbed energy voided with the faeces (Bolduc et al., 2002; Bradshaw, 2003). All variables expressed in calories per gram of dry weight (cal/g).The ratio used to transform measured aerobic metabolism into calories was the standard oxycaloric coefficient for fish which mainly excrete $N\text{-}NH_4$ (Qox = 3.20 cal mg-1 O_2). Nitrogen excretion was estimated using literary references, taking into consideration the type of fish, size, feeding habits, and physiological status [24].

Physiological experiments showed higher energy expenditures in August with values of 2500 cal/g and minimal values for December with 200 cal/g. The increased energy expenditure was found in the early hours of the day (daylight hours) and then declined, reaching minimum values at night, with the exception of October, which displayed an inverse pattern. Calorimetric analysis did not show statistically significant difference between the energy provided by sex ($p > 0.05$). Main food energy intake was 4.8 ± 0.3 Kcal g-1 of dry weight, with the total weight of the mexcalpique consisting of, on average, 85.49 ± 2.49% organic matter, and 14.50 ± 2.49% mineral matter. By replacing the caloric values in the energy balance equation, was determined that *G. multiradiatus* uses approximately 81% of the energy consumed in the production of tissue and gametes (P and G), respiration process spent 5.7% (ER) and the rest 13.3% is invested in maintenance (EU and F). Multivariate analysis of environmental factors on the metabolism, showed no significant differences, however the temperature showed the lowest value of significance ($p = 0.08$).

3. *Herinchthys cyanogutattus*

Texas Cichlids were formerly given the scientific name of *Cichlasoma cyanoguttatum*, but are now known by the name *Herichthys cyanoguttatus*. The genus *Herichthys* has been through several changes, and currently consists of nine species, native to lakes and rivers in south Texas and northern Mexico, making them the most northern naturally occurring species of cichlid in the world. It's the only native cichlid in the US and amongst the first cichlids imported to Europe, having first been imported in 1912. This species has also been introduced into areas they are not indigenous to, sometimes on purpose, but often by aquarium owners desperate to divest themselves of a fish the can no longer take care of. The areas of non-indigenous populations range from northern Texas to Florida, where it has become a popular game fish. This is due to having a tasty flavor similar to that of their distant relative, Tilapia.

In Mexico this species is called "Mojarra del norte" and It could be distinguished by a coupled of dark spots and a tiny blue circles on its sides. Adults show a olivaceus iridescent spots when viewed in the sun, there are also lines of the same color on the head, body and fins. During reproduction it is possible to see white region in the front part and a dark in the back especially in females, while males develop a prominent hump.

3.1. Biology

This fish could live in a wide range of temperature, between 5° and 30°C., [25, 26]. Trophic spectrum show variations between each population according to the region but in general is consider an omnivorous fish [27]. Many studies have described aspects of reproduction; the most relevant aspect is the monogamous behavior, when a male select a coupled it becomes aggressive and territorial [28].

3.2. Environmental conditions

The first step in this research was to looking-for a population in order to do the ecological description of its habitat. H. *cyanoguttatus* was found in several places around Queretaro State, but only one was selected due to accessibility. This place is called Taxhido river and is located at 70km E, from the Queretaro capital 20° 35′ 18″O and 99° 40′ 47″N, the climate is subhumid with summer rains (Cw1) with an average temperature of 15.1°C, the months of May and June are those with the highest temperatures. The average annual rainfall is 659.5 mm, occurring mainly during the summer and the main contribution to the dam water is from a spring. Sampling was made over a full hydrological cycle (one year, beginning in February) in which the *H. cyanoguttatus* population was monitored once every two months, at the same time physical factors were measure (T°, pH, dissolved O_2, turbidity, depth). The physicochemical parameters of water showed stable behavior during the studied hydrological cycle (pH = 7.4 to 8.2; dissolved oxygen = 4.5 to 8.3 ppm). On the other hand, the temperature was constant between 29 and 31°C., Fig. 4.

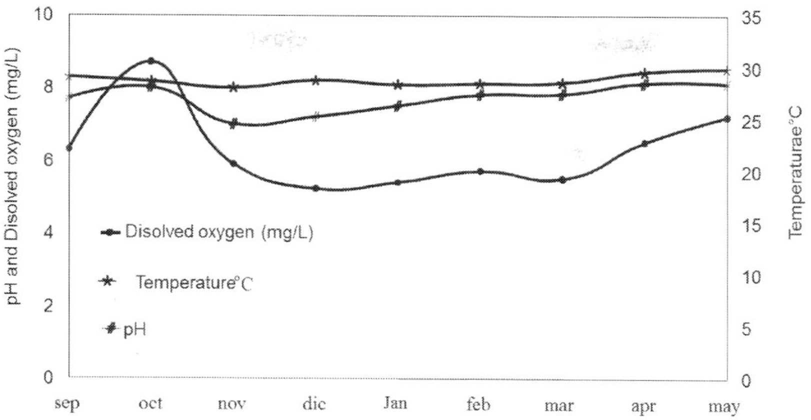

Figure 4. Environmental parameters tendency during an hydrologic cycle in Taxhido river.

3.3. Laboratory studies

Organisms were collected in Taxhido and then transport to laboratory and maintained for acclimation during a month. After this time a group was used to optimal temperature determination and other group for an optimal density experiment.

3.3.1. Optimal temperature

The fish were placed in 15 rectangular glass tanks distributed using a Latin square scheme in order to avoid spatial effects. The tanks' dimensions were of 0.4.5 m depth, 0.8 m wide and 0.3 m long, with a water storage capacity of 100 L. Five triplicated treatments with 15 organisms were applied using environmental temperature as a medium value, so the treatments were; 24, 26, 28, 30, and 32°C. The handling of tanks involves, the feces removal and partial water change (30%) weekly. The fish were feeding with a commercial diet for Tilapia (Api-Tilapia 1, maltaCleyton® with 50% protein, 12% lipid, 13% ash, 3% fiber, 12 moisture) throughout the experiment. Feeding frequency was adjusted to three provisions offered three times daily starting at 8 AM, 1 PM and 6 PM. The results show that 28°C., is the best temperature for *H. cyanoguttatus*, fig. 5.

3.3.2. Optimal density

Once the temperature was determined a similar experiment was carry out, but in this case the variable was the density. Five densities were probed, T1= 5, T2=10, T3=15, T4=20 ind/per aquarium. It is important to consider that control temperature was implemented in each aquarium in order to avoid an effect for spatial distribution. The results show that 15 individuals is the best density for *H. cyanoguttatus*, table 2.

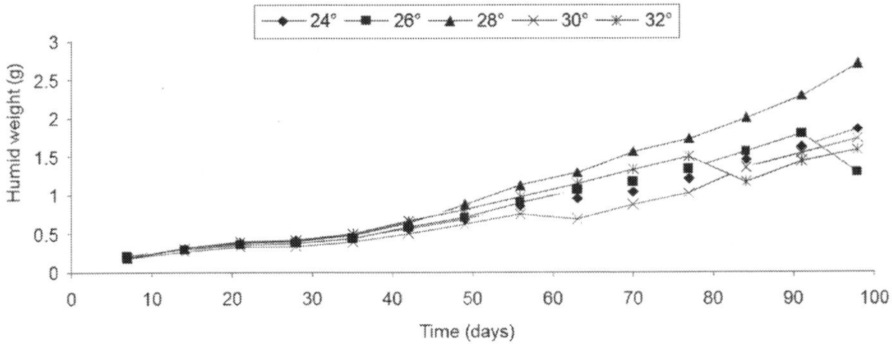

Figure 5. Humid weight behavior for the different temperature treatments

Performance parameters	T1= 5 ind	T2=10 ind	T3=15 ind	T4=20 ind
Initial number (n)	5	10	14	19
Final average number (n)	1	6	9	7
Survival rate (%)	26.66	56.66	61.7	38.88
Initial Total weight (g)	2.16	2.84	4.28	16.92
Initial individual average weight (g)	0.43	0.28	0.29	0.91
Final Total weight (g)	2.36	9.26	14.05	23.81
Final individual average weight (g)	0.59	1.09	1.49	2.05
Weight gain (%)	95.16	293.44	400.59	94.86

Table 2. add caption

3.3.3. Bioenergetics

Aerobic metabolism were determine in natural conditions with a semi closed system with (0.5L) respirometric chambers, using a method which assumes that the reduction of oxygen and the increase of nitrogen in the chamber depend on the weight of the animal, the volume of water, the period of time in which no water circulated in the chamber and the ambient temperature [22]. After each cycle, sample was oven-dried in an electric oven between 70-80°C until the samples had constant weight. From each composite sample 2 g were measured and taken as analytical sample; the samples were digested with concentrated nitric acid. The determination of the percentage proximate composition was chemically analyzed according to the method of analysis described by the Association of Official Analytical Chemist [23].

Physiological experiments showed higher oxygen consume at 15:00 hrs, while a minimum consumption was founded during the morning, Fig. 6.

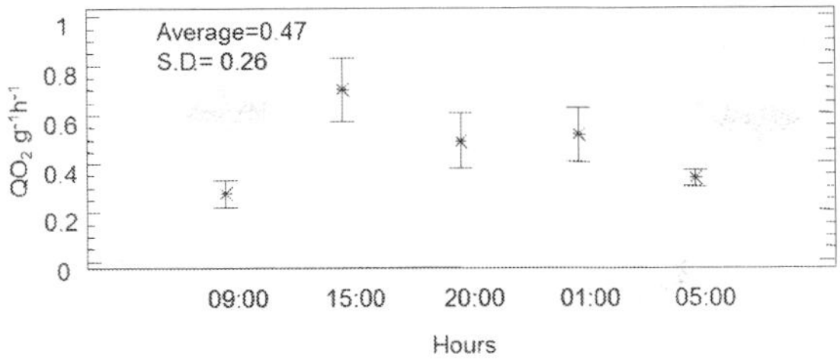

Figure 6. Oxygen consumption in natural condition for *H. cyanogutattus*

After three weeks in laboratory condition (acclimatization) oxygen consumption was measured to know if a metabolism was changed. The results show that higher consumption was found at 15:00 and minimal during the morning so the fish did not show modification in oxygen consumption.

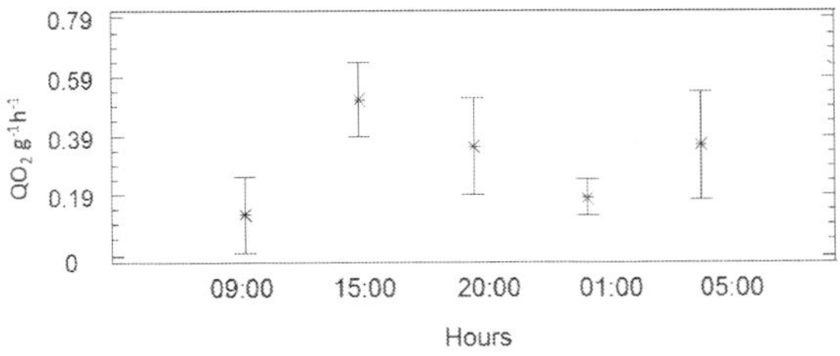

Figure 7. Oxygen consumption under laboratory condition for *H. cyanogutattus*

Finally, oxygen consumption was measured for each of the temperature treatments in order to know in which an alteration occurs. The results show that maximum values for oxygen were founded at T3= 30° minimal at T2 =26° treatments.

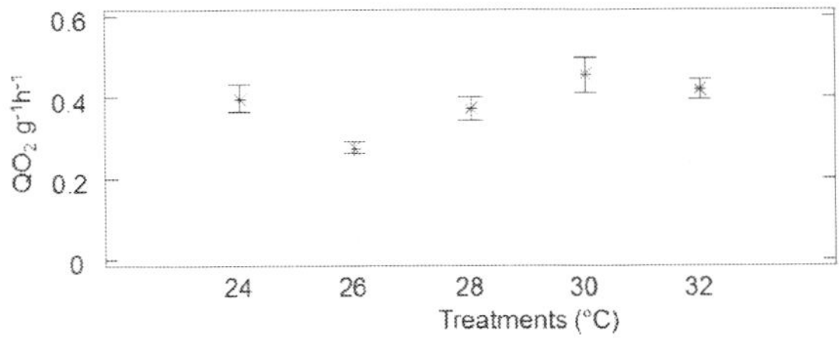

Figure 8. Oxygen consumption for each temperature treatment at the end of the experiment.

4. *Procambarus diguetti*

The cambarids, known in Mexico as acociles (in Náhuatl), makaxil (in Mayan), chapos (in Purépecha), or freshwater crayfish, according to the region, are well known, and were regularly consumed by the Aztecs and other ethnic groups based around the Valle de Mexico already during prehispanic times [29]. *Procambarus digueti* is endemic to the Mexican Central-Occidental Plateau and is found only in certain parts of the Duero and Camécuaro rivers. The Duero River is 10 km long, whist the river Camécuaro is just 2 km long. The two rivers meet at Las Adjuntas. Furthermore, this species is known to occur in Lake Chapala, though samples were collected 40 to 50 years, so is believed to have naturally disappeared from some locations, though these local extinctions may in fact be unnatural This species is threatened by over-exploitation, habitat destruction and degradation, and the introduction of exotic species.These processes have already resulted in reduction in abundance of this species, and the extinction of some populations. Much of the natural habitat of this species has been altered by chemical pollution or by human activities such as canalization, clearing, dredging and embanking of rivers, construction of reservoirs, and the regulation of water levels and stream flows. Furthermore, this species is also threatened by the introduction of exotic crayfish such as *Procambarus clarkii* and *Cherax* species from Australia, which competes for food resources and refuges, and also alters the total production of the native ecosystems (Gutiérrez-Yurrita and Latournerié-Cervera 1999). In addition to over explosion of population, a cultural pressure exists, purepecha people attributes curative properties at *Procambarus* [30].

4.1. Study area

The first part of the experiment was carry out in the national park of Camecuaro lake, which is found in Michoacan State over the municipality called Tangancícuaro, 19°54′10″N;102°12′20″O, at 1,700 meters above sea level. The national park has a spring called Camecuaro lake which is the main contribution for Camecuaro river, the depth is between 1.5 a 1.8 m; temperature 17.7 y 21°C; dissolved oxygen 7.3 y 7.5 mg/ L; visibility of 100%; hardness 138.8 y 145.5 mg/L $CaCO_3$. A handled extraction was practice in order to have a desirable sample size for the laboratory work. The organisms were transported in plastic bags with a supplemented oxygen and ice to avoid the over heat.

4.2. Laboratory studies

Acclimatization of organisms was carrying out during a month in glass aquariums considering environmental conditions. The principal problems for the maintenance of the organism were the feed and the density, so in this case these two experiments were carry out.

4.2.1. Establishing the diet

The organism were placed in 15 rectangular glass tanks distributed using a Latin square scheme in order to avoid spatial effects. The tanks′ dimensions were of 0.45 m depth, 0.8 m wide and 0.3 m long, with a water storage capacity of 100 L. Three commercial diets were probed T1= Trucha initial; T2= Camaronina, and T3=Tilapia initical, table 3. The handling of tanks involves, the feces removal and partial water change (30%) weekly. Feeding frequency was adjusted to two provisions offered three times daily starting at 8 AM and 6 PM.

Diet Compounds	**Trucha Iniciador (50:15)**	**Camaronina (35:8)**	**Tilapia Iniciador (32:4)**
Protein (%)	50	35	32
Humidity (%)	12	12	12
Grass (%)	15	8	4
Crude fiber (%)	4	5	10
Ash (%)	12	10	10
Calcium (%)	2	1.4	
Phosphorous (%)	1.2	0.9	
E. L. N.	7	30	

Table 3. Proximal Chemical Composition of the diets tested for growth

Multiple condition factor (K) was calculate for each treatment, with this factor is possible to know the relative health for organisms [31].

$$K = (10^2 * W) / L^{\ b}$$

Where;

K= Multiple condition factor

W= Weight

L= Length

b= exponent from $W = KL^b$

The type of growth was determine for the Ricker equation [32];

$$W = aL^{\ b}$$

Where;

W=weight

a = intercept

b= slope

As can be seen in

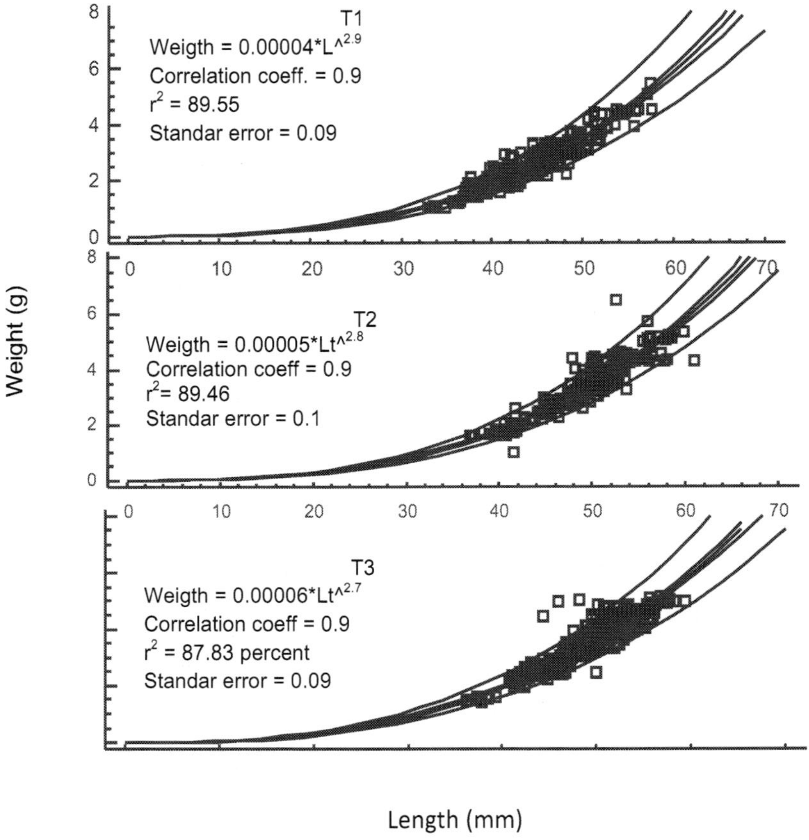

Figure 9. Regression for each one of treatments, values for the equation $W=KL^b$

Aerobic metabolism were determine in natural conditions with a semi closed system with (0.5L) respirometric chambers, using a method which assumes that the reduction of oxygen and the increase of nitrogen in the chamber depend on the weight of the animal, the volume of water, the period of time in which no water circulated in the chamber and the ambient temperature [22]. After each cycle, sample was oven-dried in an electric oven between 70-80°C until the samples had constant weight. After three weeks in laboratory under the diets treatment, oxygen consumption was measured to know if a metabolism was changed. The results show that higher consumption was found at 15:00 and minimal during the morning so the fish did not show modification in oxygen consumption, Fig. 10.

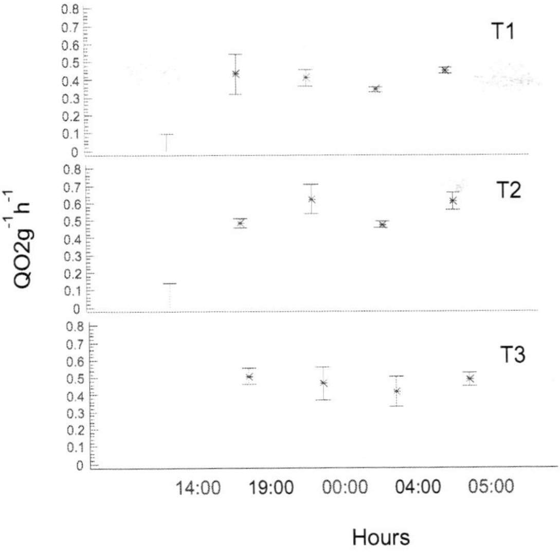

Figure 10. Oxygen consumption for each diet treatment at the end of the experiment

At the end production of total biomass was estimated, the results show that T1 is the best feed for *Procambarus digueti*, and T3 is the diet with a minor biomass production.

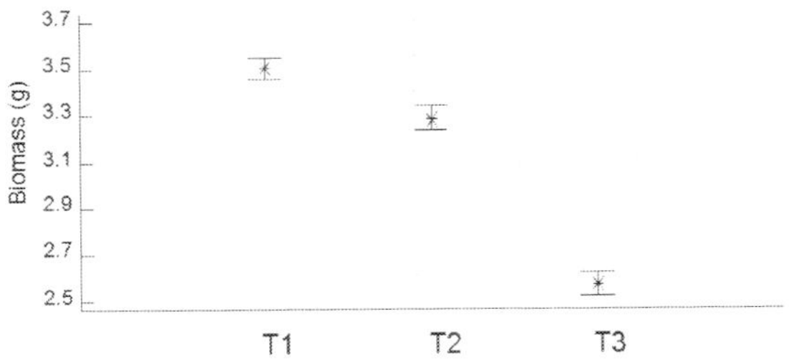

Figure 11. Biomass production at the end of experiment for each of the treatments

4.2.2. Establish an optimal density

The organism were placed in 15 rectangular glass tanks distributed using a Latin square scheme in order to avoid spatial effects. The tanks′ dimensions were of 0.4.5 m depth, 0.8 m wide and 0.3 m long, with a water storage capacity of 100 L. Five densities were probed, T1= 5, T2=10, T3=15, T4=20 ind/per aquarium. The handling of tanks involves, the feces removal and partial water change (30%) weekly. The fish were feeding with a commercial diet for trucha (50% protein, 10% humidity, 12% lipid, 12% ash, 4% fiber, 12% moisture) throughout the experiment. Feeding frequency was adjusted to three provisions offered two times daily at 8 AM, and 6 PM. The results show that a density of 15 organisms is optimal for *Procambarus* growth, table 4.

Treatment	5	10	15	20
Initial number (n)	5	10	15	20
Final average number (n)	5	8	9	6
Survival rate (%)	100	80	60	30
Initial individual average weight (g)	4.8	5.23	5.53	5.18
Final individual average weight (g)	7.12	5.67	6.36	2.82
SGR (%/día)	1.94	1.71	1.83	1.018
Weight gain (%)	612.9	467.2	536	182
Production	0.302	0.158	0.091	0.063

Table 4. Performance growth for the densities treatment in *Procambarus digueti*

4.2.3. Bioenergetics

The flow of energy that was used to determine the energy efficiency and assimilation is show in the next equation:

$$C = P + G + ER + EU + F$$

Where; C is the total energy content of food consumed, P and G are the energy equivalents of somatic and gonadal growth respectively, ER is the energy utilized in respiration, EU is the energy lost as nitrogenous and other waste compounds excreted (50% protein) is the best food for *Procambarus digueti*. The energy balance is show in the table 5.

Diet	**C**	**P**	**R**	**U**	**F**
	Consume cal/mg	**Biomass cal/mg**	**Oxygen cal/mg**	**Nitrogen cal/mg**	**Feces g**
Tilapia (32:4)	5423.44	3.664	1.309	0.450	0.014
Camaronina (34:8)	5545.03	3.845	1.175	0.525	0.009
Trucha (50:15)	5706.273	3.873	1.264	0.569	0.006

Table 5. Energetic balance for *Procambarus digueti*

http://dx.doi.org/10.5772/57215

5. Conclusion

Aquaculture has been supporting human demands for fish products for centuries and is an important industry worldwide. Global production from aquaculture has been increasing steadily, having more than doubled in the last decade; aquaculture now supplies one third of seafood consumed worldwide. With the massive increase in world aquaculture production in 1990s, the current aquaculture industry is one of the fastest growing sectors in world food production. However, the expansion of aquaculture has been accompanied by degradation of the natural environment, especially on marine aquaculture. Directly impacts of fisheries and aquaculture are habitat modification, collection of wild seedstock, changes of food webs, introduction of nonnative fish species and diseases that harm wild fish populations, and nutrient pollution. According to the FAO, major issues that need to be addressed are problems with access to proper technology and financial resources, together with environmental impacts and diseases. Another argues that further increases in aquaculture production will come mainly from further investment in biotechnology. The development of new strategies or technologies does not imply that the old one disappears; to the contrary the main idea is to use the experience and improve existing technology.

These three experiences and review of similar cases that have been developed in Mexico allow us to establish a general methodology in order to know the aquaculture potential for native species.

1. Knowledge.- The first step consist in to know and get the information above the specie. In the case of *G. multiradiatus* the knowledge practically doesn't exists so the research was oriented to the basic biology and ecology aspects. For example, with *G. multiradiatus* the main objective was to obtain reproduction but to reach these aspects it is required the maintenance under laboratory conditions and for this the knowing of food habits are essential. In the case of *H. cyanogutattus* the main problem was its taxonomic status, so the principal problem to obtain the basic knowledge is that some aspect were publish under the scientific name *Cichlasoma cyanoguttatum* and others aspects with the actual name *Herinchthys cyanogutattus.*

2. Environmental prospections.- It is necessary to know the basic physicochemical parameters (T°, pH, DO_2) in order to establish the strategy for transportation and laboratory maintenance. Field observations and ecological description is necessary in order to know the feeding habits, interspecific competence and the disposition of resources. With these parameters is possible to start the research.

3. Feed.- Under laboratory conditions feeding is the main problem in order to continue so the proofs needed is the acceptance of commercial food. The three species show an acceptance for commercial food but the problem here is the metabolism aspects such as assimilation and performance growth. This kind of problems could be solved with a bioenergetics approach.

4. Bioenergetics.- Ecophysiological basis of species should take into consideration the physiological characteristics and ecological role of the organism in question. Physiological

analysis will reflect the conditions which affect population characteristics, such as population growth, intraspecific competition, and functional and numerical responses. These studies can do more than being simply descriptive, since they enable the development of scenarios that can be tested either through strictly controlled laboratory experiments or field experiments.

The new aquaculture research must be consider to add native species, this work is an effort to get the basic information in order to development of biotechnology and a link between the basic and applied science.

Acknowledgements

This work was partially finance by CONACYT.

Author details

Fernando Garcia-Trejo[1], Silvia Hurtado-Gonzalez[1], Genaro M. Soto-Zarazua[1] and P. J Gutierrez-Yurrita[2]

1 División de Estudios de Posgrado Facultad de Ingeniería, Universidad Autónoma de Querétaro, C.U. Cerro de las Campanas, Qro, México

2 Centro Interdisciplinario de Investigaciones y Estudios sobre Medio Ambiente y Desarrollo (CIIEMAD), Instituto Politécnico Nacional, México

References

[1] FAO, editor. Review of the State of the World Marine Fishery Resources2011; Rome: FAO.

[2] Pillay TV. Aquaculture and the environment. second ed. Iowa: Blacwell publishing; 2004.

[3] Bailey C. Blue revolution: the impact of technological innovation on Third World fisheries. The Rural Sociologist. 1985;5(4): 259-66.

[4] Costa-Pierce B. Ecological Aquaculture: the evolution of the Blue Revolution second ed. Iowa: Blackwell Science; 2002. p. 383.

[5] Saborio Coze A, Flores Nava A. Review of environmental impact assessment and monitoring of aquaculture in Latin America Environmental impact assessment and monitoring of aquaculture. Rome: FAO; 2009. p. 395–454.

[6] García-Trejo F, Soto- Zarazúa GM. La nueva ecología, una perspectiva integradora hacia un paradigma de sustentabilidad. CIENCIA@UAQ. 2012;5(1).

[7] Ross LG, Beveridge MC. Is a better strategy necessary for development of native species for aquaculture? A mexican case of study. Aquaculture research. 1995[29]:539-47.

[8] Martínez-Palacios CA, Barriga-Tovar E, Taylor JF, Ríos-Durán G, Ross LG. Effect of temperature on growth and survival of Chirostoma estor estor, Jordan 1879, monitored using a simple video technique for remote measurement of length and mass of larval and juvenile fishes. Aquaculture. 2001(1-4):369-77.

[9] Ross LG, Martinez-Palacios CA, Morales E. Developing native fish species for aquaculture: the interacting demands of biodiversity, sustainable aquaculture and livelihoods. Aquaculture Research. 2008[39]:675–83.

[10] Bardach JE, Ryther JH, Mclarney WO. Acuacultura, crianza y cultivo de organismos marinos y de agua dulce. México: AGT 1986.

[11] Martinez-Palacios CA, Ross LG. The reproductive biology and growth of the Central American cichlid Cichlasoma urophthalmus (Günther) Journal of Applied Ichthyology. 1992;8:99–109.

[12] Garcia-Trejo F, Hurtado Gonzalez S, Soto Zarazua G, Alatorre-Jácome O, Rico-García E, Gutiérrez-Yurrita PJ. Ecophysiological responses to the effect of annual management on an endemic viviparous fish in central plateau of México Neotropical ichthyology 2013;11(1):117-23.

[13] Gutiérrez-Yurrita PJ, Montes C. Environmental factors controlling the crayfish Procambarus clarkii activity in Doñana National Park temporary freshwater marsh, SW. Spain Comparative Biochemistry and Physiology, Part A 1998;120:713-21.

[14] De la Vega-Salazar MY. Estado de conservación de los peces de la familia Goodeidae (Cyprinodontiformes) en la mesa central de México. Revista Biología Tropical. 2006;54(1):163-77.

[15] Macias GC. Social behavior and operational sex ratios in the viviparous fish Girardinichthys multiradiatus. Copeia. 1994;4:919-25.

[16] Álvarez del Villar J. Peces Mexicanos (Claves). In: Pesca. CNCd, editor. Instituto Nacional de Investigaciones en Biología Pesquera Mexico 1970. p. 166.

[17] Miller RR, Fitzsimmons JM. Ameca splendens, a new genus and species of goodeid fish from Western Mexico, with remarks on the classification of the Goodeidae. Copeia. 1971;1:1–13.

[18] Uyeno T, Miller R, Fitzimons J. Karyology of cyprinodontoid fishes of the Mexican family Goodeidae Copeia. 1983;2:497–510.

[19] Macias-Garcia C, Burt de Perea T. Ultraviole-based female preferences in a viviparous fish Behavioral Ecology and Sociobiology 2002;52(1):1-6.

[20] Trujillo-Jiménez P, Viveros EE. La ecología alimentaria del pez endémico Girardinichthys multiradiatus (Cyprinidontiformes: Goodeidae), en el Parque Nacional Lagunas de Zempoala, México. Revista de Biología Tropical 2006;54 1247-55.

[21] García E. Modificaciones al Sistema de Clasificación Climática de Koppen. México: Enriqueta García; 1988.

[22] Steffensen JF. Some errors in respirometry of aquatic breathers: how to avoid and correct for them Fish Physiology and Biochemistry. 1989;6:49-59.

[23] AOAC. Official Methods of Analysis of AOAC International,. sixteen ed. Gaithersburg, MD, USA.: AOAC; 1998.

[24] Klekowsky RZ, Duncan A. Physiological approach to ecological energetics. In: Grodzinzky H, Groves TDD, editors. Methods for ecological bioenergetics. Oxford: Blacwell Scientific Publication; 1975. p. 227-61.

[25] Hubbs CT, Lucier GP, Garrett RJ, Edwards SM, Dean E, Marsh E, et al. Survival and abundance of introduced fishes near San Antonio, Texas. Texas Journal of Science. 1978;4[30]:369-76.

[26] Shafland PL, Pestrak JM. Lower lethal temperatures for fourteen non-native fishes in Florida. Environmental Biology of Fishes 1982(7):149–56.

[27] Lee DS, Gilbert CR, Hocutt CH, Jen RE. Atlas of North American freshwater fishes. Raleigh, NC: North Carolina State Museum of Natural History; 1980.

[28] Itzkowitz M, Nyby J. Field Observations of Parental Behavior of the Texas Cichlid Cichlasoma cyanoguttatum. American Midland Naturalist. 1982;2[108]:364-8.

[29] Gutierrez-Yurrita PJ, Morales-ortiz A, Oviedo A, Ramirez-Perez C. Distribution, spread, habitat characterisation and conservation of the crayfish species (Cambaridae) in Querétaro (central México). Freshwater Crayfish. 2001;13:349-58.

[30] Gutiérrez-Yurrita PJ, R LCJ, editors. Preliminary studies on exploitation of two native crayfish species of the genus Procambarus (CRUSTACEA: DECAPODA) by purépecha people in Michoacán, México. Proceedings of the III International Congress of Ethnobiology; 1992.

[31] Kuri-Nivon E. Manual para la producción de trucha y salmón.. Idaho: Universidad de Idaho; 1976. Manual para la producción de trucha y salmón. Universidad de Idaho.

[32] Richter JB. Anfangsgründen der Stöchiometrie oder Messkunst chemischer Elemente (Foundations of stoichiometry or surveying chemical elements). Breslau1792.

Chapter 3

Biofloc, a Technical Alternative for Culturing Malaysian Prawn *Macrobrachium rosenbergii*

Carlos I. Pérez-Rostro, Jorge A. Pérez-Fuentes and
Martha P. Hernández-Vergara

Additional information is available at the end of the chapter

http://dx.doi.org/10.5772/57501

1. Introduction

Aquaculture is a major food-producing activity that is growing steadily, coupled with growing population density and land use needs of other industries. To maintain growth, aquaculture must shift to intensive or semi-intensive practices, effective and sustainable use of resources, and sustainable environmental stewardship. This often requires application of technologies that increase production efficiency and avoids competition for space and resources with other activities, such as agriculture and ranching. Aquacultural practices must be sustainable and minimally destructive to the environment, maintain quality and safety standards, and enable efficient use of space and natural resources and possibilities for expansion. Technology alternatives that reduce environmental impact and are efficient without affecting the health and growth of stock organisms must be incorporated into current practices. One option is to apply biofloc technology. Biofloc forms naturally in pond water as aggregates of nitrifying bacteria, organic material, inorganic flocculants, and suspended algae. These ingredients serve as food for the stock under cultivation and promote direct use of nitrogenous compounds in feces, urine, and food waste. Activity of nitrifying bacteria increases with addition of carbon sources and constant aeration, which maintains or significantly improves water quality during cultivation. Thus, the large volume of water required in intensive aquafarming is greatly reduced [1, 2, 3]. An example is using biofloc during cultivation of the Malaysian river prawn *Macrobrachium rosenbergii*. The approach led to major savings of water, without affecting the quality of the prawns.

1.1. What is biofloc?

Biofloc culture is a system where, after adding a carbon source and providing constant aeration, biofloc bacteria maintain water quality during cultivation of freshwater shrimp. The metabolic processes and biochemical transformations take place directly in the water column, which promotes overall balance of the system and the health of the farmed shrimp. The biofloc forms in the pond water naturally as aggregates of nitrifying bacteria, organic material, inorganic flocculants, and suspended algae. The algae serves as food for the pond stock and the bacterial promotes direct conversion nitrogenous waste to simpler compounds. The self-cycling process maintains or greatly improves the quality of the pond water during cultivation. Improvement in water quality drastically reduces the need to cycle large volumes of additional water in the farm pond system. This leads to a sustainable activity that is in balance with the environment and reduces the cost of water and feed for the pond stock [1, 2, 3] (Fig. 1).

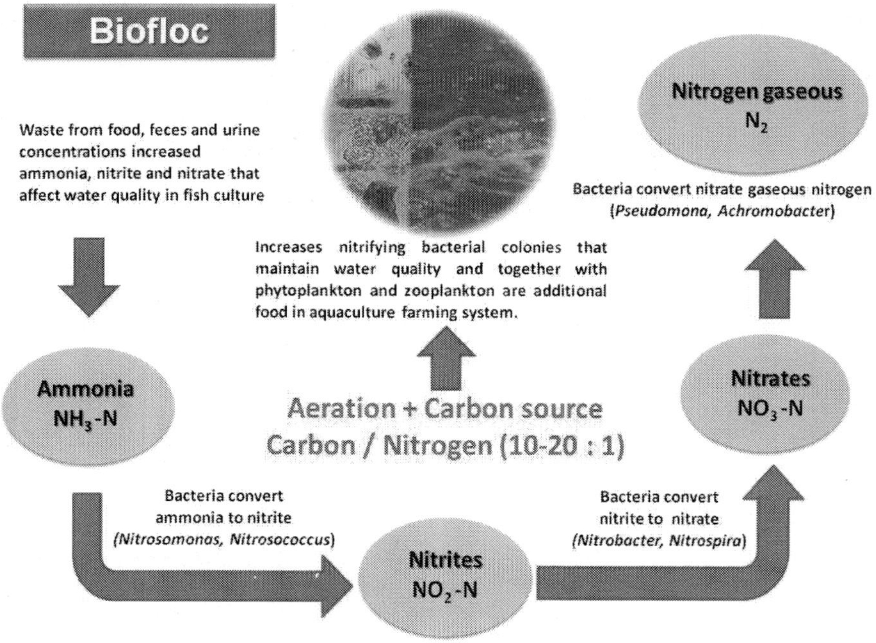

Figure 1. Biological processes in biofloc cultivation

In a biofloc system, the biological nitrification process occurs in three stages. In the first stage, bacteria of the genera *Nitrosomonas* and *Nitrosococcus* act on ammonia (NH_3/NH_4) generated by food scraps and feces and urine. The waste is oxidized to nitrite (NO_2). In the second stage, the nitrite is converted to nitrate (NO_3) by bacteria of the genera *Nitrobacter* and *Nitrospira*. The nitrate is reduced to nitrogen gas (anoxic denitrification) by bacteria mainly from the genera *Achromobacter* and *Pseudomonas* [4, 5].

Biofloc are of two main types. Classification is based on the amount and nature of organic matter and its component organisms, the latter can be bacterial or autotrophic, mainly composed of algae. The importance of this is that, in both cases, the microorganisms present in the bioflocs maintain water quality because they decrease nitrogen compounds and are also nutrients for the bacterial and algae. It is important to understand that, depending on the nature of the biofloc microorganisms, their nutritional quality can vary. This affects the supply of nutrients for the stock organisms in the ponds (Fig. 2).

Figure 2. Types of biofloc, based on predominant species.

The microorganisms that populate biofloc systems typically inhabit natural aquatic systems. Their presence depends directly on two environmental variables: intensity of solar energy absorbed by the system and the concentration of organic matter and carbon sources that enter the system. In a biofloc system, colonies of bacteria depend mostly on organic matter present in the system for survival and proliferation, and, to a lesser degree, on the intensity of sunlight. Bacteria are also directly dependent on the system to supply constant aeration because the bacteria consume large volumes of oxygen, which is, in turn, directly related to the bacteria consuming carbon from the system. The carbon is a source of power for growth and proliferation. The concentration of carbon in the system must be maintained at a C/N ratio appropriate for maintaining reproduction of bacteria with inorganic nitrogen to a maximum concentration of 200 mg L^{-1} [6]. This level is an indicator that the system is effectively and economically controling nitrogen. High concentrations of nitrogen will usually upset the balance of the system and affect the health of stock species, especially shrimp.

In a biofloc system based on bacteria and algae, nitrogen compounds are removed as bacteria increased uptake of ammonium and better control the products of waste. This does not depend on the intensity of sunlight to run efficiently. In a microalgal biofloc system, productivity depends directly on sunlight; excess illumination can generate an excess of algae, which leads to low oxygen at midday. Hence, a system using microalgae and bacteria will be a more efficient alternative bioremediation because it is possible to maintain an efficient balance between nitrifying bacteria and algae to maintain a suitable level of nitrogen, a balanced C/N ratio, and sunlight. The diversity of live food available in ponds also increases in mixed biofloc systems, which brings benefits to the stock under cultivation. This includes reducing the amount of artificial feed necessary to meet nutrient requirements under semi-intensive and intensive pond farming. It also includes nutritients not present in synthetic diets. Not to be dismissed lightly is the great savings in costs of providing fresh water and handling organic wastes in water discharge.

1.2. Source of energy for bacteria and algae in biofloc

The microorganisms that form the biofloc and process nitrogen compounds that pollute fish pond water need a source of energy for metabolism. In aquatic biofloc systems, there are three likely energy sources, depending on the nature of the organisms present in the biofloc system (bacteria–algae aggregations). Most important is sunlight, which is the main source of energy for phototrophic microorganisms, such as algae and vascular plants. Solar reception can be controlled or semi-controlled to support the needs of the biofloc crop and promote any type of biofloc system. The second source of energy is the forms of inorganic compounds that are used by the microorganisms that oxidize reduced forms of simple compounds, especially nitrogen to obtain energy. In fish farming, by metabolizing organic nitrogen and ammonia, nitrogen is oxidized to nitrite and nitrate. The third source of energy is organic compounds that are transformed by microorganisms that derive energy from the metabolic oxidation of organic carbon and transform it to carbon dioxide.

Both chemothrophic and phototrophic microorganisms naturally consume and deplete nitrogen concentrations in the water because of the relatively large quantity of energy sources that are present, but also because this is an indispensable function of the microorganisms. Transformed energy is used to synthesis proteins from the nitrogen sources.

Systems for cleaning and wastewater bioremediation, using microalgae and bacteria, is a widely known technology; however, in aquaculture systems, they should be used with caution because, with microalgae, the efficiency of the system depends directly on solar energy and intensity, which in open systems can be a risk because there is no control over productivity [7, 8]. An excess of primary production leads to constant consumption of oxygen during the night. On cloudy days, productivity will reduce water quality.

Biofloc or nitrifying colonies of bacteria in aquaculture requires incorporation of additional carbon sources into the system to adequately reproduce biofloc and maintain high density because carbohydrates in the system may be insufficient. Some of the main sources of carbon that can be used in aquaculture crops are: glycerol and sodium acetate, sugar, tapioca flour, wheat flour, and molasses [9, 10, 11, 12, 13, 14]. Use depends directly on the local costs of these products. For a biofloc system to operate efficiently, it is best to maintain a C/N ratio between 10:1 and 20:1 [15, 1, 16, 17, 10]. The amount of carbon depends on several factors, including: water quality, physiology and growing body density of the stock, quality and quantity of food to be cultivated, and solubility of the carbon source. The carbon additive must be continuously monitored to ensure that the system is functioning properly.

1.3. Ecological importance of using biofloc in aquaculture

Biofloc technology provides more efficient and sustainable aquaculture by reducing environmental impacts. One major advantage is reducing the volume of water required by the system during cultivation. Biofloc in the cultivation system uses the initial water volume throughout the production cycle and needs additional water only to replace water lost by evaporation, leakage, or to remove organic material during production. The biofloc microorganisms serve as natural food, depending on the eating habits of the stock species. This will reduce consumption of artificial food and lead to more efficient conversion of food. Biofloc is more than a supplemental source of nutrients in aquatic systems. It brings economic benefits during production and enables more efficient use of resources, given that the main source of protein during production is fish meal. Fish meal often comes from overharvesting of fisheries. Considering the rapid growth of pond farming, biofloc can directly contribute to reduced pressure on fisheries.

1.4. Physical-chemical parameters of water in the biofloc

Efficient operations with biofloc aquaculture systems depend on maintaining water physico-chemical parameters within the range of tolerance of cultivated stock because this affects yield per unit volume. This is important because biofloc pond farming is a form of simple and complete synthetic ecosystems, based on three components that interact in the same space: (1) Stock of one or more commercial species; (2) Microalgae interact and function as biofiltrates that also have oxygen demand and, like commercial stock, produce metabolites; and (3) Bacteria responsible for transforming nitrogenous metabolites that are used by the planktonic microalgae. The purpose of a complex pond ecosystem with a biofloc cultivation system is the comprehensive use of energy and biotransformed products that maintain water quality and provide natural nutrients that promote the health and quality of a commercial animal crop without negative impacts on adjacent water bodies.

Water quality in aquatic systems is directly related to biological and chemical processes that occur in the aquatic environment and depend on several factors.

1.4.1. Dissolved oxygen

Oxygen in aquatic systems should be >5 mg L–1. In a biofloc pond system, the bacteria and algae that form the biofloc also have oxygen demand, so competition can occur in the pond. It is recommended that dissolved oxygen be maintained at 7–8 mg L–1 to ensure proper functioning of the system.

1.4.2. pH

pH should range from 6.5 to 9, depending on the cultivated stock. Aslo, pH <6 and >8.5 usually affects the efficiency of the biofloc components, as well as growth and survival of the cultivated stock. In a biofloc system, pH varies during the day as concentration of carbon dioxide build up from the respiration of the stock. We recommend a range of pH 7.0–8.5, which favors functioning of biological cycles in the system. To maintain the pH balance, low pH can be adjusted with calcium hydroxide, potassium hydroxide, sodium carbonate, or sodium bicarbonate. High pH can be adjusted with carbonic acid, hydrochloric acid, sulfuric acid, phosphoric acid, or their salts.

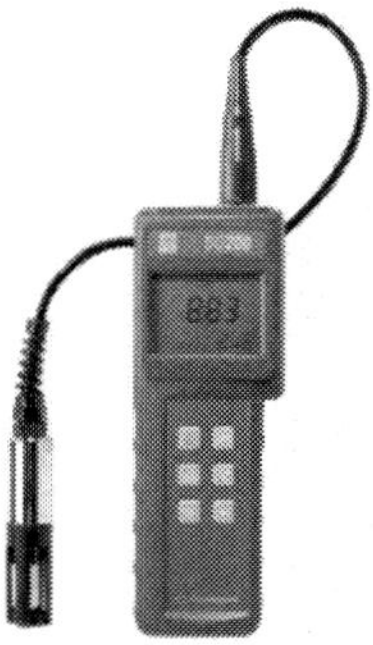

Oximeter

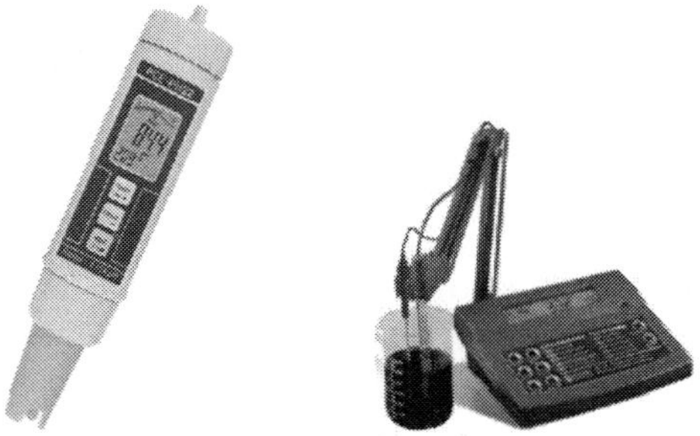

pH meter

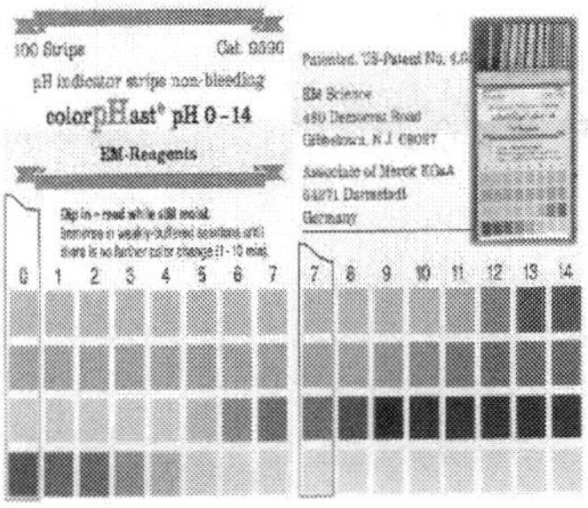

pH by color

1.4.3. Dissolved solids and volatiles

Bacteria depend on suspended solids as a substrate for adhesion and as a source of energy from carbon. We maintained concentrations of suspended matter in the range of 250–450 mg L–1, which ensures efficient bacterial activity. An excess of suspended matter can affect breathing processes in the stock species, lead to stress, or, in extreme cases, lead to death by clogging gills. Cultivation of Litopenaeus vannamei, in one biofloc system contained 453.0 ± 50.0 mg L^{-1} total suspended solids and 256.0 ± 106 mg L–1 volatile solids, which improved shrimp production and provided efficient exchange of oxygen [18].

1.4.4. Turbidity

In aquaculture systems, transparency is directly affected by the amount of organic and inorganic matter in suspension (suspended solids, phytoplankton, zooplankton, and bacteria). Turbidity is measured with a turbidimeter or nephelometer, which uses a beam of light passing through a water sample. In aquaculture, a Secchi disk is frequently used because turbidity is measured by the depth when the disk cannot be seen. Solar heating of the water is also affected by

TDS meter

transparency or turbidity. Using the Secchi disk, turbidity of 35–40 cm is acceptable. Turbidity produced by plankton in pond water should be >30 cm [29]. Higher cocentrations of plankton can increase oxygen demand of the fish stock during the night, when the same plankton community that contributes to the turbidity and dissolved oxygen during the day competes with the fish stock at night. Low oxygen not only damages the stock, but also affects the biofloc bacteria and plankton. Oxygen demand may increase up to 300% overnight. A simple method for maintaining the concentration of suspended matter at optimal levels is by sedimentation, using tanks with conical bottoms to remove solid waste in the recycling systems [19].

Secchi Disk

1.4.5. Temperature

Temperature is one of the most influential parameters in fish pond systems because it affects the metabolic rate of cold-blooded fish and microorganisms, oxygen consumption, pH, and concentrations of ionized and un-ionized ammonia during cultivation. The temperature range will depend on the stock species and the bacteria adapted to the system temperature, as well as environmental and seasonal variations. This is important because biofloc systems are more efficient when water temperature is between 28 and 30 °C. [20] Reports that nitrifying bacteria can support a range from 8–30 °C, but efficiency is reduced by 50% at 16 °C and by 80% at 10 °C.

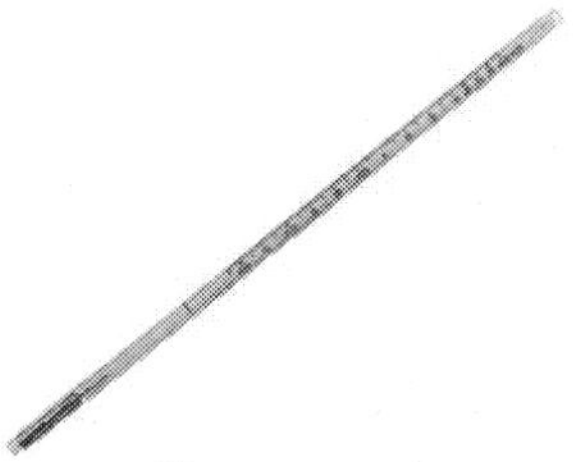

Termometer

1.4.6. Total Ammona-nitrogen

Total Ammona-nitrogen is the excretion product of feces and urine of fish, uneaten food and matter in descomposition, phytoplankton and zooplankton. Ammonia-nitrogen toxicity on aquatic organisms has been attributed to ammonia or non-ionized ammonia (NH3) (gaseous), while the ionized ammonia or ammonium ion (NH4) is considered not significantly toxic or less toxic [42].

The reaction that occurs is as follows.

$NH_3 + H_2O \rightarrow$	**$NH_4OH \rightarrow$**	**$NH^+{}_4 + OH^-$**
Nonionized form	Its rate of conjugation	Ionized form
Nontoxic form	with water depends of pH	Nontoxic form.

Ammonia-nitrogen toxicity in the unionized form (NH3), increases with a low oxygen concentration, high pH (alkaline) and a high temperature. With a low pH (acid) is less toxic. A high concentration of ammonia-nitrogen in the water has effects on the cultured organisms, causing blockage of the metabolism, affecting the balance of salts in the osmoregulation, which produces gill internal organ damage, immunosuppression and susceptibility to diseases, reduced growth and survival. In cultured crustacean as Litopenaeus vannamei, ammonia-nitrogen concentrations should be less than 1.2 and 6.5 mg / L in post-larvae and juveniles [36, 37]. The recommended concentrations less than 1.5 mg / L in cultures with biofloc.

1.4.7. Nitrite-nitrogen

They are a vital parameter for its high toxicity and for being a pollutant. The transformation process to ammonia-nitrogen to nitrite-nitrogen and their toxicity form depends on the amount of chlorides, temperature and oxygen concentration in the water. The main toxic effects of NO2 are those who have a direct effect of transport of oxygen, oxidation of important compounds and tissue damage. Nitrite-nitrogen in the larvae of M. rosenbergii tolerate concentrations of 2 mg / L, increasing their tolerance as they grow, and can support up to 16 mg / L of nitrate-nitrogen, however reduce its growth rate and culd cause their mortality [38]. Recommending nitrite-nitrogen concentration less than 2 mg / L in cultures with biofloc.

http://dx.doi.org/10.5772/57501

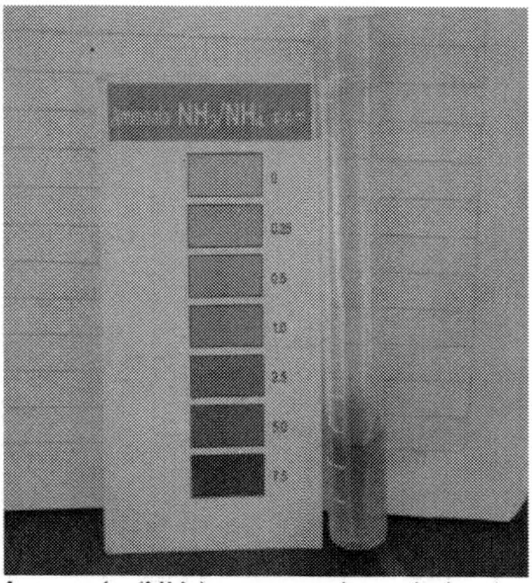

Ammonia (NH_3) measure by colorimetry

Nitrites (NO_2) measure by colorimetry

1.4.8. Nitrate-nitrogen

Nitrate-nitrogen is the end product of aerobic nitrification [32], are considered the less toxic inorganic nitrogen compounds, but can be a potential problem when its levels increase and accumulate. The toxicity of these compounds is due to its effects on osmoregulation and possibly on oxygen transport [40]. For the specie M. rosenbergii, nitrate concentrations in brackish water must be less than 20 mg / L [41]. Recommending that nitrate concentrations should not exceed 10.0 mg / L in biofloc culture.

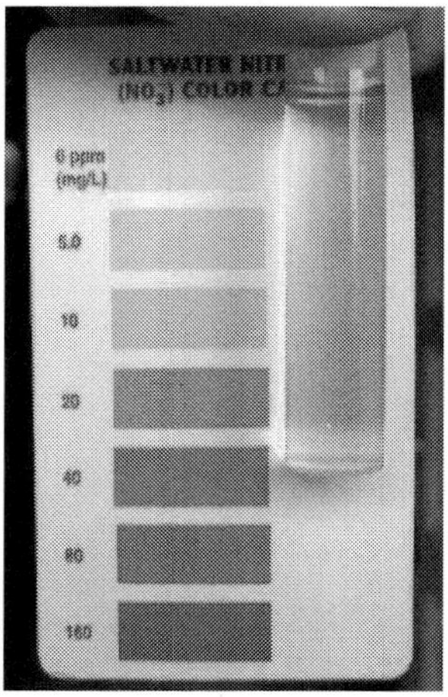

1.5. Economic benefits of crops biofloc

The economic advantages of biofloc systems to traditional pond farming are generally reflected in the profit margin, based largely on savings in feed, faster growth rates, and increased biomass during cultivation, which is related to high survival rates. However, biofloc systems have increased operating costs of the aeration system, which can be 10–40%, depending on the concentration of oxygen in ponds have to be maintained at 7–8 mg L–1, costs for the carbon source added to the system. Despite the foregoing, [21] reports savings of 14% in a shrimp biofloc system compared to traditional methods.

In our laboratory in one study, Malaysian prawn Macrobrachium rosenbergii raised in a biofloc system achieved a 13.27% saving in operating expenses. In a second study, tilapia Oreochromis niloticus raised in a biofloc system achieved a 12.90% savings in operating expenses compared with the costs in traditional pond farming of both species. The major savings in our study led to less pumping time for maintaining water quality in the system, an increase in survival from 10 to 30%, and an increase in final biomass average content from 20 to 45%. [22] report that, Nile tilapia (Oreochromis niloticus), net production was 45% higher in the biofloc tanks than in tanks without biofloc, where there was also a significant improvement in feed conversion. [23] indicates that the cultivation of Litopenaeus vannamei in biofloc systems led to a 30% decrease in the use of a commercial feed. In tilapia cultivation, producers can expect to reduce commercial feed by up to 20% [24]. Natural food produced by microalgae and bacteria in biofloc systems have high nutritional value.

2. Biofloc cultivation of Malaysian prawn *Macrobrachium rosenbergii*

Biofloc technology has been successfully applied mainly in shrimp farming [9]. Despite the positive results, few fish farms use this technique [25, 26]. The benefits associated with the

http://dx.doi.org/10.5772/57501

production of aquatic organisms under biofloc technology are apparent, so it is necessary to develop cultivation with omnivorous species in a scheme of sustainability and ecological balance to obtain the best performance with the least environmental impact.

Among economically important crustaceans in aquaculture having omnivorous eating habits, the Malaysian prawn (*Macrobrachium rosenbergii*) has successfully adapted to farming conditions, thanks to their physical endurance, fast growth, and high survival rate. This species is widely distributed in tropical and subtropical areas and, compared with similar wild shrimp [27], are a suitable candidate for biofloc practices. Despite this, there are few attempts to cultivate this shrimp in a biofloc system, making it difficult to validate the technology for application on commercial farms.

At the laboratory facilities of the breeding and production technology institute at Boca del Rio, Veracruz, Mexico, cultivation of Malaysian prawns was undertaked for six months in rectangular ponds 10 m × 2 m × 1.20 m high, with a capacity of 20 m^3, which were inside a shadehouse with shade cloth providing 90% reduction of sunlight. During cultivation, a continuous air supply was provided by a 2 hp blower connected to a 1.5 inch PVC pipe at the bottom of the ponds. Placed in the pond were four clay bricks per m^2 with 3 holes in each one. These served as dens for the prawns.The study measured the growth performance of the prawns under two conditions: biofloc shrimp farming and traditional farming, including standard water exchanges in the latter treatment. During the study the prawns were fed twice daily (9:00 and 18:00 h) with a commercial shrimp diet (El Pedregal Silver Cup with 35% protein), by an estimated 20% of the initial biomass for the first month of cultivation. Subsequently adjusted percentage monthly food supplied in connection with the consumption and increased biomass (Table 1). To promote training and biofloc production, molasses added daily diluted in water as a carbon source in ponds, in a ratio of 20:1 C: N, according to the recommendations of [10], considering feed rate.

Month	% Biomass (biofloc)	% Biomass (traditional)	Food per day (g)	Molasses per day (g)
Start	20	20	3.78	7.42
1	5	7.54	30.81	60.38
2	5	6.13	92.00	180.33
3	5	7.67	237.59	465.67
4	3	3.18	167.75	328.79
5	3	3.19	228.33	447.52

Table 1. Percentage of biomass per month to provide same amount of food in two treatments.

2.1. Physicochemical parameters of water during cultivation

During cultivation, physicochemical parameters were similar among treatments and within the tolerance range for growing Malaysian shrimp [28]. The average temperature was 25.90 ± 0.78 °C, dissolved oxygen 5.8 ± 0.55 mg L^{-1}, pH 8.77 ± 0.18. Transparency in both treatments was within the range recommended by [29] for aquaculture crops (minimum visibility of 30 to 40 cm). If turbidity is greater, there is a substantial increase in oxygen demand.

The average concentration of ammonia during the study remained at 0.1 mg L^{-1}, N-nitrite was 0.5 mg L^{-1}, and N-nitrate was 10 mg L^{-1} in both treatments, concentrations below what is considered toxic. Stability of the parameters in the biofloc system results from bacterial activity, which according to several authors, transform bacteria metabolites to the advantage of the shrimp because they are nutrients, as well as prevent accumulation of toxic products in the production system [1, 10]. By nitrification, where ammonia-nitrogen (N-NH_3/N-NH_4) is transformed by oxidation to nitrite-nitrogen (N-NO_2) by bacteria of the genera Nitrosomonas and Nitrosococcus, and others. Nitrite-nitrogen is converted to nitrate-nitrogen (N-NO_3) for nitrite-oxidizing bacteria of the genera *Nitrobacter* and *Nitrospira*. Ultimately, nitrate-nitrogen is reduced to nitrogen gas (denitrification) by bacteria of the genera *Pseudomonas* and *Achromobacter* and others [4, 5]. Unlike biofloc systems, water quality in traditional systems is maintained by continuous dilution of metabolites by influx of fresh water.

2.2. Response variables

Survival of the prawns in the two contrasting treatments, at the end of the study, was similar (85%) and is largely attributed to maintaining water quality. In the biofloc system, there was an increase of the contact surface for bacteria, which allows increased prawn density, compared to the traditional density (8–10 org m^{-2}) and dens that increased the area of protection during molting (Pineda, 2005). High survival further suggests that the biofloc does not affect the health of the prawn. Prawns grown in the biofloc system reached a higher weight (15.17 ± 8.2 g) than prawns rose by the traditional method (12.57 ± 7.89 g; Fig. 3).

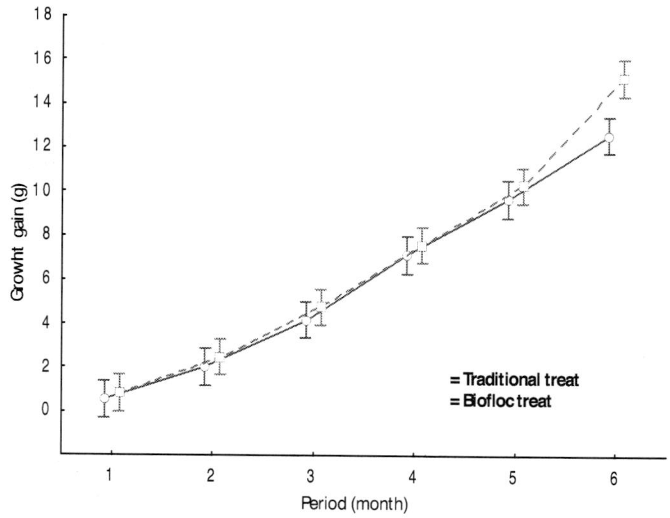

Figure 3. Weight gain of prawns raised under biofloc conditions (red) and traditional conditions (blue). The latter required large inputs of fresh water.

[30] raised Malaysian prawns for 182 days at a density of 10 prawns m^{-2}, obtaining an average weight gain similar to our biofloc system; however, we raised 37 prawns m^{-2}, and used organic and inorganic fertilized during they study. Prawns in biofloc showed a feed conversion rate that was significantly lower (2.27 ± 0.99), compared to traditional cultivation (2.74 ± 0.91), indicating that the biofloc system with the increased contact area and holes, can increase production of shrimp, along with a saving in consumption of commercial feed, which according to [24] comes from biofloc microorganisms that have high nutritional value and promote growth because the microorganisms contain up to 49% protein [31].

2.3. Saving water during biofloc cultivation

One of the major advantages of biofloc systems is reducing the volume of water required for maintaining good water quality. The biofloc system only recycles the water. It does not replace water with fresh water. Only losses from evaporation need replacement. In traditional treatments, 30% of the water is replaced every third day and 60% every two weeks (Table 2) to maintain water quality during cultivation.

Cultivation with biofloc			Traditional cultivation		
Initial fill		20 m^3	Initial fill		20 m^3
6 top-offs at 20%		24 m^3	84 replacements at 30%		504 m^3
			11 replacements at 60%		132 m^3
	Total	**44 m^3**		**Total**	**656m^3**

Table 2. Water consumption in prawn cultivation.

Additional water to maintain biofloc water quality was 24 m3, while water to maintain traditional cultivation was 636 m3. The biofloc system saved about 96% of the water needed to maintain nontoxic conditions during production. An additional large saving in electrical expenses was achieved, estimated at about 96% during production time. This is similar to the findings of [1, 2, 3].

3. Conclusion

The potential of biofloc technology applied to shrimp farming to promote good aquaculture practices is manifold resource sustainability and environmental care and in reduction in energy consumption. This is important if we expect to maintain current growth rates of aquaculture. Aquaculture is now competing for space and water with other food-producing activities, so that properly designed and improved systems to maintain high biological load in a relatively small space is essential. Intensive biofloc system is a strategy that will promote the growth of aquaculture [32, 33].

Expanding systems of semi-intensive and intensive production aquatic animals will ead to increasing volumes of waste nitrogen and solids that foul the water [35]. Therefore, reducing effluents and effluent pollution to near zero can only benefit the downstream quality of water in rivers, estuaries, lagoons, and nearshore environments [35].

While closed recirculation systems increase the costs of installation of equipment and operation of a farm (pumps, clarifiers, biological filters), nitrifying bacteria to maintain water quality and reduce environmental impact of biofloc systems lead to a large increase the density of the fish and shrimp and their final biomass, which more than compensates for the initial investment. In a closed, recirculating system, the biological treatment is within the water. Despite being efficient, recirculating systems require auxiliary equipment (pumps, filters, settlers) that increase installation costs, may limit production volumes, and increases operating costs resulting from continous pumping during the crop cycle.

Biofloc, as a culture, is a closed system that works in the same cultivations tanks to largely natural maintainance of the quality of water. In turn, the environmental impact is greatly reduced. Another advantage of biofloc systems is that the naturally occurring organisms in the system are used as complementary food, which reduces consumption of commercial feed, which usually contain products from marine fisheries. This helps to reduce the pressure on fisheries to provide ingredients for diets used in aquaculture.

Author details

Carlos I. Pérez-Rostro[1], Jorge A. Pérez-Fuentes[1] and Martha P. Hernández-Vergara[2]

1 Laboratory of Genetic Improvement and Aquaculture Production, Technological Institute of Boca del Río, Veracruz, Mexico

2 Laboratory of Native Crustacean Aquaculture, Technological Institute of Boca del Río, Boca del Río, Veracruz, Mexico

References

[1] Avnimelech Y., Lacher M. Carbon/nitrogen ratio as a control element in aquaculture systems. Aquaculture 1999; 176: 227–235.

[2] Crab R., Defoirdt T., Bossier P., Verstraete W. Nitrogen removal techniques in aquaculture for a sustainable production. Aquaculture 2007; 270: 1-14.

[3] Schryver P., Crab R., Defoirdt T., Boon N., Verstraete W. The basics of bio-flocs technology: The added valuefor aquaculture. Aquaculture 2008; 277: 125-137.

[4] Schramm A., D. de Beer A., Van den Heuvel S., Ottengraf S. Microscale distribution of populations and activities of Nitrosospira and Nitrospira spp. along a macroscale

gradient in a nitrifying bioreactor: quantification by in situ hybridization and the use of microsensors. Applied and Environmental Microbiology 199; 65: 3690-3696.

[5] Ferrer, J., Seco A. Tratamientos Biológicos de Aguas Residuales. Editorial UPV. Valencia, Spain, 2007. Pp.184.

[6] Shan H., Obbard J. Ammonia removal from prawn aquaculture water using immobilized nitrifying bacteria. Appl Microbiol Biotechnol 2001; 57:791–798.

[7] Avnimelech Y., Kochva M., Diab S. Development of controlled intensive aquaculture systems with a limited water exchange and adjusted carbon to nitrogen ratio. Israel Journal of Aquaculture- Bamidgeh 1994; 46: 119–131.

[8] Chamberlain G., Avnimalech Y., Mcintosh RP., Velasco M. Advantages of aerated microbial reuse systems with balanced C/N. In: Nutrient transformation and water quality benefits. Global Aquaculture Allliance Advocate 2001; 4: 53–56.

[9] Burford MA., Thompson PJ., McIntosh PR., Bauman RH., Pearson DC. The contribution of flocculated material to shrimp *(Litopenaeus vannamei)* nutrition in a high-intensity, zeroexchangesystem. Aquaculture 2004; 232: 525–537.

[10] Asaduzzaman M., Wahab MA., Verdegem MCJ., Huque S., Salam MA., Azim M. C/N ratio control and substrate addition for periphyton development jointly enhance freshwater prawn *Macrobrachium rosenbergii* production in ponds. Aquaculture 2008; 280: 117–123.

[11] Kurup BM., Prajiht KK. Application of biofloc technology in the semi intensive culture system of giant prawn *Macrobrachium rosenbergii*. World Aquaculture Society: Aquaculture 2010 San Diego, California 2010. https://www.was.org/.../meetings/SessionAbstracts.aspx

[12] Megahed, ME. The Effect of Microbial Biofloc on Water Quality, Survival and Growth of the Green Tiger Shrimp (*Penaeus Semisulcatus*) Fed with Different crude Protein Levels. Arabian Journal of the Aquaculture 2010; Vol.5 No. 2, p. 119-142.

[13] Emerenciano M., Ballester ELC., Cavalli RO., Wasielesky, W. Biofloc technology application as a food source in a limited water exchanger nursery system for pink shrimp *Farfantepenaeus brasiliensis* (Latreille, 1817) 2011. Aquaculture Research,doi: 10.1111/j.1365-2109.2011.02848.x

[14] Pérez-Fuentes JA., Pérez-Rostro CI., Hernández-Vergara MP. Pond-reared Malaysian prawn *Macrobrachium rosenbergii* with the biofloc system. Aquaculture 2013; 400-401:105-110.

[15] Goldman JC., Caro D.A., Dennett MR. Regulation of gross growth efficiency and ammonium regeneration in bacteria by substrate C:N ratio. Limnology and Oceanography 1987; 32 (6), 1239–1252.

[16] Avnimelech Y., Panjaitan P. Effects of carbon:nitrogen ratio control on water quality and shrimp growth in zero water exchange microcosms. Abstracts. World Aquaculture, Florence, Italy; 2006.

[17] Hargreaves JA. Photosynthetic suspended-growth systems in aquaculture. Aquac. Eng. 2006; 34: 344–363.

[18] Ray JA., Lewis LB., Browdy LC., Leffler WJ. Suspended solids removal to improve shrimp (Litopenaeus vannamei) production and an evaluation of a plant-based feed in minimal-exchange, superintensive culture systems. Aquaculture 2010; 299:89-98.

[19] Rakocy JE. Tank Culture of Tilapia. Southern Regional Aquaculture Center (SRAC) 1989. Publication Number 282. Texas Agricultural Extension Service, Texas A&M University, College Station, Texas, USA. Pp. 4.

[20] Gerardi M. Nitrification and denitrification in the activated sludge process 2002; Wiley-Interscience. Nueva York. Estados Unidos.

[21] Hanson TR., Posadas BC., Samocha, T.M., Stokes AD., Losordo TM., Browdy, C.L. Economic factors critical to the profitability of suer-intensive biofloc recirculating shrimp production system for marine shrimp Litopeneus vannamei. Pages 267-283 in C.L. Browndy and D.E. Jory, eds. The resing Tide,Proceedings of the Special Session on Sustainable Shrimp Farmin, World Aquaculture 2009.The World Aquaculture Society, Baton Rouge, Louisiana.

[22] Azim ME., Little, DC. The biofloc technology (BFT) in indoor tanks: Water quality, biofloc composition, and growth and welfare of Nile tilapia (*Oreochromis niloticus*). Aquaculture 2008; 283, 29–35.

[23] Avnimelech Y. Biofloc Tecnology – A practical guide book, 2nd edition. The World Aquaculture Society, Baton Rouge, LA; 2012.

[24] Ekasari J., Crab, R., Verstraete W. Primary Nutritional Content of Bio-flocs Cultured with Different Organic Carbon Sources and Salinity. *HAYATI Journal of Biosciences 2010;* 17(3):125-130.

[25] Milstein A., Avnimelech Y., Zoran, M., Josep D. Growth performance of hybrid bass and hybrid tilapia in conventional and active suspension intensive ponds. Israeli Journal of Aquaculture – Bamidgeh 2001; 53 (3–4): 147-157.

[26] Serfling SA. Microbial flocs. Natural treatment method supports freshwater, June 2006, marine species in recirculating systems. Global Aquaculture Advocate 2006; 34–36.

[27] Martínez-Córdova LR., Porchas-Cornejo MA., Villarreal-Colmenares H., Calderon-Perez JA. Winter culture of yellow leg shrimp *Penaeus californiensis* in aerated ponds with low water exchange. *Journal of the World Aquaculture Society* 1998; 29(1): 120-124.

[28] New, M.B., Singholka, S. Cultivo de camarón de agua dulce. Manual para el cultivo de *Macrobrachium rosenbergii*. FAO, Doc. Tec. Pesca 1984; (225): 118 pp.

[29] Arboleda-Obregón DA. Limnología aplicada a la acuicultura. Revista Electrónica de Veterinaria 2006, Vol. VII, No. 11. http://www.veterinaria.org/revistas/redvet/n111106.html. 23/05/2010.

[30] García-Rodríguez J., Granados-Ramírez JG., Quiroz-Castelán H., Molina-Astudillo FI., Díaz-Vargas M. Utilización de fertilizantes y desechos agrícolas para el crecimiento del langostino *Macrobrachium rosenbergii* (de man), en estanques rústicos. REDVET Revista electrónica de Veterinaria 2005; Vol. VI, Núm. 8, agosto-sin mes, pp. 1-10.

[31] Kuhn DD., Boardman GD., Lawrence AL., Marsh L., Flick-Junior GJ. Microbial floc meal as a replacement ingredient for fish meal and soybean protein in shrimp feed. Aquaculture 2009; 296:51-57.

[32] Twarowska JG., Westerman PW., Losordo TM. Water treatment and waste characterization of an intensive recirculating fish production system. Aquacult. Eng. 1997; 16: 133–147.

[33] Thoman ES., Ingal ED., Davis DA., Arnold CR. A nitrogen budget for a closed, recirculating mariculture system. Aquacultural Engeneering 2001; 24: 195-211.

[34] Goddar S. Feed Management in Intensive Aquaculture. Chapman & Hall,N.J.;1996. p. 189.

[35] Páez-Osuna F. Retos y perspectivas de la camaronicultura en la zona costera. Revista Latinoamericana de Recursos Naturales 2005; 1: 21-31.

[36] Frías-Espericueta MG., Harfush-Meléndez M., Osuna-López JI., Páez-Osuna F. Acute toxicity of ammonia to juvenile shrimp *Penaeus vannamei* Boone. Bulletin of Environmental Contamination and Toxicology 1999; 62: 646-652.

[37] Frías-Espericueta MG., Harfush-Meléndez M., Páez-Osuna F.Effects of ammonia on mortality and feeding of postlarvae shrimp *Litopenaeus vannamei*. Bulletin ofEnvironmental Contamination and Toxicology 2000; 65: 98-103.

[38] Mallasen M., Valenti WC. Effect of nitrite on larval development of giant river prawn Macrobrachium rosenbergii Centro Avançado do Pescado Continental, Instituto de Pesca, Caixa Postal 1052, 15025-970, São José do Rio Preto, SP, Brazilb São Paulo State University, Aquaculture Center (CAUNESP) and College of Agriculture and Veterinarian Sciences,14884-900, Jaboticabal, SP, Brazil. Aquaculture 2006; 261:1292–1298. http://www.caunesp.unesp.br/publicacoes/artigos/valenti/RP_MALLASEN_Effect%20of%20nitrite%20on%20larval.pdf

[39] Pierce RH., Weeks JM. Nitrate toxicity to five species of marine fish. Journal of the World Aquaculture Society 1993; 24: 105-107. Colt JE., Armstrong DA. Nitrogen toxicity to crustaceans, fish and molluscs. Proceedings of the Bio-Engineering Symposi-

um for Fish Culture. Fish Culture Section of the American Fisheries Society 1981; (FCS Publ. 1): 34-47.

[40] New MB. Farming freshwater prawns: a manual for the culture of the giant river prawn (*Macrobrachium rosenbergii*). FAO Fisheries Technical Paper No. 428. FAO, Rome, Italy, 2002. 212p.

[41] Emerson K, Russo RC., Lund RE., Thurston RV. (1975). Aqueous ammonia equilibrium calculations: effect of pH and temperature. Journal of the Fisheries Research Board of Canada 1975; 32: 2379-2383.

Chapter 4

Sustainable Technique for Selected Live Feed Culture

Zaleha Kassim, Akbar John, Lim Keng Chin,
Nur Farahiyah Zakaria and Nur Hidayah Asgnari

Additional information is available at the end of the chapter

http://dx.doi.org/10.5772/57212

1. Introduction

Sustainability in the aquaculture industry depends on several factors including the minimum production cost in comparison to the yield, unexpected environmental conditions which affect the farm and practices in the farm management itself. These factors are inter-connected and always incur a synergistic effect on the issue of sustainability. Live feeds as the fundamental needs for larval rearing and fry production have to be prioritised for sustainable farming activity. Dependency on imported sources of live feeds or inert feed will increase the production cost. Thus, the continued activity of screening, stocking and maintaining some local species as an option for live feed production is economically necessary.

Live feeds are an important basic diet for newly-hatched fish and shrimp larvae as they still have an incomplete digestive system and are lacking in enzymes. They are still at a very young stage to generate their own required nutrients or convert them from any pre-cursor obtained from a diet. They need a ready-made diet with readily available nutrient to be absorbed through their digestive system. There have been many species suggested or tested for their potential as live feed. All test animals were mostly zooplankton in nature and must meet the requirement as live feed. They must be in a compatible size with the mouth size or gape of the larvae predator, or they cannot be swallowed. Since larvae are still weak to track down the food, the wave created by the prey will be a great help, thus 'active' swimming prey is preferred. The most important role of a prey is the ability to supply energy and other nutrients which are essential for the larval survival and growth. Live feeds, as the starter diet in larval rearing and fry production must be continuous in supply. Good, nutritious and compatible-size prey must be able to reproduce fast to meet the requirement and adaptable to a simple mass-production technique.

1.1. Copepods as live feeds

The conventional live feed, brine shrimp and rotifers, are considered unsuitable as live feed if compared to copepods in term of nutritional value. *Artemia* sp. is deficient in polyunsaturated fatty acids (PUFAs), thus it needs to be enriched before feeding [1, 2]. Similarly, rotifer have poor nutritional value and are small in size [3]. On the other hand, copepod diets were proven to increase the growth of larval marine fish compared to diet of rotifers *Brachionus plicatilis*, [4, 5] or *Artemia* [4]. The potential use of copepods as live feed due to their excellent fatty acid content has been highlighted by using an example, a paracalanid [6]. They improved the quality of the cultured organism, particularly the larval stages. The superiority of copepods over other live feed such as brine shrimp and rotifers was further confirmed [7]. They have the appropriate ratio of docosahexaenoic acid (DHA) to eicosapentaenoic acid (EPA) which will improve the growth and survival of fish larvae if compared to the conventional live feeds [8]. Nonetheless, it is reminded that copepods could be better used as supplement rather than a sole diet in larval rearing, unless they are used for some high-valued commercial reef fish due to the high operation cost [9]. The possibility of using copepods, particularly the harpacticoids as alternative live feed in fish larval rearing has been stressed [10] and it is confirmed that the nutritional quality of harpacticoid copepods as live feed is extremely high [11,12]. *Macrosetella gracilis*, a planktonic harpacticoid copepod, is also reported to have better diet quality when compared to *Artemia* [13].

Despite these positive findings, rotifers and *Artemia* continue to be the live feeds of choice in commercial hatcheries, because copepods are not currently cultured at sufficient densities to be economically efficient on a commercial scale [14, 15, 16]. In term of culture condition, it was found that the optimum condition for the high production of a tropical harpacticoid copepod, *Pararobertsonia* sp., was at salinity 35psu and temperature of 25°C [17]. The fluctuation in salinity, pH and temperature in the culture vessel would definitely influence the reproduction and population growth of copepods such as the harpacticoids if kept in small containers [18]. Nonetheless, a strategy to produce harpacticoid in large quantities for hatchery use by using a tray-culture method has been suggested [19].

Another copepod group, a Cyclopoida, *Apocyclops dengizicus* was found to increase the survival and growth of *Panaeus monodon* postlarvae when used as live feed [20]. Cyclopoids are omnivorous, and can be fed a mixture of feeds, mainly phytoplankton or a combination of phytoplankton, yeast or other feeds [7]. As for *Apocyclops panamensis*, there is a report on a successful technique for outdoor ponds [15]. Information on the use of copepods in aquaculture, particularly from the tropical *Apocyclops* sp., is still scarce. The species reported in abundance and potentially exploited as live feeds for shrimp post-larvae in Malaysia for example is *A. dengizicus*. A new species, *Apocyclops ramkhamhaengi*, has been described [21] and added to the present report of 3 species of *Apocyclops* recognised from Asia: *A. dengizicus* (Lepeshkin), *A. royi* (Lindberg), and *A. borneonensis* (Lindberg). This new species is found in abundance in eastern Thailand water and has yet to be determined for its potential in aquaculture. Planktonic copepods such as cyclopoids feed on other plankton including planktonic microalgae. To maintain planktonic copepods in the hatchery or aquaculture ponds, a continuous supply of their diet, particularly the microalgae, will definitely be required.

1.2. Microalgae

Microalgae are a diverse group of unicellular autotrophs inhabiting almost all aquatic water bodies. Microalgae are rich in many specific and attractive compounds [22] and their nutritional values for aquaculture had been highlighted [23]. Production of microalgae is mandatory in the hatchery as it is a basic and nutritious diet for live feed, specifically the zooplankton. However, its mass production is generally costly due to huge manpower, space requirements and operation which usually related to the cost of the energy used. A good strategy in manipulating the culture environment, particularly during the production process of microalgae would scale down the operational cost.

Light plays a fundamental role in the development of microalgae through photosynthesis. It is one of the major environmental factors which control the performance of microalgae phototrophic growth and productivity [24, 25, 26]. Light may either be natural or supplied by fluorescent tubes giving the maximum effective radiation which can be absorbed by the pigments of the microalgae. Light intensity plays a vital role, but the requirements vary with the culture depth or volume as well as the density of the algae in the culture. At a higher volume, light intensity must be increased to enable it to penetrate through the culture. However, an extreme light intensity may result in photo-inhibition which reduces the photosynthetic rates and growth [27,28]. Furthermore, overheating due to artificial or natural illumination should be avoided in microalgal culture. The most often employed light intensity is 1000 lux which is suitable for Erlenmeyer flasks but 5000-10000 lux is needed for a greater volume of microalgal culture [29, 30]. The duration of illumination can be varied where photosynthesis of microalgae can be enhanced or increased in the light/dark (LD) cycle (usually 12:12 or 14:10 LD, maximum 16:8 LD). For some microalgae, photosynthesis rate could also be increased exponentially with increasing light/dark frequencies where a longer period of dark in relation to the light period can further increase photosynthetic efficiencies but not vice versa [31]. The illuminations also affect the nutrient utilisation in the culture vessel [32].

A cost-effective and nutritionally-adequate alternative to costly maintenance of live microalgae is the production of moist-microalgae concentrates. It is seen to simplify hatchery procedures and has shown promising potential in the aquaculture industry [33,34].The storing of microalgae concentrates in moist form under low temperature can preserve their high nutrient composition and excellent cell viability [35,33]. Juvenile pacific oyster (*Crassostrea gigas*) fed with different algal pastes had shown significant improvement in growth rate than oyster fed with other diets [36]. Concentrates of *Chaetoceros muelleri* and *Tetraselmis pseudonana* refrigerated for 6 weeks at 4°C were found to promote similar survival rates of the tiger prawn *Penaeus monodon* larvae fed with live microalgae [37]. Supplementation of microalgae concentrates to bivalves, oysters and scallops have also recorded the same extent of growth rates as live microalgae [38,34]. It has been documented that most of the demand for mariculture feed in Japan is supplied with live and fresh microalgae which is thickly concentrated and readily stored at 2-4ºC for 1-8 weeks with good shelf life [39].

Preparation of concentrated condition of microalgae usually involves centrifugation technique. Nonetheless, although this technique has been successfully applied and utilised for preparing microalgae concentrates, it poses some limitations. First, the process involves exposure of cells to high gravitational and shear forces deteriorating the cell structure with the leaking of nutritional contents. Second, centrifuging large volumes of cultures is time-consuming and requires expensive equipments. Several alternative procedures, less damaging to the cells, which can be applied are filtration [40], foam fractionation [41] and flocculation [33, 34, 38]. Previous studies have observed the excellence of ultrafiltration technique in preserving and retaining the cellular structure and properties of fragile algal cells with little loss of material [42, 43].

The level of natural resources exploitation for aquaculture purposes is commonly high. Coastal land and mangroves forests always become the target area for brackish-water aquaculture ponds. The water source of this area, which is always from the nearby river estuary and lagoon, is also used as the live feeds (zooplankton and microalgae) source. Nonetheless, the supply is always seasonal and could become unavailable unexpectedly due to many factors and natural phenomena. This chapter aims to discuss the possible ways to produce local live feeds, a marine microalgae species and a planktonic copepod, sustainably using a simple technique for larval-rearing purposes. Maintaining local species is hypothesised to be more economical and practical. The usage of the microalgae as an enrichment element for live feed copepods will be proved.

2. Methodology

Experiment 1: Production of *Chlorella vulgaris* Concentrate Isolated from Bidong Island and Assessment as Copepod Diet

Seawater samples were obtained from Bidong Island, Terengganu. The collection was made by lowering a Niskin water sampler to a required depth, following the light-penetration depth. Concentrated water samples were then transferred into chilled, white-plastic containers and brought back to the laboratory for microalgae isolation process. Successive plating out on agar plates was performed in order to select the desired marine *Chlorella* colonies. Monospecific colonies were then transferred into trial culture tubes before scaling up into larger volumes of Erlenmeyer flasks.

The microalgae was then cultured for the preparation of moist concentrates in the temperature controlled room (20±2°C) using the standard batch culture method. Triplicate of actively-growing starter cultures were inoculated into 30 litres acrylic tanks enriched with Conway medium under constant illumination (cool-white type; 110 watts). All cultures were started with an initial inoculum of $2x10^6$ cells mL^{-1}. Cultures were aerated continuously using humidified filtered air. Evaporation in the culturing tanks was kept to a minimum by covering the top of the tanks. Cellular density of microalgae cultures was monitored daily using a Neubauer haemocytometer [29]. Scanning electron microscopic observation was also done to determine the ultra structure of the cell. Measurement of radius and height of the target microalgae cells was done under the advanced research microscope (Model Nikon Eclipse 80-i, Japan) and twenty individual cells were measured for the calculation of cell biovolume to

http://dx.doi.org/10.5772/57212

avoid biasing results. Cell biovolume was calculated as assumed round-shape volume with the following formula proposed by Sun and Liu [44]:

$$\text{Cell volume} = 4/3\ \pi R^3 \tag{1}$$

Where, π= 3.142, R= radius of cell

Specific growth rate was calculated from the expression as proposed [45] which is shown below:

$$\text{Specific growth rate } (\mu) = \ln(F_1/F_0)/t_1\text{-}t_0 \tag{2}$$

Where, μ = specific growth rate, F_1= biomass at time harvest, t_1 and F_0= biomass at time zero, t_0.

Doubling time was computed based on the formula as proposed [45] which is shown below:

$$\text{Doubling time}(\grave{o}) = \log(2)/\mu \tag{3}$$

Where, T = doubling time, μ = specific growth rate.

All microalgae cultures were grown to late-logarithmic phase for the preparation of concentrates via ultrafiltration technique. The concentrated aqueous suspensions of microalgae were filtered through a membrane filter (0.1 μm pore size) to remove access water from the suspension without rupturing the microalgae, thereby obtaining the microalgae concentrate or paste. Cell viabilities of microalgae concentrates were assessed using Eosin dye as a viability assay on the basis of its penetration into non viable-cells based on the expression as proposed [46]:

$$\text{Cell viability } (\%) = \frac{\text{Viable cells x 100}}{\text{Total cells}} \tag{4}$$

The harvesting efficiency or percentage recovery (%) was evaluated by comparing the remaining total number of cells in the concentrate with the total number of cells before filtration with the following expression:

$$\text{Harvesting efficiency/Percentage recovery } (\%) = C_B/C_A \text{x } 100 \tag{5}$$

Where, C_B = total number of cells before filtration, C_A = total number of cells after filtration

Microalgal concentrates were compared to live cultures of the same algae as food for marine copepods. Copepods were obtained from existing culture in UMT's laboratory. Two different sets of cultures were done using a Petri dish where each of them was fed with live and microalgae concentrate respectively. Individual copepods were counted daily under the Leica stereo microscope before being fed (1 drop). The maximum specific growth rate (K) was calculated [47] as shown below:

$$K=\ln\left(X_1/X_0\right)/t_1\text{-}t_0 \quad (6)$$

Where, K = specific growth rate, X_1= the number of copepods at harvest time, t_1 and X_2= the number of copepods at time zero, t_0

The doubling time was computed as:

$$\text{Doubling time}\left(\grave{o}\right)=\log\left(2\right)/K \quad (7)$$

Where, τ = doubling time, K = specific growth rate.

Experiment 2: Effects of Photoperiod and Culture Size on *Chlorella vulgaris* Stock Growth

Pure strains of *Chaetoceros* sp. and *C.vulgaris* were obtained from the microalgae maintenance laboratory at Institute of Tropical Aquaculture, Universiti Malaysia Terengganu, Malaysia, while the pure strain of *Nannochloropsis* sp. was obtained from the Fisheries Research Centre, Pulau Sayak, Kedah, and was maintained in autotrophic conditions in liquid and semisolid agar with Conway media [48]. This axenic culture was stored at 25±2^{0}C for 3 days which served as an inoculum for further experiments.

Microalgae were grown in autotrophic conditions as a monospecific axenic culture in different volumes (250mL, 500mL and 2000mL) containing Conway media. 25mL of pure strain with the cell density of ~2 x 10^6 cells mL^{-1} were transferred to each Erlenmeyer culture flask and kept at complete illumination provided by luminescent tubes (1000 Lux). Carbon source was provided by bubbling sterile 2% (v/v) CO_2 in air through the cultures. Culture flasks were maintained at a constant temperature (22°C± 1°C) with the pH range of 7-8 and salinity of ~35ppt in an air-conditioned laboratory over 2-3 weeks. Daily cell count was calculated using a haemocytometer. To determine the effect of different photoperiods, microalgae cultured in a 2 litre flask containing Conway media was treated at different photoperiods (light/dark) (24:0, 12:12 and 8:16) in replicates and cell count was achieved as mentioned above. Growth curve for each species of algae was constructed and One-way ANOVA with Dunnett's post-test was performed using Graph-Pad Prism.

Mean cell count and specific growth rate were calculated using the formula $X=\frac{\Sigma X_i}{n}$ and $SGR=Ln\,\frac{W_2/W_1}{t_2-t_1}$ respectively (where, X = mean cell count; χi = total number of cells; 'n'= number of cell counts; SGR = specific growth rate; W_1= Initial cell density, $cell^{-1}$; W_2= Cell density at late exponential phase, $cell^{-1}$; t_1 = Time at initial cell density, $cell^{-1}$; t_2 = Time at late exponential phase, $cell^{-1}$).

http://dx.doi.org/10.5772/57212

Experiment 3: Low-Cost Commercial Fertiliser for Mass Culture of Marine *Chlorella vulgaris*: Manipulation of N:P:K Ratio

An investigation was made to see the adaptability of the local marine *C. vulgaris* to the natural conditions in an aquaculture farm. This means that they need to adapt to different fertilisers other than Conway media, different salinity regimes and uncontrolled temperatures. Preparation of NPK-based fertiliser was made by manipulating the ratio of nitrogen, phosphorus and potassium source as summarises in Table 1. Each of the different N:P:K ratio treatments was prepared in triplicate. Source of nitrogen was obtained by using urea fertiliser.

Culture containers were well-cleaned with bleach and rinsed thoroughly before filling up with 1L of the farm water (salinity of between 20-25ppt). The marine *C.vulgaris* concentrate was prepared and 1mL of it was inoculated into the container and 1mL of the fertilizer was added. The containers were vigorously aerated to provide required quantity of oxygen and to keep cells and media in suspension. The containers were kept in the open under 100% outdoor light exposure.

N:P:K ratios			Type and fertiliser used		
N	P	K	Urea (g)	P* (mL)	Potash* (g)
1	1	1	0.98	0.98	0.98
15	15	15	14.7	14.7	14.7
8	8	2	7.84	7.84	1.96
16	8	6	15.68	7.84	5.88
12	6	4	11.76	5.88	3.92
12	8	4	11.76	7.84	3.92

Table 1. Type and fertiliser used in N:P:K ratio for mass culture of marine *Chlorella vulgaris*

Sampling of microalgae cells was done daily and counting was carried out using a Neubauer Hemocytometer covered with glass slide under a compound microscope.

The growth rate, divisions per day, and generation time or doubling time was calculated following [49]

$$\text{Growth rate; } K' = \text{Ln}\left(N_t/N_o\right)/\left(t_2 - t_1\right) \quad (8)$$

$$\text{Divisions per day; } \text{Div.day}^{-1} = K'/\text{Ln}2 \quad (9)$$

$$\text{Generation time}\left(\text{days}\right);\ \text{Gen't} = 1/\text{Div.day}^{-1} \quad (10)$$

$$\text{Generation time } (\text{hours});\ \text{Gen't} = 24\left(1/\text{Div.day}^{-1}\right) \tag{11}$$

Where, No and Nt = final and initial populations at time t1 and time t2, respectively.

Since sample was collected daily, therefore, t2 – t1 = 1.

Experiment 4: Egg Production, Growth and Development of *Apocyclops ramkhamhaengi* Fed on Marine *Chlorella vulgaris*

Detailed observation on the reproduction performance of a zooplankton depending solely on a *C.vulgaris* diet was planned to prove the important role played by this local microalgae in live-feed production. Samples of copepods were collected from Sungai Semerak (N 05° 51.737, E 102° 30.809'), Tok Bali, Kelantan using a zooplankton net. This area receives sea water from the South China Sea, which is near to the Thailand coast where the copepod species was first identified and reported. Live copepods were maintained and adapted to the laboratory environment. Sand-filtered sea water from the Marine Hatchery, Universiti Malaysia Terengganu was diluted with deionised water to be at salinity of 25ppt and was further filtered through a GFC membrane filter and then autoclaved at 121°C for 15 minutes [11]. Salinity was measured using a portable hand-refractometer (ATAGO, Japan). Microalgae diet for the copepod was prepared from the marine algae *C.vulgaris* stocked at the Marine Hatchery. The microalgae were cultured in 29-31ppt Conway medium with 24h-light, room temperature of 25-27°C and continuous aeration for 7days. The cell concentration in each 500ml conical flask was determined by using Neuber haemocytometer (0.25mm^2 x 0.1 mm) under a compound microscope. The algal production was done weekly and supplied to *A. ramkhamhaengi* culture.

The investigation on the reproduction performance started with fifteen gravid females of *A. ramkhamhaengi* placed into two sets of triplicate of 250mL beakers. The diet constituted,1mL of Baker's yeast (0.02g/L) and 1mL of *C.vulgaris* at density 1x10^{6}cells/mL which were introduced into both sets of the beakers and covered with parafilm layer to avoid contamination. Three subsamples (approximately 1mL) from each beaker of the cultures were observed daily. The number of the copepods at all stages, including nauplii, copepodite, adult and gravid female, were counted under a dissecting microscope (Leica ZOOM 2000) and then returned to the culture. Changing of approximately 80% of the culture medium was done every alternate day by passing the copepods culture through 100 and 40 microns nylon net which would retain all stages of copepods (the smallest size of 60 microns) but remove most of the waste.

The population growth of *A. ramkhamhaengi* was studied for 30 days. The specific growth rates (K) of all stages of the copepods in both diets given were calculated by using the formula [50]:

$$K = \frac{\ln Nt - \ln No}{t} \tag{12}$$

Where, t is the culture days, No and Nt is the number of copepods at the initial and final selected time interval. The doubling time (Dt) was calculated by dividing $\log_e 2$ by the population growth rate (K) of all stages of *A. ramkhamhaengi* in both diets given:

$$Dt = \frac{\log_e 2}{K} \tag{13}$$

Although cyclopoid copepods are known to suspend in water column, *A. ramkhamhaengi* showed its adaptability to swim on the near bottom of its culture vessel. The culture for this experiment was started by introducing a gravid female on the experimental petri dish. The adult was removed after the eggs hatched and the nauplii were monitored until they reached copepodite-v stage and were ready to mate. Adult females and males from the culture were prepared for the experiment. A pair of male and female was put into each set of glass Petri dish filled with 15mL sea water. The use of a Petri dish instead of a beaker eased the daily observation of different stages of the copepod in the population. The cultures were maintained at room temperature of 25-27°C without additional oxygen supply or aeration. Observation was done twice per day under a dissecting microscope (Leica ZOOM 2000) before feeding to avoid the disturbance of the diet materials during individual or population counting. The culture medium was changed approximately 80% daily, and culture containers were subsequently changed every 4days. Daily feeding was done in the morning and evening by dropping 1mL of 1 x 10^6 cells/ml *C.vulgaris* into the culture. The time taken for the females to become gravid was based on the observation recorded twice per day (morning/evening). Once the females become gravid, the male broodstocks were removed, and the female were left alone inside the Petri dish in order to determine the number of eggs per female from its first copulation. Observation on the development time from nauplii to adult, maturation time and generation time of *A. ramkhamhaengi* were recorded coupling with the numbers of offspring produced and percentage of hatching.

3. Result and discussions

Experiment 1: Production of *Chlorella vulgaris* Concentrate Isolated from Bidong Island and Assessment as Copepod Diet

The ultra-structure of the *C.vulgaris* isolated and cultivated in this study is shown in Figure 1. The scanning electron micrographs displayed the characteristic features of green single cells with spherical shape and possession of rigid cell wall. There are some differences found in the present specimen if compared to some other established species. The outer shell is rough if compared to the latest SEM of *C.sorokiana* [51].The feature is almost the same as found in SEM of *C.vulgaris* [52]. In terms of size, the specimen was found to be in between the size of marine *C.vulgaris* (2.1μm) and estuarine *C.vulgaris* (2.3μm) from Korean waters [53].

Kingdom PLANTAE
Phylum CHLOROPHYTA
Class CHLOROPHYCEAE
Order CHLORELLALES
Family CHLORELLACEAE
Genus *CHLORELLA*

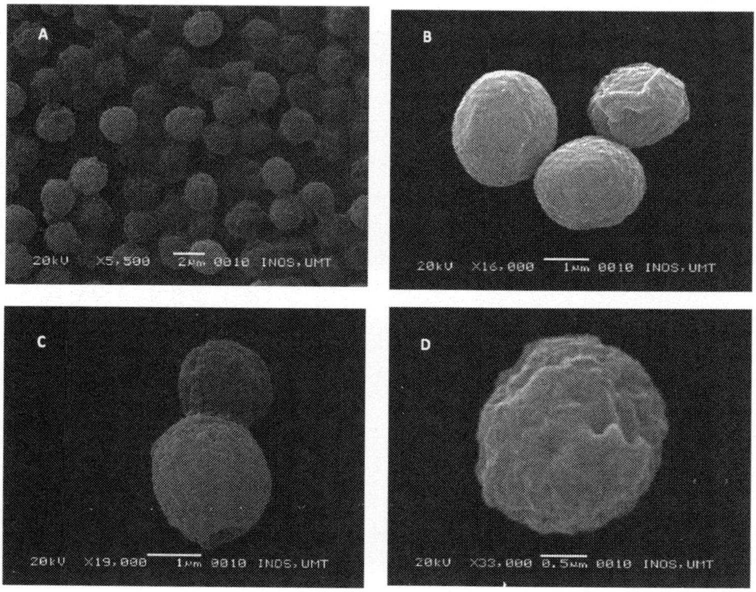

Figure 1. The scanning electron micrographs of *C.vulgaris* under different views and various magnifications. A, Cell aggregation. B, *C.vulgaris* cells under different magnification C. Cell divison in *C.vulgaris*. D, Single cell and cell wall structures.

The cell has an average cell biovolume of 5.26±0.87 μm^3. The cell densities changed following the culture period in both culture of concentrates (paste) and live condition (although they both started at the same density). Nonetheless, they followed more or less the same growth patten. The variation in cell densities during the experimental period is shown in Figure 2. Cell density of *C.vulgaris* increased rapidly to 227.22±0.87 $\times 10^6$ cells mL^{-1} prior to stationary phase. After that, the cell densities maintained at this point for ten days before decreasing significantly thereafter (Figure 2). The average specific growth rate (SGR, μ) achieved during the exponential phase was 0.660±0.001 day^{-1} with the doubling time (T) of 0.580±0.004 hour which then decreased drastically to 0.126±0.001 day^{-1} during the retardation phase with the doubling time of 2.420±0.019 hour before the death phase. Based on cell density and growth rate observed, the following growth phase is described:

i. Exponential (log) phase (days 0-6),

ii. Declining of relative growth rate phase (days 6-12),

iii. Stationary phase (days 12-22),

iv. Death phase (days 22-26).

C.vulgaris paste was successfully concentrated from the pure culture isolated from Bidong Island. This concentrate contains cell density of approximately 58.46±2.44 x10^9 cells mL^{-1} - 227.22±0.82 x10^6 cells mL^{-1}. The present result also showed that this *C.vulgaris* concentrate can still be inoculated after refrigeration for a duration of six weeks and exhibited similar growth characteristics as the live culture (Figure 2). The cells had very high viability even after 6 weeks of storage in chilling conditions (4ºC) as shown in Figure 3. It is interesting to note that the paste had recorded the highest cell viabilities of 99.51±0.57% and continued to display slow and steady decrement of cell viabilities to 83.28±0.58% on the sixth week of storage. Microscopic examination also indicated that the cells were in single forms without any aggregation occurring and can be readily dispersed in seawater medium as single suspension of cells upon inoculation (Figure 1A). The harvesting efficiency of the ultrafiltration technique using membrane filter had recorded a very high percentage recovery of 93.14±1.35% showing the effectiveness of this technique for harvesting and concentrating the microalgae biomass.

Copepod species, *Apocyclops* sp., showed a higher population density when fed with *C.vulgaris* paste (60±4.36 individual mL^{-1}) than with the live culture (14.33±0.58 individual mL^{-1}) (Figure 4). In addition, the copepod populations fed with this concentrate exhibited a higher instantaneous growth rate, K=0.455±0.008 day^{-1} and faster doubling time (0.662±0.012 hour) than live culture which recorded an instantaneous growth rate of 0.296±0.005 day^{-1} and doubling time of 1.108±0.016 hour. Results from one-way ANOVA test has shown that there is a significant difference (P<0.005) between the copepod population densities fed with the *C.vulgaris* concentrate and live culture.

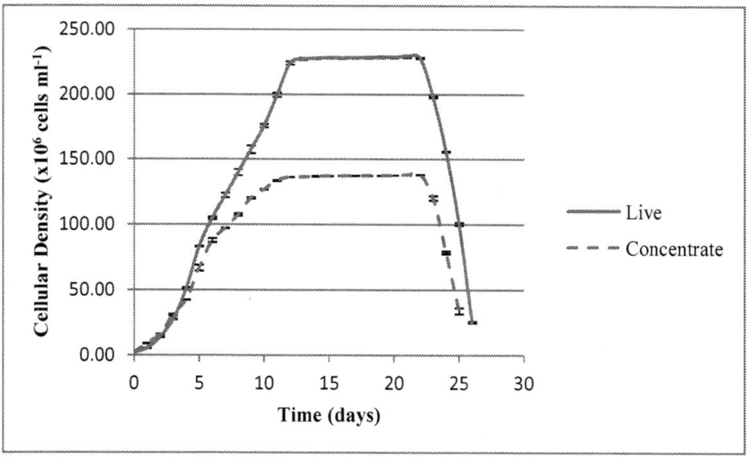

Figure 2. Cellular densities of live *C.vulgaris* (30L) and concentrate/paste (after reinoculation in 5L) cultured under laboratory conditions with Conway medium. Data are mean value and standard deviation of 3 repetitions.

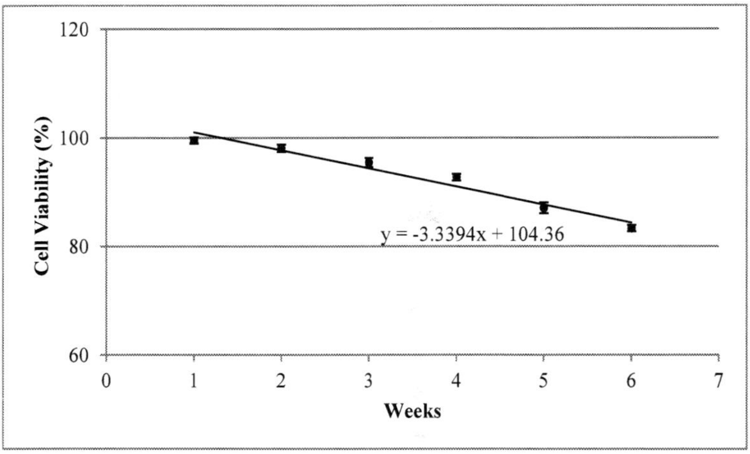

Figure 3. Variations in the cell viabilities of *C.vulgaris* microalgae concentrate over 6 weeks of storage in chilling condition at 4° C. Data are mean value and standard deviation of 10 repetitions.

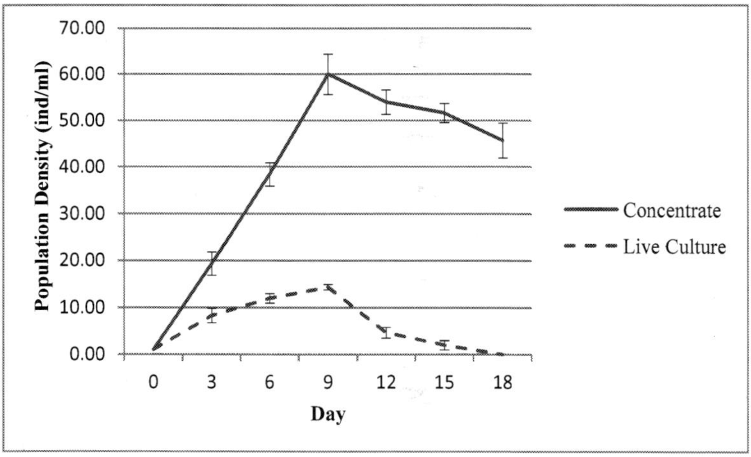

Data are mean value and standard deviation of 3 repetitions.

Figure 4. Population densities of *Apocyclops* sp. fed with live culture and microalgae concentrate of *C.vulgaris*.

C.vulgaris isolated from Bidong Island had exhibited an extremely rapid growth rate. This might be attributed to its relatively small cell volume with an average of 5.26±0.87 μm^3. This is indeed much smaller than the cell sizes recorded for some other species; *Chaetoceros calcitrans* (8μm diameter, volume 276.95 μm^3) and *Isochrysis galbana* (4μm diameter, volume

33.49 μm^3) [54]. Small size species of microalgae grow faster with a rapid growth rate. This is due to the reason that the greater surface-to-volume ratio of smaller size cells facilitates assimilation of nutrients at a relatively faster rate. In addition, the smaller size cells may achieve high density because they occupy less space. Apart from that, it was also cultured with optimal values for all environmental factors in the laboratory, thus promoting favourable environmental conditions for the cells to grow to extremely high density. In this high-density culture, the possibilities of contamination were excluded. The sudden collapse in the growth rate after day six could be mainly the result of the depletion of the nutrient in the culture. Growth rate declines and growth of microalgae ceases when the nutrient in shortest supply relative to the metabolic needs of algal population [27,55]. The populations of *C.vulgaris* cells then entered the stationary phase of the growth cycle and collapsed after day twenty two. The long stationary phase of this culture might indicate that contamination was absent during the culture period. It has been reported that this stationary phase can last for several weeks if there is no contamination in the culture [30].

The ultrafiltration technique which was used to concentrate the *C.vulgaris* cells in this study can be applied to concentrate a range of other microalgae species used as aquaculture feeds. Concentrating and storing the microalgae concentrate in moist form preserves its high nutritional value through maintaining excellent cell viability [33, 35]. The cells were readily re-suspended upon dilution in sea water with high cell viability which was proven by their ability to be inoculated even after storage for a duration of 6 weeks. The efficiency of ultrafiltration through this study was ≥90% which is very comparable to the reported efficiency of ≥80% for flocculation technique by Knuckey *et al.* [34]. There has been no comparative assessment of concentrates prepared by ultrafiltration with those prepared by centrifugation. However, from a practical and theoretical point of view, it is proven that the centrifugation method possesses some disadvantages due to its exertion of shear gravitational forces rupturing the microalgae cell structure during harvesting procedure. This reduces their nutritional values due to leaking of nutritional contents. On the other hand, microalgae concentrates prepared by ultrafiltration are not subjected to the same gravitational forces during harvesting. As reported earlier [38], the major production cost of centrifuged concentrates may exceed US$10,000 (RM32,620) which is unaffordable for small-scale hatcheries and is likely to be limited to larger hatcheries with specialised equipments or facilities specifically set up to produce microalgae concentrates to hatcheries. Advantages of the ultrafiltration technique used in this study is that it is a relatively simple, inexpensive and volume-independent process which can be readily adopted by small-scale hatcheries to prepare their own microalgae concentrates on site.

The use of *C.vulgaris* concentrate as diet for cyclopoid copepods increases population density, instantaneous growth rate as well as doubling time and it was proven as a better diet than the *C.vulgaris* live cultures. This might be possibly due to the significantly higher cellular density of the microalgae concentrates. The rates of ingestion and egg production in copepods are dependent on the quantity of the provided microalgae [56, 57] implying that quantity of food is the most important factor regulating the productivity of copepod culture. Other studies have also demonstrated that the rate of egg production of calanoid copepod, *Acartia tonsa,* increases with increasing food concentrations [58, 59]. Essential substances such as cholesterol, HUFA and PUFA are present or exist abundantly in microalgae, and, copepod production is positively related to the lipid levels or DHA: EPA ratio in the diet [60]. Thus, microalgae concentrate could be useful as a replacement for live or fresh microalgae. This is extremely important as a stable and continuous supply of live feed for aquaculture hatcheries must always be provided.

Experiment 2: Effects of Photoperiod and Culture Size on *C.vulgaris* Stock Growth

It is very important for hatcheries to be able to maintain the stock for microalgae for their sustainable live-feeds supply. Batch cultures need to be maintained under optimal environmental conditions and in a suitable culture vessel which will not affect the cell density and quality. Comparison on the effect of photoperiod and culture sizes between *C.vulgaris* and other microalgae was made to investigate the adaptability of the species to simple stock handling in the laboratory or hatchery conditions. No significant difference in the cell density was noted in *Nannochloropsis* sp. (Figure 5A) and *C.vulgaris* (Figure 5C) cultures grown in different volume flasks while *Chaetoceros* sp. Figure 5B showed significant variation in cell-density level at similar culture conditions ($P < 0.001$). However, stationary phases of all cultured species were achieved earlier in 250mL flask compared to the cultures in 500mL and 2L flasks. All cultures showed greater response towards daylight variations whereby higher cell density was noted in culture flasks exposed to continued illumination (24:0 L/D), and it was followed by 12:12 L/D and 6:18 L/D condition. *Nannochloropsis* sp. (Figure 5D) responded less towards the treatment compared to the other 2 species (Figure 5E & F) which clearly showed a specific response towards culture conditions.

It is well-documented that, in natural conditions, microalgae growth is not curtailed by ambient environmental conditions because the growth rate is just enough for species survival. However, their multiplication rate is highly influenced by various environmental parameters. In an *In vitro* setup, the proper maintenance of optimum culture condition triggers the metabolic pathway of target species in a unidirectional fashion to achieve high cell density. In this study, a higher cell density of microalgae in low volume flask culture together with early stationary phase was observed could be used to obtain continuous harvest of selected micro-algae.

Highest cell density and specific growth rate were recorded in selected species cultured in 250mL culture flask compared to the cultures in 500mL and 2000mL flasks (Table 2). The highest cell density was achieved during the end of the log phase. Cell density of early stationary phase, which is the end of the log phase for *Nannochloropsis* sp., *Chaetoceros* sp. and *C.vulgaris* was achieved on the 10th day of culture in 250mL. After the 10th day, density decreased and the lowest level was different for different species and different culture volumes. Similar results were also noted for the specific growth rate values. Significant variation in cell density and specific growth rate was observed between the cultures in different sizes of culture flask ($P < 0.05$ or $P < 0.001$). This observation might probably be due to the light-penetration efficiency in the culture flask. Similar observation was noted for the culture of *Nannochloropsis* sp. in 2000mL flasks which produced greater cell density compared to the culture in 20L carboys [61]. The effect of light saturation could decrease in the denser culture and the average irradiance in the culture reduced due to absorption from other cells [62]. The large volume culture needed higher light intensity to allow light penetration while the smaller volumes were less affected by the light penetration.

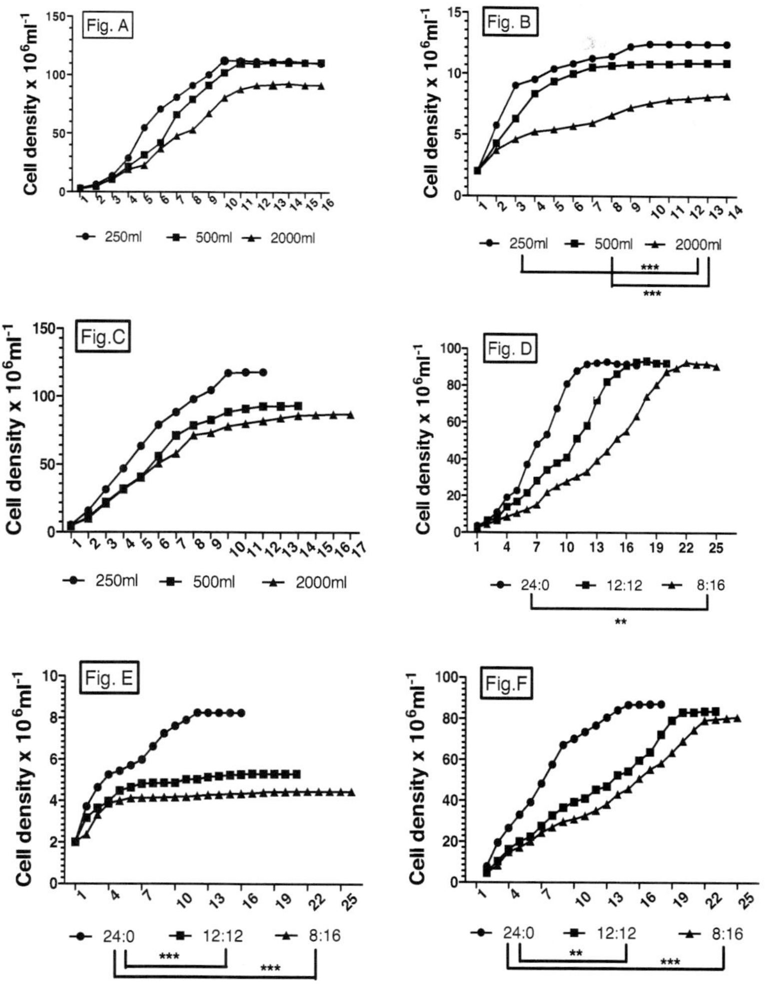

Figure 5. Influence of different photoperiods and culture flask volumes on the cell density of *Nannochloropsis* sp. (Fig A & D), *Chaetoceros* sp. (Fig B & E) and *C.vulgaris* (Fig C & F) respectively. X-axis shows the days of culture. Data represented as 250ml, 500ml and 2000ml are culture flask volumes while 24:0, 12:12 and 8:16 are photoperiods (Light : Dark phase). [***] and [**] are significantly different at $P < 0.001$ and $P < 0.05$ level in the culture conditions respectively.

Microalgae	Flask volume (mL)	Early stationary phase	Cell density ($x10^6$cells mL^{-1})	Specific growth rate (K')
Nannochloropsis sp.	250	Day 10	112.5 ± 2.36^{a}	0.52 ± 0.01^{a}
	500	Day 11	110.17 ± 1.77^{a}	0.46 ± 0.04^{b}
	2000	Day 13	92.2 ± 0.87^{a}	$0.34\pm0.01^{c*\,(ac)}$
Chaetoceros sp.	250	Day 10	12.460 ± 0.018^{a}	0.203 ± 0.002^{a}
	500	Day 12	$10.889\pm0.013^{b**\,(ab)}$	$0.145\pm0.001^{b*\,(ab)}$
	2000	Day 14	$8.225\pm0.001^{c**\,(ac\,\&\,bc)}$	$0.037\pm0.003^{c**\,(ac),\,*\,(bc)}$
C.vulgaris	250	Day 10	117.53 ± 0.84^{a}	0.4749 ± 0.0007^{a}
	500	Day 11	91.0 ± 0.55^{a}	0.4081 ± 0.0002^{b}
	2000	Day 14	86.13 ± 0.81^{a}	$0.3166\pm0.0007^{c*\,(ac)}$

Note: Data represented in Mean ± SD. [*] and [**] indicates significant difference at $P < 0.05$ and $P < 0.001$(respectively) level between different superscripts depicted for each species.

Table 2. Cell density and specific growth rate of selected microalgae cultured at different flask volumes.

Microalgae	Photo period (Light : Dark phase) in hours	Early stationary phase	Cell density ($x10^6$cells ml^{-1})	Specific growth rate (K')
Nannochloropsis sp.	24:0	Day 13	112.5 ± 2.36^{a}	0.34 ± 0.01^{a}
	12:12	Day 17	110.17 ± 1.77^{a}	0.25 ± 0.01^{b}
	8:16	Day 23	92.2 ± 0.87^{a}	$0.19\pm0.02^{c*\,(ac)}$
Chaetoceros sp.	24:0	Day 12	8.225 ± 0.001^{a}	0.129 ± 0.003^{a}
	12:12	Day 17	$5.293\pm0.009^{b**\,(ab)}$	$0.061\pm0.002^{b*\,(ab)}$
	8:16	Day 22	$4.453\pm 0.003^{c**\,(ac)}$	$0.037\pm0.003^{c**\,(ac)}$
C.vulgaris	24:0	Day 14	86.60 ± 0.17^{a}	0.3170 ± 0.0001^{a}
	12:12	Day 19	$83.04 \pm 0.19^{b**\,(ab)}$	0.2313 ± 0.0001^{b}
	8:16	Day 21	$79.23 \pm 0.21^{c**(ac)}$	$0.2010\pm0.0001^{c*\,(ac)}$

Note: Data represented in Mean ± SD. [*] and [**] indicates significant difference at $P < 0.05$ and $P < 0.001$ (respectively) level between different superscripts depicted for each species.

Table 3. Cell density and specific growth rate of selected microalgae cultured at different photo periods.

Highest cell density and specific growth rate were recorded in all cultured species that were exposed to continued illumination (24:0. L/D) followed by 12:12 and 6:18 L/D respectively. Early stationary phases differed for *Nannochloropsis* sp. (day 13), *Chaetoceros* sp. (day 12) and *C.vulgaris*. (day 14) respectively while the corresponding specific growth rate was also highest under 24hours illumination. Significant variation in both growth parameters was observed between the cultures exposed to different photoperiods ($P < 0.05$ or $P < 0.001$) (Table 3). Photosynthetic efficiency of microalgae can be enhanced by sudden alteration between light and dark phase [63]. During this process, the fast reduction of e-acceptors, Qa and Qb, associated to photosystem II (PSII) followed by their oxidation in the dark period will take place that will ultimately maximise the proton-accepting capacity of PSII during sudden irradiant of light [64].

C.vulgaris proved its adaptability to different culture volumes and lighting periods with good growth performance comparable to *Nannochloropsis* sp. and better than *Chaetoceros* sp. The cells responded positively towards continuous illumination of light by producing higher cell density and specific growth rate in the culture media. It was also noted that the culture in the low-volume flask produced an early stationary phase due to high penetration of light and continuous sharing of available nutrients in the media for faster growth and survival. On the other hand, *C.vulgaris* consistently grew at significant cell densities even in larger volume containers and shorter period of illumination than dark condition (comparable to *Nannochloropsis* sp. and better than *Chaetoceros* sp.). In another study to analyse the effect of photoperiod to the cellular essential fatty acid in these species, the photoperiod of 12:12h L/D regime is recommended for the fast and economical technique for batch culture production [65]. A better ratio of essential fatty acid accumulated in *C.vulgaris* exposed in the 12:12h if compared to 24:0 or 8:16 L/D photoperiod.

Experiment 3: Low-cost Commercial Fertiliser for Mass Culture of Marine *Chlorella vulgaris*: Manipulation of N:P:K Ratio

C.vulgaris showed its adaptability to grow well when fertilised with a low-cost commercial N:P:K plant fertiliser (Figure 6). Duration of the log phase for *C.vulgaris* varied among treatments. The 12:6:4 and 12:8:4 ratios had a result of 3 days while the longest period was in the 15:15:15 treatment. Combined applications of urea, P+ and K+ (N:P:K; 12:6:4) produced the highest cell number (4.0×10^6 cells mL^{-1}) during log period at 5 days while N:P:K; 15:15:15 (control) produced highest cell number (4.16×10^6 cells mL^{-1}) at 7 days of log period. Different ratios of N:P:K, 12:8:4, 8:8:2 and 16:4:6 resulted in decrease of cell density, 3.3×10^6, 3.0×10^6 and 2.7×10^6 cells mL^{-1}, respectively.

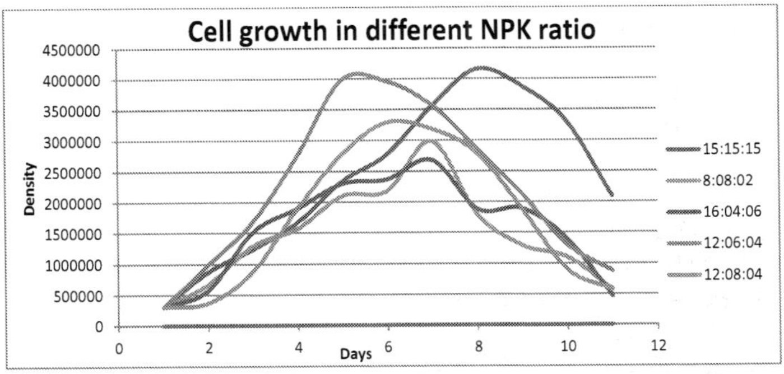

Figure 6. Density (cells mL^{-1}) of *C.vulgaris* cultured with commercial fertiliser of different N:P:K ratios

			Days in Log Phase				
NPK ratio	2nd	3rd	4th	5th	6th	7th	
		Growth rate; K'					Average
15:15:15	1.07	0.35	0.30	0.34	0.17	0.25	0.41
8:08:02	0.81	0.65	0.20	0.29			0.49
16:04:06	0.65	0.96	0.23	0.19			0.51
12:06:04	1.19	0.53	0.50				0.74
12:08:04	0.20	0.87	0.77				0.61

Table 4. Growth rate of *C.vulgaris* cultured with commercial fertiliser of different N:P:K ratios

			Days in Log Phase				
NPK ratio	2nd	3rd	4th	5th	6th	7th	
		Division/ Day; Div. day-1					Average
15:15:15	1.54	0.50	0.44	0.48	0.25	0.36	0.59
8:08:02	1.16	0.94	0.28	0.42			0.70
16:04:06	0.94	1.39	0.33	0.28			0.73
12:06:04	1.72	0.77	0.72				1.07
12:08:04	0.29	1.25	1.11				0.89

Table 5. Division per day of *C.vulgaris* cultured with commercial fertiliser of different N:P:K ratios

The 12:6:4 ratios showed the best average (74%) growth rates of natural increase at log phase. The second was 12:8:4 with 61% average growth rate.15:15:15 NPK ratio showed the lowest average growth rate of 41% (Table 4). The *C.vulgaris* cell in 12:6:4 NPK ratio recorded an average division per day by 107% which was the best compared to others. In 12:8:4 ratios the average cell division was 89% and decreasingly followed by 16:4:6 and 8:8:2 for 73% and 70% respectively. The control ratio which was 15:15:15 showed the lowest average division which was 59% (Table 5).

Measurement of generation time for *C.vulgaris* is summarised in Table 6 and Table 7. *C.vulgaris* cultured with fertiliser of the ratio 12:6:4 only took 1.09 days (26.22 hour) to complete one generation of replication, the shortest time compared to other treatments. The longest generation time was when using the15:15:15 ratio which was completed in 2.31 days (55.49 hour). The other three intermediate treatments recorded 1.70 (40.82 hour), 1.96 (47.11 hour) and 2.1 days (50.63 hour) for 12:8:4, 8:8:2 and 16:4:6 respectively. When comparing the performance by using all of the growth parameters, N:P:K; 12:6:4 ratio gave the best result with average

http://dx.doi.org/10.5772/57212

		Days in Log Phase					
NPK ratio	2nd	3rd	4th	5th	6th	7th	
		Generation time (days); Gen't					Average
15:15:15	0.65	2.00	2.28	2.06	4.07	2.80	2.31
8:08:02	0.86	1.06	3.55	2.38			1.96
16:04:06	1.06	0.72	3.04	3.62			2.11
12:06:04	0.58	1.31	1.39				1.09
12:08:04	3.41	0.80	0.90				1.70

Table 6. Generation time (days) for *C.vulgaris* cultured with commercial fertiliser of different N:P:K ratios

		Days in Log Phase					
NPK ratio	2nd	3rd	4th	5th	6th	7th	
		Generation time (Hour); Gen't					Average
15:15:15	15.56	47.99	54.84	49.55	97.66	67.32	55.49
8:08:02	20.60	25.55	85.22	57.05			47.11
16:04:06	25.44	17.28	73.01	86.79			50.63
12:06:04	13.97	31.34	33.37				26.22
12:08:04	81.76	19.14	21.54				40.82

Table 7. Generation time (hour) for *C.vulgaris* cultured with commercial fertiliser of different N:P:K ratios

growth rate per day (74%), maximum growth rate day^{-1} (107%), maximum cell density (4.0×10^6 cell/mL), division's day^{-1} (107%) and generation time (1.09 day; 26.22 hour). *C.vulgaris* in control treatment (15:15:15) exhibited the poorest growth performance. Nonetheless, it is interesting to note that they experienced longer log period which could give more time for reproduction activity, thus the density did not decrease drastically as when cultured using other ratios. The fluctuation of temperature and different salinities could be the reason why cell densities were not as high as the first and second experiment.

Numerous nutrient media have been use for the culture of pure *Chorella* sp. Most of those were for laboratory use and/or for low-grade production of algae. Majority of these media are composed of pure nutrients (N-8). Commercial fertilisers are least considered for *Chlorella* culture because of the conception that they do not provide required nutrients for algal growth and are mostly suitable for crop (land) agriculture. Nevertheless, it has been proved that the commercial plant fertiliser could support a freshwater *Chlorella* [66]. The use of N:P:K fertiliser could be a better choice if compared to the organic fertiliser. Organic matter has its own limitations and depends on the microbial activity to release the inorganic nutrient and it cannot be compared to the performance of pure nutrients. Despite good growth performance, the short period of the log phase when *C.vulgaris* is cultured using N:P:K; 12:6:4 need specific and efficient up-scaling or harvesting method, indicating that other ratios such as 15:15:15 could be a better choice.

Experiment 4: Egg Production, Growth and Development of *Apocyclops ramkhamhaengi* Fed on *C.vulgaris*

Different diets gives significantly (P<0.05) different densities of *A. ramkhamhaengi*. The mean gravid production of *A. ramkhamhaengi* fed on *C.vulgaris* and Baker's yeast was highest on 23rd day with 1.11ind./ml and 0.67ind./ml respectively. The production peaked on 11th, 23rd, and 26th day and on 17th, 20th and 23rd day when fed with *C.vulgaris*. and Baker's yeast respectively (Figure 7). In this 30 days culture, the highest mean population density of *A. ramkhamhaengi* fed with *C.vulgaris* was recorded on the 9th day with 3.31ind./ml and when fed on Baker's yeast was on day 19th with 1.83ind./ml (Figure 8). *A. ramkhamhaengi* fed on *C.vulgaris* showed the higher instantaneous growth rate (K) than when fed on Baker's yeast (Table 8). The period taken to double their population (Dt) was shorter in *A. ramkhamhaengi* fed *C.vulgaris* (8days) than Baker's yeast (11days).

Diets	Instantaneous growth rate (K)	Doubling Time(Dt)(day)
C.vulgaris	0.1150	8
Baker's yeast	0.0756	11

Table 8. The instantaneous growth rate and doubling time of *A.ramkhamhaengi* fed on different diets.

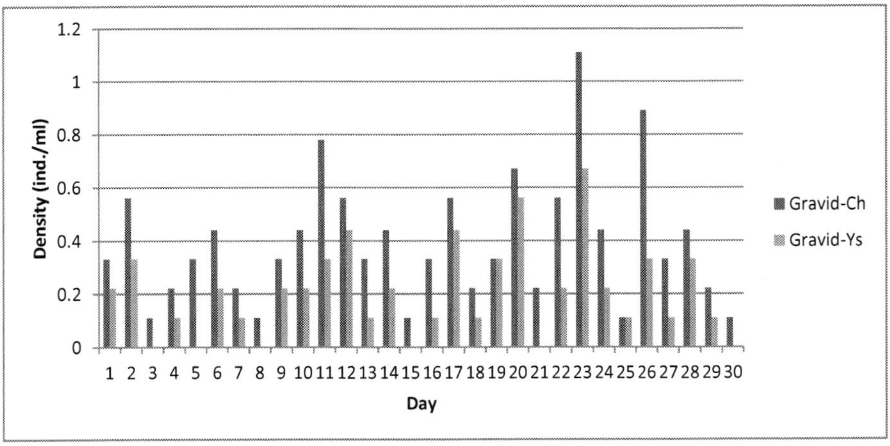

Figure 7. Mean density of gravid female of *A.ramkhamhaengi* fed with *C.vulgaris* (Gravid-Ch) and Yeast (Gravid-Ys) in 30 days.

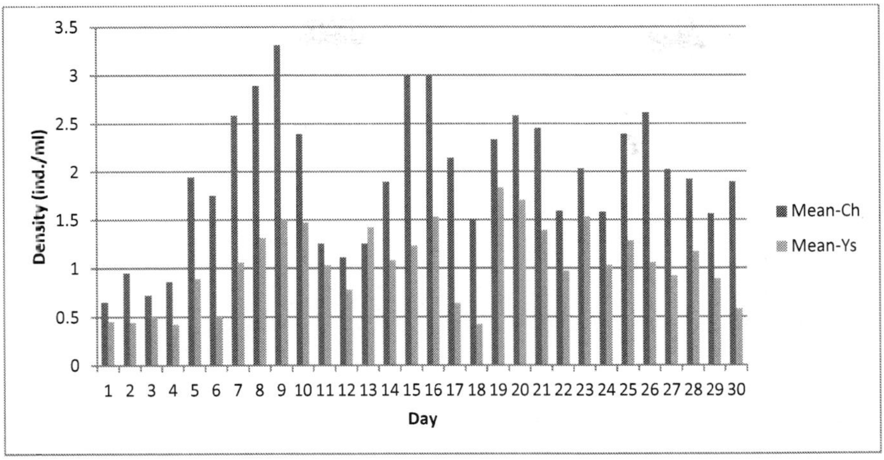

Figure 8. Mean total density of *A.ramkhamhaengi* fed with *C.vulgaris* (Mean-Ch) and Yeast (Mean-Ys) in 30 days.

The development times for nauplii, copepodite, adult and gravid female were observed separately using the copepod culture fed on *C.vulgaris*. The longest period was at copepodite stage (7.33 ± 2.08days) and the shortest period was the naupliar stage which needed only 1.33 ± 0.58 days (Figure 9). The mean number of eggs produced was 21.33 ± 1.53. Hatching percentage of the three individuals of *A. ramkhamhaengi* was 96.82 ± 2.77 % (Table 9). Maturation time which is the time between the appearance of eggs and their hatching time was 1.33 ± 0.58 days. The time taken to become gravid female from the produced nauplii was about 20.67 ± 3.51days and it is known as generation time.

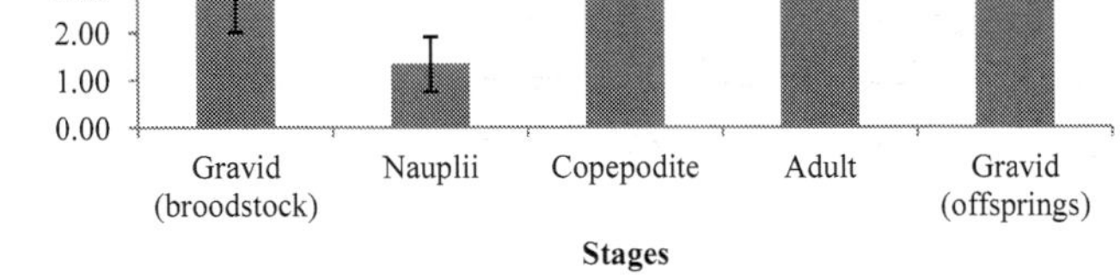

Figure 9. Mean development time in each stage in *A.ramkhamhaengi* life cycle

Parameter	N	Mean ± SD	Minimum	Maximum
Number of eggs	3	21.33 ± 1.15	20	22
% hatching	3	96.82 ± 2.77	95	100
Maturation time(days)	3	1.33 ± 0.58	1	2
Generation time(days)	3	20.67 ± 3.51	17	24

Table 9. The number of observation (N), mean and standard deviation (SD), minimum and maximum of total number of eggs, percentage of hatching of *A.ramkhamhaengi*, maturation time and generation reared in laboratory-controlled condition.

Production of the gravid females and the population density obviously increased when *A. ramkhamhaengi* was fed on *C.vulgaris* as compared to Baker's yeast. This finding is in agreement with the previous study [20] on the population growth and production of *A. dengizicus* fed on different diets. Population reached its peak in term of total density for several periods in the 30day culture condition indicates the existence of different populations. These populations reached their peak density in accordance with the diet taken where *Chlorella*-fed population were found to grow faster than those fed on Baker's yeast. It seemed that *A. ramkhamhaengi* has the potential to become more nutritional when enriched, thus growing faster when fed on a good-quality diet such as microalgae if compared to Baker's yeast. The nutritional value has been shown to increase when cyclopoid nauplii stage such as in *A. panamensis* was offered an enriched diet [15]. Although other microalgal diets such as *Tetraselmis* sp. and *Isochrysis* sp. could be a better choice for *Chlorella* sp. [67], at least the present finding is able to prove the potential of the species to reproduce and grow when fed on the marine *C.vulgaris* as used in this study.

A female of *A. ramkhamhaengi* fed on *C.vulgaris* could produce between 20 and 22 eggs with about 97% hatching success. This is more than what has been reported before for *A. panamensis* [68], and it could be related to many environmental factors and culture procedure. Environmental parameters such as temperature, food availability and predation were reported to influence the life-history strategy in copepods [69]. Binary diet of *Nannochloropsis* sp. and T-ISO improved the hatching rate by 88.1 ±2.1 % in a calanoid copepod, *Acartia sinjiensis* [70]. The brackish water cyclopoid, *A. ramkhamhaengi* has shown its potential to be cultured and reproduced under controlled conditions. The population adapted very well to the introduced diet, a marine *C.vulgaris* and a common Baker's yeast. *Chlorella*-fed population of *A. ramkhamhaengi* grow faster and need fewer number of days to double its population than those fed on Baker's yeast.The number of eggs produced was 21.33 ± 1.15 eggs at the maturation time of 1.33 ± 0.58 days and generation time was 20.67 ± 3.51 days. The species show great potential to be cultured together with *C.vulgaris* for hatchery and farm use. A more comprehensive study is essential to investigate the reproductive biology of this species, particularly in a large-scale production system to verify its suitability in aquaculture.

http://dx.doi.org/10.5772/57212

4. Conclusion

C.vulgaris isolated from Bidong Island exhibited a rapid growth rate under optimum environmental conditions in the laboratory culture and was able to achieve an extremely high density when cultured in bigger containers. The photoperiod of 24:0 proved to be the best condition for cells growth but 12:12 L/D photoperiod could be the more economical. The high-density culture could be harvested using a relatively cheap, inexpensive and simple ultrafiltration technique for other use or reinoculation. This will save the space and long period of maintaining live algae for unexpected use. The cells collected using a ultrafiltration technique showed high viability and long shelf life when kept in 4°C refrigerator. The product is called as *C.vulgaris* paste or concentrate which could be used to enrich or maintain the zooplankton live feeds for aquaculture purposes. The *C.vulgaris* also showed its best growth performance when cultured using a common commercial plant fertiliser with certain ratio of N:P:K. This was shown by their ability to perform cell division and grow and easily adapted to certain ratio such as 15:15:15 and 12:6:4. Nonetheless, the cells density is very much lower than those cultured with the specific chemical fertilizer, Conway media. This problem could be overcome by further investigation on their ion requirement when cultured openly in hatchery or ponds. The suitability of C.vulgaris as enriched diet for a zooplankton potentially used as live feed, A. ramkhamhaengi was proved by the population increase and reached high individual density with good reproduction performance. Maintaining local species of microalgae and zooplanktons in hatchery and ponds will definitely support the continued supply of live feeds for larval rearing and the aquaculture industry.

Acknowledgements

The authors are grateful to the Ministry of Science, Technology and Innovation (MOSTI) for the funding given under the ABI-MOSTI grant (2009-2011), "Mass fry production technology for grouper (Epinephelus sp.)", the Ministry of Education Malaysia for the Knowledge Transfer Programme grant (2012-2013) Using Microalgae and Copepod Live Feeds for Brackish Water Aquaculture Farm, UMT-PPKJBS.

Author details

Zaleha Kassim[1*], Akbar John[3], Lim Keng Chin[2], Nur Farahiyah Zakaria[4] and Nur Hidayah Asgnari[1]

*Address all correspondence to: zaleha@umt.edu.my

1 Department of Fisheries, Faculty of Fisheries and Aqua-Industry, University Malaysia Terengganu, Terengganu, Malaysia

2 Institute of Tropical Aquaculture University Malaysia Terengganu, Terengganu, Malaysia

3 Kuliyyah of Science, International Islamic University Malaysia, Kuantan Campus, Malaysia

4 Unit of Farmer's Organisation of South Johor Bahru, Johor, Malaysia

References

[1] Sorgeloos, P.The use of brine shrimp Artemia in Aquaculture. The brine shrimp, Artemia. 1980. Proceedings the International Symposium on the Brine shrimp, *Artemia salina*. Universa Press, Wetteren, Belgium. Pp. 25- 46.

[2] Leger, P., Bengtson, D. A., Simpson, K. L. & Sorgeloos, P. The use and nutritional value of Artemia as a food source. Marine biology and oceanography: an annual review 1986; 24, 521–623.

[3] Chu, K. H. & C. K. Shing. Feeding behaviour of the shrimp, *Metapenaus ensis*, on Artemia nauplii. Aquaculture 1986; 58, 175-184.

[4] Kuhlmann, D. G., Quantz and Witt, U. Rearing of turbot larvae (*scopthalmus maximus*) on cultured food organism and postmetamorphosis growth on natural and artificial food. Aquaculture 1981; 23: 183-196

[5] Watanabe, T., and Kiron, V. 1994. Prospects in larval fish dietetics. Aquaculture 1994; 124: 223-251.

[6] McKinnon, A.D., Duggan,S., Nichols, P.D., Rimmer, M.A., Semmens, G. and Robino, B.The potential of tropical paracalanid copepods as live feeds in aquaculture. Aquaculture 2003; 223(1-4), 89-106.

[7] Stottrup, J.G. Review on status and progress in rearing copepods for marine larviculture and advantages and disadvantages among calanoid, harpacticoid and cyclopoids copepods. 2006. VIII Symposium Internacional Acuicola 15-17 Noviembre, Universidad Autonoma de Nuevo Leon, Mexico.

[8] Rajkumar & Kumaraguru vasagam. Suitability of the copepod, *Acartia clausi* as a live feed for Seabass larvae (*Lates calcarifer* Bloch): Compared to traditional live-food organisms with special emphasis on the nutritional value. Aquaculture 2006; 261(2), 649-658.

[9] Schipp, G. 2006. The use of Calanoid Copepods in semi-intensive, tropical marine fish larviculture. VII Simposium Internacional de Nutricion Acuicola. 15-17 Noviembre. Universidad Autonoma de Nuevo Leon, Monterrey, Nuevo Leon Mexico.

[10] Zaleha,K., Ibrahim, B., Akhbar John, B., Kamaruzzaman, B.Y. Generation times of some marine harpacticoid species in laboratory condition. Journal of Biological Sciences 2012; 12(8), 426-432. DOI: 10.3923/jbs.2012.433.437

[11] Carli, A., Mariottini, G. L. & Pane, L. Influence of nutrition on fecundity and survival in *Tigriopus fulvus* Fischer (Copepoda, Harpacticoida). Aquaculture 1995; 134,113-119.

[12] Cutts, C.J. Culture of harpacticoid copepods: potential as live feed for rearing marine fish. Advances in Marine Biology 2003; 44:295-316.

[13] Ananth, S. and Santhanam, P. Laboratory culture and biochemical profile of marine copepod *Macrosetella gracilis* (Dana). Aquaculture 2011; 12(1): 49-55.

[14] Rhodes. A. 2003. Methods for high density batch culture of *Nitokra lacustris*, a marine harpacticoid copepod.The Big Fish Bang. In: Howard I. Browman, A., Skiftesvik, B. (eds.) Proceedings of the 26th Annual Larval Fish Conference. Institute of Marine Research, Bergen, Norway.

[15] Sumiarsa, G.S. Cyclopoid copepod nauplii Apocyclops panamensis: ~~mass production~~ method in outdoor ponds and their fatty acid profiles.VDM Verlag Dr.muller GmhH & Co. KG. Germany. 2011.

[16] Drillet,G., Frouël,S., Sichlau, M.H., Jepsen, P.M., Jonas K., Højgaard, J.K., Joarder,A.K., Hansen, B.W. Status and recommendations on marine copepod cultivation for use as live feed. Aquaculture 2011; 315:155-166

[17] Zaleha, K., Farahiyah, I. J. Culture and growth of a marine harpacticoid, *Pararobertsonia* sp in different salinity and temperature. Sains Malaysiana 2010; 39(1): 135–140

[18] Kassim, Z.,Ibrahim, B. Culture of Harpacticoid Copepods: Understanding the Reproduction and Effect of Environmental Factors. In: Muchlisin, Z. (ed.) Aquaculture. Rijeka: Intech; 2012.p343-360.

[19] Rhodes. A. Methods for high density batch culture of *Nitokra lacustris*, a marine harpacticoid copepod.The Big Fish Bang. In: Browman, H.I., Skiftesvik, A.B. (eds.). Proceedings of the 26th Annual Larval Fish Conference. Institute of Marine Research, Postboks 1870 Nordnes, N-5817, Bergen, Norway. ISBN 82-7461-059-8. 2003

[20] Farhadian, O., Yusoff, M.F., Arshad, A. Population growth and production of *Apocyclops dengizicus*(Copepoda: Cyclopoida) Fed on Different Diets. Journal of the World Aquaculture Society, 2008; 39(2), 384-396. DOI: 10.1111/j.1749-7345.2008.00172.x

[21] Chullasorn, S., Kangtia, P., Pinkaew, K.,Ferrari, F.D. *Apocyclops ramkhamhaengi* sp. nov. (Copepoda: Cyclopoida) in a Culture Originating from Brackish Waters of Chang Island, Trat Province,Thailand. Zoological Studies 2008; 47(3),326-337.

[22] Apt, K. E., Behrens, P. W. 1999. Commercial developments in microalgal biotechnology. *Journal of Phycology, 35*(2), 215-226. doi: 10.1046/j.1529-8817.1999.3520215.x

[23] Geudes, A.C., Malcata, F.X. Nutritional values and uses of microalgae in aquaculture. In: Muchlisin, ZA (Ed); InTech Open Acess Publisher

[24] Muller-Feuga, A., Moal, J. and Kaas, R. 2003. The microalgae of aquaculture. In: Stottrup, J.G and Mcevoy, L.A (eds.) Live Feeds in Marine Aquaculture. Blackwell Publishing Ltd., USA, p206-252.

[25] Mata, T.M., Martins, A.A., Caetano, N.S. 2010. Microalgae for biodiesel production and other applications: a review. Renewal Sustainable Energy Review 2010; 14 : 217-232.

[26] Chen, C.Y., Yeh, K.L., Aisyah, R., Lee, D.J. and Chang, J.S. 2011. Cultivation, photobioreactor design and harvesting of microalgae for biodiesel production: a critical review. *Bioresource Technology*, 102, pp. 71-81.

[27] Lee, Y.K. and Shen, H. 2004. Basic culturing techniques for microalgae. In: *Handbook of Microalgal Culture, Biotechnology and Applied Phycology* (ed. A. Richmond). Blackwell Publishing Ltd., USA, pp. 40-56.

[28] Chen, C.Y., Yeh, K.L., Aisyah, R., Lee, D.J. and Chang, J.S. 2011. Cultivation, photobioreactor design and harvesting of microalgae for biodiesel production: a critical review. *Bioresource Technology*, 102, pp. 71-81.

[29] Becker, E.W., 1995. Microalgae biotechnology and microbiology. *Cambridge University Press*, Great Britain, pp.56- 62.

[30] Borgne, Y.L. 1996. Culture of microalgae. In: *Aquaculture* (ed. G. Barnabe). *Ellis Horwood Ltd.*, England, pp. 197-206.

[31] Grobbelaar, J.U., Nedbal, L. and Tichy, V. 1996. Influence of high frequency light/dark fluctuations on photosynthetic characteristics of microalgae photoacclimated to different light intensities and implications for mass algal cultivation. *Journal of Applied Phycology*, 8, pp. 335-343.

[32] Meseck S.,L., Alix J.,H., Gary, H., Wikfors G.,H., 2005. Photoperiod and light intensity effects on growth and utilization of nutrients by the aquaculture feed microalga Tetraselmis chui (PLY429). *Aquaculture* 246: 393–404.

[33] Heasman, M., Diemar, J., O'Connor, W., Sushames, T. and Foulkes, L. 2000. Development of extended shelf-life microalgae concentrate diets harvested by centrifugation for bivalve mollusks- a summary. *Aquacult. Res.*, 31, pp. 637-659.

[34] Knuckey, R., Brown, M., Robert, R. and Frampton D. 2006. Production of microalgal concentrates by flocculation and their assessment as aquaculture feeds. *Aquacult. Eng.*, 35, pp. 300-313.

[35] Montaini, E., Zitelli, G.C., Tredici, M.R., Molina Grima, E.M., Fernandez Sevilla, J.M. and Sanchez Perez, J.A. 1995. Long-term preservation of *Tetraselmis suecica*: influence of storage on viability and fatty acid profile. *Aquaculture*, 134, pp. 81-90.

[36] McCausland, M.A., Brown, M.R., Barrett, S.M., Diemar, J.A. and Heasman, M.P. 1999. Evaluation of live and pasted microalgae as supplementary food for juvenile Pacific oyster (*Crassostrea gigas*). *Aquacult. Res.*, 174, pp. 323-342.

[37] D'Souza, F.M.L., Knuckey, R.M., Hohmann, S. and Pendrey, R.C. 2002. Flocculated microalgae concentrates as diets for larvae of the tiger prawn *Penaeus monodon* Fabricius. *Aquaculture Nutrition*, 8, pp. 113-120.

[38] Brown, M. and Robert, R. 2002. Preparation and assessment of microalgal concentrates as feeds for larval and juvenile Pacific oyster (*Crassostrea gigas*). *Aquaculture*, 207, pp. 289-309

[39] Iwamoto, H. 2004. Industrial production of microalgal cell-mass and secondary products-Major industrial species: *Chlorella*. In: *Handbook of Microalgae culture, Biotechnology and Applied Phycology* (ed. Richmond, A.). Blackwell Science Ltd., Oxford, UK, pp. 255-263.

[40] Rossingol, N., Vandanjon, L., Jaouen, P. and Quemeneur, F. 1999. Membrane technology for the continuous separation microalgae culture medium: compared performances of cross-flow microfiltration and ultrafiltration. *Aquacult. Eng.*, 20, pp. 191-208.

[41] Csordas, A. and Wang, J.K. 2004. An integrated photobioreactor and foam fractionation unit for the growth and harvest of *Chaetoceros* sp. in open systems. *Aquacult. Eng.*, 30, pp. 15-30.

[42] Petrusevski, B., Bolier, G., Van Bremen, A.N. and Alaerts, G. J. 1995. Tangential flow filtration: a method to concentrate freshwater algae. *Water Res.*, 29, pp. 1419-1424.

[43] Rossingol, N., Vandanjon, L., Jaouen, P. and Quemeneur, F. 1999. Membrane technology for the continuous separation microalgae culture medium: compared performances of cross-flow microfiltration and ultrafiltration. *Aquacult. Eng.*, 20, pp. 191-208.

[44] Sun, J. and Liu, D.Y. 2003. Geometric models for calculating cell biovolume and surface area for phytoplankton. *J. Plankton Research.*, 25, pp. 1331-1346.

[45] Muller-Feuga, A. 2000. The role of microalgae in aquaculture: situation and trends. *J. Appl. Phycol.*, 12, pp. 527-534.

[46] Baraniak, B., Niezabitowskab, M., Pieleckic, J. and Wojcika, W. 2004. Evaluation of usefulness of Magnafloc M-22S flocculant in the process of obtaining protein concentrates from peas. *Food Chem.*, 85, pp. 251-257.

[47] Omori, M. and Ikeda, T. 1984. Methods in Marine Zooplankton Ecology. *Wiley*, New York, 322 pp.

[48] Usha, T.L., Sarada, R., and Ravishankar, G.A. 2002. Effect of culture conditions on growth of green alga *Haematococcus pluvialis* and astaxanthin production. Acta Physiologiae Plantarum. 24 (3): 323-329

[49] Fogg and Thake 1987. Algal Cultures and Phytoplankton Ecology. University of Wisconsin Press.

[50] Abu-Rezq, T. S., Yule, A. B. & Teng, S. K. 1997. Ingestion, fecundity, growth rates and culture of the harpacticoid copepod, *Tisbe furcata,* in the laboratory. *Hydrobiologia, 247,* 109-118.

[51] Imase, M., Ohko, Y., Takeuchi, M., Hanada, S. 2013. Estimating the viability of Chlorella exposed to oxidative stresses based around photocatalysis. International Biodeterioration and Biodegradation. 78: 1-6

[52] Dwaish, A.S., Mohammed, D.Y., Jawad, A.M., Al-kubaicy, A.A. Determine the Uptake of Lead in Chlorella vulgaris Isolated from Tigris River in Baghdad City. International Journal of Scientific and Engineering Research, 2011, 2(9): 1-4

[53] Bae, J.H. and Hur, S.B. 2011. Selection of suitable species of Chlorella, Nannochloris and Nannochloropsis in high- and low temperature seasons for mass culture of the rotifer *Brachionus plicatilis*. Fisheries and Aquatic Sciences, 14(4):323-332.

[54] Phatarpekar, P.V., Sreepada, R.A., Pednekar, C. and Achuthankutty, C.T. 2000. A comparative study on the growth performance and biochemical composition of mixed culture of *Isochrysis galbana* and *Chaetoceros calcitrans* with monocultures. *Aquaculture,* 181, pp. 141-155.

[55] Barsanti, L. and Gualtieri, P. 2006. Algae: anatomy, biochemistry and biotechnology. *CRC Press,* New York, 301 pp.

[56] Stottrup, J.G. and Jensen, J. 1990. Influence of algal diet on feeding and egg-production of the calanoid copepod *Acartia tonsa* Dana. *J. Exp. Mar. Biol. Ecol.,* 141, pp. 87-105.

[57] Stottrup, J.G. 2003. Production and nutritional value of copepods. In: *Live Feeds in Marine Aquaculture* (eds. Stottrup, J.G and Mcevoy, L.A.). *Blackwell Publishing Ltd.,* USA, pp. 145-205.

[58] Kiorboe, T. 1997. Population regulation and role of mesozooplankton in shaping marine pelagic food webs. *Hydrobiologia,* 363, pp. 13-27.

[59] Dam, H.G., Peterson, W.T. and Bellantoni, D.C. 1994. Seasonal feeding and fecundity of the calanoid copepod *Acartia tonsa* in Long Island Sound: is omnivory important to egg production? *Hydrobiologia,* 292/293, pp. 191-199.

[60] Payne, M.F. and Rippingale, R.J. 2001. Intensive cultivation of the calanoid copepod *Gladioferens imparipes*. *Aquaculture,* 201, pp. 329-342.

[61] Hernandez, D.E. and Martinez. S.F. 2001. The effect of the culture conditions on the growth and lipid contents of two strains of Nannochloropsis sp. to be used in aquaculture. Hidrobiologia 11(2):163-168.

[62] Sandnes, J.M., K°allqvist, T., Wenner, D. and Gislerød. H.R. 2005. Combined influence of light and temperature on the growth rates of *Nannochloropsis oceanica*: linking

cellular responses to large-scale biomass production. Journal of Applied Phycology 17: 515–52.

[63] Nedbal, L., Tichyacute, V., Xiong, F. and Grobbelaar. J. U. 1996. Microscopic green algae and cyanobacteria in high frequency intermittent light. Journal of Applied Phycology, 8(4/5), 325-333.

[64] Matthijs, H.C.P., Balke, H., van Hes, U.M., Kroon, B.M.A., Mur, L.R. and Binot. R.A. 1996. Application of light emitting diodes in bioreactors: Flashing light effects and energy economy in algal culture (Chlorella pyrenoidosa). Biotechnology and Bioengineering, 50(1), 98-107.

[65] Lim, K.C and Zaleha, K. 2013. Effect of Photoperiod on the Fatty Acid Composition of Three Tropical Species of Marine Microalgae. *Malaysian Applied Biology Journal,* 42(1): 41-49

[66] Ashraf, M., Javaid, M. Rashid, T., Ayub, M., Zafar, A., Ali, S., Naeem,M. 2011. Replacement of expensive pure nutritive media with low cost commercial fertilizers for mass culture of freshwater algae, *Chlorella vulgaris. International Journal of Agriculture and Biology,* 13: 484–490

[67] Lee, K.W., Park, H.G., Lee, S.& Kang, H. 2006. Effects of diets on the growth of the brackish water cyclopoid copepod *Paracyclopina nana* Smirnov. *Aquaculture, 256,* 346-353. http://dx.doi.org/10.1016/j.aquaculture.2006.01.015

[68] Phelps, R.P., Sumiarsa, G.S., Lipman, E.E., Lan, H.-P., Moss, K.K., & Davis, A.D. 2005. Intensive and extensive production techniques to provide copepod nauplii for feeding larval red snapper *Lutjanus camperchanus.* In: Copepods in Aquaculture (Ed. By C.-S. Lee, P.J. O'Bryen, N.H. Marcus), pp. 151-168. Blackwell Publishing Ltd. Oxford, UK.

[69] Da Grac, M., Melao, A.G., & Rocha, O. 2004. Life history, biomass and production of two planktonic cyclopoid copepods in a shallow subtropical reservoir. *Journal of Plankton Research, 26(8),* 909-923.

[70] Milione, M., & Zeng, C. 2008. The effects of temperature and salinity on population growth and egg hatching success of the tropical calanoid copepod, *Acartia sinjiensis. Aquaculture,275,*116-123.

Chapter 5

Use of Yeasts as Probiotics in Fish Aquaculture

Paola Navarrete and Dariel Tovar-Ramírez

Additional information is available at the end of the chapter

http://dx.doi.org/10.5772/57196

1. Introduction

According to the UN Food and Agriculture Organization (FAO), "aquaculture, probably the fastest growing food-producing sector, now accounts for nearly 50 percent of the world's food fish" [1]. However, production is hampered by unpredictable mortalities that may be due to the negative interactions between fish and pathogenic bacteria. Intensive fish farming has resulted in a problematic growth in bacterial diseases, prompting the necessary and intensive use of antimicrobials for their treatment.

Because of the rapid expansion of aquaculture, a limited supply of fishmeal has the potential to impede the future growth of this industry. Consequently, much effort has been given to studying other protein and oil sources, but finding a suitable alternative has proved to be challenging. Among the alternatives, plant-based formulations are the least expensive, and many such formulations have a suitable protein profile and long-term availability. Oilseeds, in particular soybean and grain products, have great potential as alternative sources of fish feed. Soybeans are rich in protein and represent the most commonly used plant protein source on the world market. Soybean meal (SBM) has already become an important protein source in fish feed. However, the inclusion of some vegetable proteins, such as SBM, in the diets of fish at levels of >20% may induce intestinal disorders including pathomorphological changes in the distal intestinal epithelium accompanied by diarrhea [2, 3], sometimes caused by the anti-nutritional factors that are present in SBM. The addition of probiotics (acid lactic bacteria) to starter diets appeared to improve SBM utilization in first feeding rainbow trout [4].

In this context, two of the major challenges in fish aquaculture facilities are 1) the control of diseases, especially during the earliest life stages, and 2) the improvement of nutrition by optimizing food utilization, especially for new fish species.

It is well recognized that the bacterial microbiota of fish is beneficial to the host and affects important biological processes, including nutrient processing and absorption, the development of the mucosal immune system, and angiogenesis, as was demonstrate in gnotobiotic mice. In larval gnotobiotic zebrafish studies, was shown that the microbiota also influences enterocyte morphology and epithelial renewal, host-transcriptional responses to the microbiota regarding epithelial proliferation, nutrient and xenobiotic metabolism, and immune responses [5].

Yeast have been identified as part of the normal microbiota of both wild and farmed fish, and their role in fish health and nutrition has been addressed in the literature, as yeast have been used either alive to feed live food organisms or after processing as a feed ingredient after demonstrating an artificial colonization of the intestinal host.

Even when accounting for less than 1% of the total microbial isolates in the host, yeast can represent a major physiological contribution beyond what has been observed for probiotic bacteria; in fact, cell volumes from yeast may be larger than those of bacteria by a hundredfold [6]. In contrast to bacteria, yeast cells utilize a wide spectrum of simple and more complex organic compounds. This phenomenon results from the extensive metabolic potential of yeast, which is reflected by the production of diverse enzymes. Polyamines secreted by yeasts are also involved in the maturation of the digestive tract of fish larvae. Furthermore, some yeast species and their components, such as β-glucans and mannoproteins, can stimulate the immune and antioxidant systems of the host. Understanding the participation of yeast microbiota in fish health and nutrition may improve both the sanitary conditions and the production performance of fish.

The aim of this chapter is to describe the current knowledge regarding the use of yeasts as probiotics in aquaculture systems. The chapter will include a recent review on the presence and diversity of yeast in marine and aquaculture systems, focusing on the yeast diversity found in the fish gut microbiota. The chapter will also include basic information on the molecular methods used for yeast identification. Finally, the chapter will emphasize topics related to the essential role of probiotic yeasts used in disease control and nutritional improvements in aquaculture, with a special focus on the beneficial effects of yeast β-glucans.

2. Yeast identified in the marine and other aquatic environments

Yeasts are unicellular eukaryotic microorganisms that are taxonomically placed within the phyla Ascomycota and Basidiomycota within the Kingdom Fungi [7]. Ascomycete yeasts comprise approximately 1,000 phylogenetically diverse species that have recently been assigned to 14 different lineages on the basis of multigene sequence analysis [8]. The other species of yeasts are classified as the basidiomycetes [9]. Some of the general characteristics and ecological properties of each phylum include the following: 1) the cell wall polysaccharide composition is dominated by chitin in the basidiomycetes and by β-glucans in the ascomycetes; 2) the guanine + cytosine (G + C) composition of the nuclear DNA tends to be higher than 50% in basidiomycetes and lower than 50% in ascomycetes; 3) ascomycetes yeasts are generally more fermentative, more copiotrophic (but at the same time nutritionally specialized), more fragrant, and mostly hyaline, while basidiomycete yeasts more often form mucoid colonies,

http://dx.doi.org/10.5772/57196

display intense carotenoid pigments and tend to use a broader range of carbon compounds more efficiently at lower concentrations [7]; and 4) ascomycete yeasts are often found in specialized niches involving interactions with plants and insects or other invertebrate animals that they rely upon for dispersal, while basidiomycete yeasts seem to be adapted to the colonization of nutrient-poor solid surfaces and may not rely to the same extent on animal vectors for their dispersal [10].

Marine samples	**Identified yeast**	**Identification method**	**Reference**
Northern Biscayne Bay	*Candida tropicalis* *Candida guilliermondii* *C. parapsilosis* *Rhodotorula rubra*	identification of cultivated yeasts by the methods described by Lodder and Kreger-van Rij [23] and Wickerham [24].	[25]
Marine grass flats (Soldier Key)	*Rhodotorula pilimanne* *R. rubra* *Cryptococcus*	identification of cultivated yeasts by the methods described by Lodder and Kreger-van Rij [23] and Wickerham [24].	[25]
Gulf Stream 15 miles east of the coast of South Florida.	*R. graminis* *R. glutinis*	identification of cultivated yeasts by the methods described by Lodder and Kreger-van Rij [23] and Wickerham [24].	[25]
Marine vegetation	*Cryptococcus albidus*	identification of cultivated yeasts by the methods described by Lodder and Kreger-van Rij [23] and Wickerham [24].	[25]
Suwannee Florida estuary (water)	*Candida guilliermondii* *Candida krusei* *Candida valida* *Cryptococcus laurentii var. laurentii* *Cryptococcus laurentii var. flavescens* *Hansenula saturnus var. saturnus* *Hansenula spp.* *Rhodotorula marina* *Rhodotorula minuta var. minuta* *Rhodotorula rubra* *Rhodotorula spp.* *Torulopsis candida* *Trichosporon cutaneum*	identification of cultivated yeasts by the methods described by Lodder [26].	[27]

Marine samples	Identified yeast	Identification method	Reference
Suwannee Florida estuary (sediment)	*Brettanomyces intermedius* *Candida boidinii* *Candida diversa* *Candida glaebosa* *Candida ingens* *Candida krusei* *Candida lambica* *Candida maritima* *Candida melibiosica* *Candida silvae* *Candida solani* *Candida valida* *Candida spp.* *Cryptococcus dimennae* *Cryptococcus laurentii var. laurentii* *Cryptococcus laurentii var. flavescens* *Debaryomyces cantarellii* *Debaryomyces phaffii* *Hansenula beijerinckii* *Hansenula saturnus var. saturnus* *Kluyveromyces polysporous* *Leucosporidium capsuligenum* *Pichia membranaefaciens* *Pichia ohmeri* *Rhodotorula glutinis* *Rhodotorula graminis* *Rhodotorula lactosa* *Rhodotorula marina* *Rhodotorula rubra* *Rhodotorula spp.* *Saccharomyces spp.* *Torulopsis candida* *Torulopsis inconspicua* *Torulopsis mogii* *Torulopsis spp.* *Trichosporon aculeatum* *Trichosporon cutaneum* *Trichosporon penicillatum*	identification of cultivated yeasts by the methods described by Lodder [26].	[27]

Marine samples	Identified yeast	Identification method	Reference
Western coast of Baja California Sur, Mexico	*Sporobolomyces roseus* *Sporobolomyces puniceus* *Sporobolomyces hosaticus*	morphological and biochemical identification	[28]
Suruga and Sagami Bay (sediments, crab and *Calyptogena*)	*Kluyveromyces nonfermentans*	18S rDNA, 5.8S rDNA and ITS sequencing	[29]
Northwest Pacific Ocean (benthic animals)	*R. aurantiaca* *R. glutinis* *R. minuta* *R. mucilaginosa* *Sporobolomyces salmonicolor* *S. shibatanus*	5.8S-ITS rDNA sequencing of cultivated yeasts	[30]
Northwest Pacific Ocean (sediments)	*R. glutinis* *R. mucilaginosa*	5.8S-ITS rDNA sequencing of cultivated yeasts	[30]
Sagami bay (deep-sea tubeworm)	*R. lamellibrachii*	sequencing of ITS, 5.8S rDNA, and D1/D2 of the 26S rDNA	[31]
Sagami Bay and Iheya Ridge (deep-sea tubeworm)	*R. bentica*	sequencing of ITS, 5.8S rDNA, and D1/D2 of the 26S rDNA	[32]
Sagami Bay (deep-sea clam)	*R. calyptogenae*	sequencing of ITS, 5.8S rDNA, and D1/D2 of the 26S rDNA	[32]
Deep-sea hydrothermal systems of the Mid-Atlantic Rift	*C. atlantica* *C. atmosphaerica* *C. lodderae* *C. parapsilosis* *D. hansenii* *P. guilliermondii* *Rhodosporidium babjevae* *R. diobovatum* *R. kratochvilovae* *R. sphaerocarpum* *R. toruloides* *Rh. Mucilaginosa* *Rh. minuta* *S. dacryoides*	26SrRNA gene sequencing of cultured yeasts	[33]
Northwest Pacific ocean	*R. pacifica*	sequencing of ITS, 5.8S rDNA, and D1/D2 of the 26S rDNA	[34]
11 deep-sea samples	*Pichia fermentans* *Saccharomyces cerevisiae* *Debaryomyces hansenii*	Cloning and sequencing	[35]

Marine samples	Identified yeast	Identification method	Reference
Japan Trench (deep-sea sediments)	*Dipodascus tetrasporeus*	Sequencing of 18S rDNA, ITS1, 5.8S rDNA, ITS2 and D1/D2 of the 26S rDNA	[36]
Coastal waters of northeastern Taiwan	*Candida tropicalis* *Pichia anomala* *Issatchenkia orientalis* *C. glabrata* *Saccharomyces yakushimaensis* *Kodamaea ohmeri* *Hanseniaspora uvarum* *Kazachstania jiainicus* *Torulaspora delbrueckii*	5.8S-ITS rDNA sequencing of cultivated yeasts	[37]
Deep-sea shrimps Deep-sea mussels	*R. mucilaginosa*	26SrRNA gene sequencing of cultured yeasts	[14]
R. exoculata exuviae in decomposition on smoker rocks, *B. azoricus* and a sponge	*Rhodosporidium diobovatum* *Sporobolomyces roseus*	26SrRNA gene sequencing of cultured yeasts	[14]
carbonate colonization module	*Cryptococcus uzbekistanensis*	26SrRNA gene sequencing of cultured yeasts	[14]
B. azoricus mussel	*Leucosporidium scottii*	26SrRNA gene sequencing of cultured yeasts	[14]
R. exoculata, *M. fortunata*, a deep-sea coral and the gills of the gastropod *Ifremeria nautilei*	*Debaryomyces hansenii*	26SrRNA gene sequencing of cultured yeasts	[14]
R. exoculata exuviae in decomposition *B. azoricus*	*Candida atlantica*	26SrRNA gene sequencing of cultured yeasts	[14]
Deep-sea sponge	*Pichia guilliermondii* *Candida viswanathii*	26SrRNA gene sequencing of cultured yeasts	[14]
Deep-sea coral	*Candida sp.*	26SrRNA gene sequencing of cultured yeasts	[14]
B. azoricus	*Phaeotheca triangularis* *Hortaea werneckii*	26SrRNA gene sequencing of cultured yeasts	[14]
Arabian sea (200 m depth)	*Candida* *Lipomyces* *Yarrowia*	morphological and biochemical identification	[38]

Marine samples	Identified yeast	Identification method	Reference
	Rhodotorula *Pichia*		
Arabian sea (500 m depth)	*Candida* *Yarrowia* *Lipomyces* *Rhodotorula* *Debaryomyces* *Pichia* *Wingea* *Dekkera*	morphological and biochemical identification	[38]
Arabian sea (1000 m depth)	*Lypomyces* *Candida* *Wingea* *Dekkera* *Rhodotorula*	morphological and biochemical identification	[38]
Bay of Bengal (200 m depth)	*Wingea* *Candida* *Cryptococcus* *Rhodotorula* *Bullera* *Lipomyces* *Oosporidium* *Dekkera*	morphological and biochemical identification	[38]
Bay of Bengal (500 m depth)	*Candida* *Rhodotorula* *Cryptococcus* *Yarrowia* *Pichia* *Bullera* *Wingea* *Dekkera* *Oosporidium*	morphological and biochemical identification	[38]
Bay of Bengal (1000 m depth)	*Candida* *Wingea* *Rhodotorula* *Bullera* *Lipomyces* *Trichosporon*	morphological and biochemical identification	[38]

Table 1. Yeast identified in the marine and other aquatic environment

Yeasts are widely distributed in several natural environments such as soil, freshwater, and seawater. Their numbers and species distributions are dependent on the concentrations and types of available organic materials. Nearshore environments are usually inhabited by 10 to 1000 of yeast cell/L of water, whereas low organic surface to deep sea oceanic regions contain 10 or fewer cells/L. Marine yeasts are divided into "obligate" and "facultative" groups. When yeast are able to grow on a marine substrate and are frequently collected from the marine environment, they are called "obligate marine" yeasts; in contrast, "facultative marine" yeasts can also be recovered from terrestrial habitats. Marine yeasts participate in several ecological processes in the sea, especially in estuarine and nearshore environments, such as the decomposition of plant substrates [11], nutrient recycling [12], and the biodegradation of oil/ recalcitrant compounds [13]; they are also part of the microbiota of marine and aquaculture animals [6, 14]. This functional diversity is due, in part, to the fact that yeasts have extraordinary metabolic potential. This potential is available for exploitation [15-20], but notably, the vast majority of this potential has yet to be discovered. Several yeast compounds have significant biological value as reagents, cell proteins, vitamins, pigments, and enzymes. Different yeast species have been identified in several marine locations (Table 1). For excellent reviews on marine yeasts, see [21, 22]. The ascomycete yeasts *Debaryomyces hansenii,* and *Candida* spp. are typical ubiquitous species found in oceanic, and other aquatic environments. Basidiomycete yeasts often account for the majority of the total yeast population found in oligotrophic oceanic water. Among the basidiomycete yeasts, some species of *Cryptococcus, Rhodotorula,* and *Sporobolomyces* are widespread across various oceanic regions [22].

3. Yeast as part of the gut microbiota of fish

Most of the literature on the yeast microbiota of fish is based on the identification of cultivable yeast (Table 2). Yeast have been isolated from the gills, skin, mouths, feces and guts of different fish species. The occurrence of yeast in the fish gut is variable and can fluctuate from non-detectable levels to 10^7 CFU/g of intestinal content [6]. Both ascomycete and basidiomycete yeasts have been isolated from fish intestines (Table 2): among ascomycetes, *Saccharomycetaceae* (which include *Candida, Pichia, Saccharomyces,* and *Debaryomyces*) is likely the most important family, while basidiomycetes, include the genera *Rhodotorula, Cryptococcus, Sporobolomyces,* and *Trichosporon* [6]. The yeasts *Metschnikowia zobelii, Kloeckera apiculata,* and *Debaryomyces* sp. dominate in some marine fish (*Tachurus symmetricus* and *Atherinopis affinis littoralis*) [39], and in these fish species, the yeast concentration was significantly higher inside the fish than in the surrounding sea water, suggesting that the yeast may grow inside the fish intestine [39]. The ascomycetes *Debaryomyces hansenii, Candida* sp., and *Saccharomyces cerevisiae,* the basidiomycete *Leucosporidium* sp., and *Rhodotorula* have been frequently isolated as the dominant yeast found in the rainbow trout intestine [6]. Yeast can also be isolated from the waters of fishponds with different abundance and diversity depending on the season of the year. Fishponds from the Záhorie Lowlands in Slovakia, sampled in summer, harbor the most heterogeneous yeast species, with *Aureobasidium, Sporobolomyces, Candida,* and *Cryptococcus* as the most frequently isolated species [40]. In autumn, the yeast numbers were higher than in

http://dx.doi.org/10.5772/57196

summer, with *Candida, Hyphopichia burtonii, Aureobasidium pullulans, Hansenula anomala* and *Cryptococcus laurentii* being most frequently identified [40].

It has been reported that yeasts isolated from the intestine of rainbow trout may adhere to and grow in intestinal mucus [41]. Some yeast cells can colonize the intestine of fish after dietary introduction [42], and this ability to colonize may be related to cell surface hydrophobicity [43] and the ability of the strains to grow on mucus [41, 44]. Some experiments have shown that high levels of yeast intestinal colonization can be achieved when a pure culture of yeast is inoculated into fish. Rainbow trout and turbot were inoculated with *Rhodotorula glutinis* and *D. hansenii* HF1, and up to 3.8×10^4, 3.1×10^6, and 2.3×10^9 viable yeast cells/g of intestine or feces were recovered in three separate colonization experiments [45]. It is important to note that the majority of the studies published until 2007, on the yeast species identified from the aquaculture fish gut, were published by the Gatesoupe Lab [6]. Later, studies focused more on the probiotic effects of different yeast strains in aquaculture fish (as described below), and less on the actual yeast species isolated from the fish gut.

Location	Fish intestine	Identified yeast	Identification method	Reference
Estuarine and coastal areas (Biscayne Bay, Florida)	*Haemulon* *Carana* *Anisotremus*	*T. cutaneum* *C. parapsilosis* *C. guilliermondi* *C. tropicalis* *R. rubra* *R. pilimanae* *H. anomala* *D. kloeckeri* *Il. valbyensis*	identification of cultivated yeast by the methods described by Lodder and Kreger-van Rij [23] and Wickerham [24].	[25]
Tropical island-(Bimini, The Bahamas, 1960)	*Haemulon* *Stenotomus* *Ocyurus* *Anisotremus* *Lachnolainus*	*R. minuta* *C. parapsolosis* *R. glutinis*	identification of cultivated yeast by the methods described by Lodder and Kreger-van Rij [23] and Wickerham [24].	[25]
Tropical island-(Bimini, The Bahamas, 1961)	*Ilaemulon* *Lutjanus* *Sphyraena* *Seriola* *Balistes* *Malacanthus* *Halichoeres* *Holocentrus* *Carana*	*C. tropicalis* *R. pilimanae* *Torulopsisspp.* *C. parapsilosis*	identification of cultivated yeast by the methods described by Lodder and Kreger-van Rij [23] and Wickerham [24].	[25]

Location	Fish intestine	Identified yeast	Identification method	Reference
	Anisotremus			
La Jolla coast	*Atherinopis affinis littoralis* *Trachurus symmetricus*	*Metschnikowia zobellii* *Kloeckera apiculata* *Debaryomyces sp.*	identification of cultivated yeast by the methods described by Lodder and Kreger-van Rij [23], Wickerham [24], and van Uden and Farinha [46].	[39]
Clyde estuary North Sea	*Herring* *Haddock* *Whiting* *Skate* *Halibut* *Flounder* *Lemon sole*	*Candida* *Cryptococcus* *Debaryomyces* *Rhodotorula* *Torulopsis* *Trichosporon pullulans*	identification of cultivated yeast by the methods described by Lodder and Kreger-van Rij [23] and Kreger-van Rij [47].	[48]
Sweden farm	*Salmo gairdneri*	*S. cerevisiae* *D. hansenii* *Cryptococcus* *Leucosporidium* *Rhodotorula rubra* *R. glutinis*	Identified by the CBS Yeast Collection	[45]
Swedish west coast	*P. platessa* *P. flesus*	*Rhodotorula*	Identified by the CBS Yeast Collection	[45]
Experimental fish farm at Sizun (France)	*Oncorhynchus mykiss*	*Debaryomyces hansenii* *D. hansenii var. fabryi* *Trichosporon* *Rhodotorula mucilaginosa*	ITS 1, 5.8S rRNA gene, ITS 2, and partial sequencing of 26S rRNA gene	[49]
Experimental fish farm at Sizun (France)	*Oncorhynchus mykiss*	*Debaryomyces hansenii*	ITS 1, 5.8S rRNA gene, ITS 2, and partial sequencing of 26S rRNA gene	[42]
Cabras (Oristano, Sardinia)	*Mugil auratus* *M. chelo* *M. capito* *M. saliens* *M. cephalus*	*Candida* *Metschnikowia* *Sporidiobolus* *Clavispora* *Sporobolomyces*	morphological, physiological and biochemical tests of the isolated yeast strains according to [50]	[51]

Table 2. Yeast identified in the fish intestine

4. *Debaryomyces hansenii*, an ubiquitous yeast frequently associated with fish and the marine environment

Debaryomyces hansenii is a halotolerant, non-pathogenic ubiquitous yeast capable of growing in a variety of environments, such as the marine environment and the fish gut. *D. hansenii* has been described as one of the most frequently isolated yeast associated with fish (Table 2). This species is prevalent in seawater, which may explain its high incidence in the fish gut. One study reported the presence of sequences affiliated with *D. hansenii* in hydrothermal sediments [35]. A major biotechnological advantage of *D. hansenii* over *Saccharomyces cerevisiae* is that *D. hansenii* is osmotolerant and can grow in media containing up to 4 M NaCl, whereas the growth of *S. cerevisiae* is restricted to media containing less than 1.7 M NaCl. *D. hansenii* has been extensively studied because of its significant enzymatic potential [52]. For example, several enzymes of biological and biotechnological interest have been identified and characterized in this yeast, including inulinase [53], protease [54], superoxide dismutase (SOD) [55], lipase [56], catalase [57] and α-galactosidase [58-62]. Additionally, β-glucosidase from *D. pseudopolymorphus* [63] and phytase from *D. castellii* [64] has also been identified. The ability of *D. hansenii* to synthesize α-galactosidase has been useful in the treatment of soybean products to reduce raffinose oligosaccharides [59], which are recognized as anti-nutritional factors for mammals and fish. Interestingly, *D. hansenii* SOD has been proposed as a therapeutic anti-inflammatory agent in animal models (Wistar rats) [55]. Because of these characteristics, *D. hansenii* is one of the yeast species that has been selected for complete sequencing [65]. The beneficial effects of this yeast species in cultured fish are described below.

5. Methods to analyze the yeast microbiota

In the past, the identification of yeast species was a tedious and labor-intensive process that was generally based on the morphological and physiological properties of the isolated yeasts. To resolve these problems, the identification of cultivated yeasts is now based on DNA sequence analysis, which is a faster and more accurate process. Sequence-based approaches to the study of yeast biodiversity have resulted in a two-fold increase in the number of described species over the past decade, and a 100-fold increase is predicted in the coming decades. A previous work [8] has studied the phylogenetics of the Saccharomycetales by performing DNA sequence analysis based on five loci: 1) the nuclear small subunit (SSU) ribosomal RNA gene, 2) the D1/D2 region of the nuclear large subunit (LSU) 26S rDNA, 3) the elongation factor 1α gene (EF-1α), and 4) the largest and 5) the second largest subunits of the RNA polymerase II gene (RPB1 and RPB2). Based on the availability of the sequence data in the GenBank and AFTOL databases, the LSU rDNA genes were found to be more reliable for yeast identification. Sequencing the 400-650 bp D1/D2 region or a wider LSU region is extremely useful and distinguishes yeast rapidly at a near-species level. The D1/D2 LSU rDNA region has been sequenced for almost all known yeast, both as an identification tool and as a means for estimating phylogenetic relationships among the Saccharomycetales. A previously study [66] published the sequences of the fungal primers ITS1 and ITS4, which amplify the internal transcribed spacer (ITS) ITS1-5.8S-ITS2 region that has also been used for yeast identification. Some specialized databases are available on the web in order to help with yeast identification. The Centraal bureau voor Schimmelcultures (CBS) database aids BLAST analysis by allowing pairwise identification of LSU and SSU rDNA, ITS [67] and miscellaneous sequences [68].

It is generally accepted that in every ecosystem, there are cultivable organisms as well as viable organisms that cannot be cultivated in the laboratory. Less than 1% of the microbial species from the marine environment can be cultivated [69]. Similarly, only approximately 1% of yeast species has been described thus far [70]. In the last decades, several molecular methods have been developed to study natural samples. These methods allow for the identification of microorganisms without isolation and for the determination of the phylogenetic affiliation of community members, thus revealing the enormous extent of microbial diversity. Methods based on the amplification of fragments coding for the 16S rRNA gene have emerged as a powerful tool for studying the bacterial diversity. Denaturing or temperature gradient gel electrophoresis (PCR-DGGE/TTGE) techniques have been introduced into molecular microbial ecology to determine the genetic diversity of the bacterial communities found in the fish gut [71-76]. These techniques were also applied to characterize the dominant active bacteria in the intestine of different rainbow trout families using RNA that was extracted directly from the samples [77]. One important limitation of PCR-based methods, however, is low sensitivity, which can identify approximately 1% of the total number of species [78]. On the other hand, the use of rRNA gene fingerprinting requires sequencing of the cloned bands to identify the community's members accurately. Sequencing is necessary because the amplicons from different species may migrate to the same positions, or one species may give multiples bands because of multiple gene copies with intra-gene differences [74, 79]. The use of molecular approaches to study yeast communities has been scarce and generally limited to the study of food matrices. Yeast communities have been studied using PCR-DGGE, and amplification of a portion of the 26S rRNA gene of yeast [80-85], while PCR-TTGE has been applied to establish phylogenetic relationships of species of the genus *Saccharomyces* [86].

Recently, high-throughput sequencing methods, such as pyrosequencing, have been shown to be fast and very efficient tools for identifying members of the complex populations. In general, two approaches can be taken: diversity studies based on the sequencing of ribosomal gene (rRNA gene) amplicons, and metagenomic studies where whole-community DNA is subjected to shotgun sequencing [87]. While sequencing ribosomal amplicons is much cheaper because only one gene is being sequenced, the metagenomic approach sequences all of the DNA genes, thus revealing the functions of the microbial community [87]. A useful innovation for these two approaches is to analyze multiple samples at the same time (multiplexing), which can be accomplished using barecoded pyrosequencing or by physically separating the samples in the sequencing plates. In the barecode technique, the sequences in each sample are tagged with a unique barecode using barcoded primers during PCR amplification. The result of these high-throughput sequencing methods is several thousand sequences per sample in just few days, which must necessarily be analyzed using bioinformatics tools. Although the costs associated with these new technologies are less than for the Sanger method (considering the cost for one sequence), high-throughput sequencing methods remain an expensive approach. To date, these methods have been applied to the study of the microbial diversity of several communities, especially from human and animal guts [88-93]. Diversity analysis targeting the D1/D2 domain of the 26S rRNA gene or the internal transcribed regions (ITSs) allow yeast species to be distinguished [8], and have recently been applied to the study of some fungal communities [94, 95], and to the identification of some clinical yeast isolates [96-98]. These methods appear to be very suitable for studying the yeast biodiversity in the fish gut. This new knowledge, together with the information available in various databases, will allow both the accurate identification of new yeast isolates and the application of molecular strategies to characterize the yeast population in the fish gut.

6. Use of yeast as probiotics in aquaculture: stimulation of the immune and antioxidant systems, gut maturation and fish growth

The natural occurrence of numerous yeast species in the gastrointestinal tract of healthy fish has been well described, and yeast have been shown to constitute an important part of the microbiota of the fish gut [6]. In addition to *S. cerevisiae*, the halotolerant yeast *D. hansenii* has been considered an excellent probiotic candidate in fish aquaculture. Because, the number of experiences reporting the use of *D. hansenii* is increasing, this allows us to know the capacity of this yeast to enhance growth, survival, and gut maturation and to improve of the immune and antioxidant systems in fish larvae and juveniles. Yeast cells can be a hundred times larger than bacterial cells, which may explain the fact that the introduction of a low yeast population (10^4 CFU/g) through feed can induce beneficial effects in the host.

The effects of probiotics differ greatly depending on the microbial species, isolation source, experimental concentration and finally, the rearing conditions of the fish. However, the improvement of the immune response is one of the most encountered side effects in the host because immune system stimulation or immunomodulation are considered important mechanisms supporting probiosis. Yeast have immunostimulatory properties because they possess components such as β-glucan, mannoproteins, chitin (as a minor component) and nucleic acids [99].

Recent studies have shown the beneficial effect of dietary administered *Saccharomyces cerevisiae* in fish. Yeast supplemented diets stimulate growth, feed efficiency, blood biochemistry, survival rate, and non- specific immune responses in *Uronema marinum*-infected olive flounder (*Paralichthys olivaceus*) [100]. A diet supplemented with *S. cerevisiae* treated with beta-mercaptoethanol was better than whole cell yeast and n-3 highly unsaturated fatty acids (HUFA)-enriched yeast as an immune system and growth stimulator in juvenile rainbow trout challenged with *Yersinia ruckeri* [101]. Similarly, the dietary administration of the probiotic *S. cerevisiae* P13 at a minimum level of 10^5 CFU/kg enhanced the growth, innate immune responses and disease resistance of grouper (*Epinephelus coioides*) [102]. Cellular yeast components also stimulate the immune system: an improvement in gut mucus lysozyme activity was observed in European sea bass (*Dicentrarchus labrax*) fed mannan oligosaccharides (MOS) derived from the outer cell wall of a select strain of *S. cerevisiae* (Bio-Mos, Alltech Inc, USA) [103]. Furthermore, channel catfish (*Ictalurus punctatus*) juveniles fed diets supplemented with whole-cell *S. cerevisiae* (Levucell SB20®) or yeast subcomponents such as commercial preparations of β-glucan (MacroGard ® and Betagard A ®) had a significantly higher survival rate after *Edwardsiella ictaluri* challenge than did catfish fed with a controlled diet [104].

As previously described, most published studies have been performed using *S. cerevisiae*; however, promising results have also been obtained with *Debaryomyces hansenii*. A diet supplemented with *D. hansenii* stimulates the immune system of juvenile leopard grouper, *Mycteroperca rosacea*, by increasing IgM and superoxide dismutase (SOD) activity and enhances the resistance of the fish to infection by the dinoflagellate, *Amyloodinium ocellatum* [105]. Additionally, other studies have demonstrated the immune system improvement of *M. rosacea* when the fish were fed for 4 weeks with a compound diet enriched with 1.1% *D. hansenii*. After the 4 weeks, the fish were challenged with the pathogenic bacteria *Aeromonas hydrophila* strain Ah-315, resulting in an increase in IgM levels as well as catalase (CAT) and SOD activities in those fish fed yeast diets. Improvements were also observed at the molecular level, where CAT and HSP70 expression levels were enhanced in *M. rosacea* fed with *D. hansenii*, and challenged with *A. hydrophila* [106].

D. hansenii administration to the gilthead seabream (*Sparus aurata* L.) significantly enhances leukocyte peroxidase and respiratory burst activity by week 4 of feeding with yeast. Yeast feeding causes an up-regulation in the expression of the immune-associated genes Hep, IgM, TCR-ß, NCCRP-1, MHC-IIa, CSF-1R, C3, TNF-α and IL-1ß in the head-kidney: C3 expression was only stimulated in the liver, whereas the expression of TCR-ß, TNF-α and C3 was stimulated in the intestine of *S. aurata* [107].

When *D. hansenii* was administered at 1.1% in a compound diet to *D. labrax* larvae, the yeast stimulated the antioxidant status [108]. The group fed with yeast showed lower glutathione peroxidase (GPX) and SOD activity compared to fish fed the control diet, suggesting a possible involvement of superoxide anion retention in fish larvae, which could represent importance to the host to increase cell or tissue responsiveness to growth- and/or differentiation-enhancing factors [109]. The group fed the control diet showed oxidative stress represented by an increase in GPX activity at 48 days post hatching (dph) and gene expression levels for both GPX and SOD at 23 dph.

The ontogeny of the digestive tract of fish larvae has been the subject of many studies in the last decades with the purpose of increasing production rates by reducing the bottlenecks in larviculture. In this sense, the number of reports concerning the use of yeast to enhance gut maturation and digestive enzyme activity in fish are also increasing. The activity and expression of digestive enzyme-related genes during fish development provides an excellent marker of digestive development in fish larvae. The enzymes secreted from the pancreas (trypsin, lipase, and amylase) as well as those encountered in the intestinal brush border membranes (BBM) (leucine aminopeptidase N, alkaline phosphatase, maltase, γ-glutamyl-transpeptidase, and the cytosolic leucine-alanine peptidase), are the most common indicators of digestive system maturation in fish larvae. The degree of enterocyte maturation is described by increasing ratios of activities of BBM enzymes *vs* cytosolic leucine alanine peptidase; in the case of pancreatic enzymes, a decrease in amylase activity or expression level with the concomitant increase in trypsin or lipase activities characterizes the maturation of the exocrine pancreas.

Given this previous work, the effect of *D. hansenii*-enriched microparticulated diets were tested in European sea bass larvae. The authors observed an increase in intestinal brush border and pancreatic enzyme activities at 27 dph, indicating an achievement of gut morphology in this larvae stage compared with the control diet lacking yeast [110]. In a second feeding trial, where the introduction of yeast to the microparticulated diet was improved, larvae fed 1% *D. hansenii* matured earlier than fish fed a control diet after day 26, as revealed by lower amylase expression, higher lipase and trypsin expression, and high levels of the BBM enzymes, aminopeptidase N, maltase and alkaline phosphatase [111].

In another study [42], two yeast strains: *Saccharomyces cerevisiae* and *S. boulardii*, were evaluated as probiotics for rainbow trout (*Oncorhynchus mykiss*) fry to compare the cross effects of the two rearing conditions, with the intestinal microbiota and the brush border enzyme activities. Intestinal maturation at 10 dph was observed in trout fed *S. boulardii*, and kept in spring water, and these fish displayed the highest ratios of BBM leucine aminopeptidase N *vs* leucine-alanine peptidase, compared with those fish fed *S. cerevisiae*, and kept in river water.

Overall, yeast has been added directly to the water, administered as an additive in microparticulated diets, and has been used alive to feed live food (rotifers or *Artemia*) as a possible vector to deliver yeast into the gut of fish larvae. Rotifers have been established as the most common live prey to feed larval fish in hatcheries around the world, and baker's yeast (*S. cerevisiae*) is the most common nutrient source for culturing rotifers in addition to algae, emulsified oils or bacteria. Currently, efforts are being made to introduce *D. hansenii* into *Brachionus rotundiformis* to deliver yeast into the intestine of *L. guttatus* to accelerate digestive maturation [112].

In this regard, the use of Levucell® (*S. boulardii*), Bactocell® (*Pediococcus acidilactici*) and live yeast (*S. cerevisiae*) produced no significant effect on trypsin, lipase, and leucine aminopeptidase activities in California halibut larvae, *Paralichtys californicus*, at 46 dph. Contrary to this, an increase in pepsin and chymotrypsin activity was only observed in fish larvae fed Bactocell® at the final endpoint of the experiment (46 days), suggesting a potential use of these probiotics once metamorphosis is completed [113].

In fish aquaculture, the most utilized growth-promoting additives are hormones, antibiotics, ionophores, and salts [114]. The use of probiotics as growth-promoters has been recognized in the last decade with a number of studies related to this topic being published. Probiotics can be used as an alternative to avoid the use of antibiotics for growth promotion, thus eliminating the possibility of generating antibiotic-resistant bacteria in the aquaculture systems. When yeast probiotics have been used in the earliest developmental stages of fish larvae, enhanced growth and survival have been observed. Several yeast species have been documented to enhance growth following artificial colonization, particularly *S. cerevisiae* and *D. hansenii* either alone, or in synergic association with bacteria. One study [115] observed that pollock (*Pollachius pollachius*) larvae grew better when *Artemia nauplii* was first treated with *S. cerevisiae* var. *boulardii* CNCM I-1079 and then with *Pediococcus acidilactici* MA185 M than did larvae fed with one or no probiotic. When *S. cerevisiae* was used alone in feeding trials, it improved feed efficiency in Israeli carp and Nile tilapia [116, 117]. In addition to growth enhancement, an improvement in the conformation of the larvae was also observed in *D. labrax* fed with *D. hansenii*; a reduction in spinal deformity from 13.6% to 1.1% in fish fed yeast *vs* the control group was observed [111].

The use of probiotics in well-established fish industry is easier of because the existence of many individuals to experiment upon (*S. salar, D. labrax, S. aurata, S. maximus, O. mykiss*, etc). Nevertheless, it is advisable to use probiotics in species with the potential for exploitation to optimize the results for growth, survival and health. In this context, several studies have been performed in fish species with emerging aquaculture potential to contribute to the establishment of a continuous production line for experimental purposes. A commercial preparation of live yeast (*S. cerevisiae* and *Lactobacillus coagulans*) was tested on Indian carp fry, *Labeo rohita*, with no conclusive effects on growth [118]. *D. hansenii* has been observed to function as a growth promoter in *Mycteroperca rosacea* juveniles because after 4 weeks of continuous feeding, a weight gain (33%) and condition factor were observed in those fish fed a microparticulated diet enriched with yeast, compared with those fish fed without yeast [106]. In spotted sand bass larvae, *Paralabrax maculatofasciatus*, the highest survival (13.0%) was obtained with *D. hansenii* enriched microparticulated diets, but no effects on growth were observed with the use of probiotics [119].

At present, much of the existing evidence indicates that yeast promotes growth, and survival because of gut maturation, conformation of the larvae, and stimulation of the immune system by a possible involvement of endoluminal yeast-secreted polyamines in the host. As was earlier demonstrated, *D. hansenii* produces more polyamines (putrescine, spermidine and spermine) than *S. cerevisiae* and *S. boulardii* [111]. Polyamines participate in several physiological processes such as cell proliferation and differentiation and appear to have a broad influence on digestive tract maturation. In particular, the roles of dietary spermine and spermidine have been previously described [120]. These molecules enter enterocytes, where they induce a hormonal cascade that affects organs such as the pancreas and liver. Recently, the production of polyamines in 13 strains of *D. hansenii*, isolated from different sources, using high-pressure liquid chromatography (HPLC) has been reported [121]. In this study, they found that the L2, and CBS004 strains isolated from citrus fruit and marine water, respectively, were the main polyamine-secreting yeasts. Later, L2 strain was shown to have a probiotic effect because it enhanced the immune status, and intestinal function of gilthead seabream, *Sparus aurata* [122].

Finally, evidence of polyamine contribution to larviculture performance was reported when the spotted sand bass larvae, *P. maculatofasciatus* fed with *D. hansenii* with un-inhibited ornithine decarboxylase (ODC) activity had precocious digestive maturation compared to those larvae fed ODC-inhibited (with α-difluoromethylornithine (DFMO)) yeast [119]. Ornithine decarboxylase, which catalyzes the formation of putrescine, is the rate-limiting enzyme in the biosynthesis of polyamines in cells.

7. Yeast β-glucans: Structure, mechanisms of action and its application as immunostimulant in aquaculture

The glucose polymer β-glucan is a major structural component of the cell wall of some plants (such as the cereals oat and barley), seaweeds, and the outer cell wall of bacteria, fungi and yeast. Different β-glucans vary in structure, size, branching frequency, structural modifications, conformation and solubility, which may influence their physiological functions. Glucose molecules, in all β-glucan polymers, are linked together by a β-(1→3) linear β-glycosidic chain core, but differ in their length and branching structures. For example, the β-glucans from oat and barley are linear with β-(1→4) linkages, and shorter stretches of β-(1→3) linkages, while the structure of yeast β-glucans is composed of β-(1→3)-D-glucans with β-(1→6)-glycosidic linked branches, which apparently corresponds to the most active form of β-glucan. The relationship between structure and biological activity is controversial, but it appears that large molecular weight β-glucans are the most active compared with small β-glucans below 5,000-10,000 Da that are generally inactive. The solubility of β-glucans also influences their biological activity, with soluble β-glucans appearing to be more active.

The consumption of β-glucans has been associated with beneficial health effects in humans, including anticancer properties [123], metabolic syndrome prevention [124, 125], cholesterol-lowering effects [126], anti-atherogenic properties [127] and skin health promotion [128]. *In vitro* and *in vivo* studies in animals and humans show that the β-glucans derived from fungi and yeasts in particular, have interesting immune modulating properties [129-132]. This immune stimulation can be achieved when β-glucans are administered by a parenteral or an oral (dietary) route.

Despite their structural versatility, β-glucans are highly conserved structural components and belong to a group of physiologically active compounds called biological response modifiers [133]. Because of their large molecular weight, they cannot penetrate the cell membrane and therefore they must interact with cell-surface receptors; it has been shown that β-glucans are recognized by several receptors found on neutrophils, macrophages, and dendritic cells [129, 134]. Additionally, β-glucans belong to the group of non-self-molecules called pathogen-associated molecular patterns (PAMPS), which are recognized by pattern recognition receptors (PRRs) on the cell surface [135]. The principal β-glucans PRRs are dectin-1 and the toll-like receptors (TLRs), but other receptors are suggested to be involved, such as scavenger receptors, complement receptor 3, and lactosylceramide [136, 137]. Dectin-1 specifically recognizes β-(1→3)(1→6) glucans from fungi, plants, and bacteria [138], but it is not reactive toward β-(1→4) glucans or α-mannan [139]. The stimulation of dectin-1 activates the innate immune response, ROS and inflammatory cytokine production [140] through the activation of phospholipase C [141], the PI3K/Akt pathway, MAPK, NFAT, and NF-κB [142]. The interaction of β-glucans with TLRs results in activation of NF-κB and MAPK signaling [143]. Zymosan (β-glucans from the yeast *Saccharomyces cerevisiae*) binds to TLR2 and TLR4 (as well as dectin-1) found on macrophages, leading to an increase in cytokine production, such as TNF-α and IL-12, through the NF-κB pathway [143]. The β-glucan (β-(1→3) (1→6)-D-glucan) from *Aureobasidium pullulans* ADK-34 stimulates intestinal Peyer's patch cells both *in vitro* and *in vivo* as reflected by an increase in IL-5, IL-6, and IgA [129]. The interaction of β-glucans with specific receptors on macrophages and dendritic cells can induce the production of several cytokines, which in turn activate other immune cells such as B and T cells, thus activating the systemic immune response.

Yeast β-glucans have been applied in aquaculture to modulate the innate immune system of fish to improve their survival until adaptive immune responses are sufficiently developed to mount effective responses against pathogens [144, 145]. If the β-glucans are administered as feed additives, they can exert their primary effects at the intestinal level through the induction of cytokines, which in turn affect the systemic immune response in fish. Different sources of β-glucans have been evaluated, although the most frequent sources are those obtained from the baker's yeast, *Saccharomyces cerevisiae.* Some commercial preparations of β-glucan (MacroGard ®, Betagard A ®, EcoActiva™, Nutriferm ™, BG, Fibosel®, etc) are available on the market that can be used in aquaculture. Many studies have explored the *in vitro* response of the macrophage to β-glucan [146, 147], while other studies have addressed the *in vivo* effect of β-glucans in different fish species (Table 3). The β-glucans from several sources have been administered to fish via the oral or intra-peritoneal route with different effects. Many studies have shown the immune effects of β-glucans specifically on antibody production, expression of immune system genes, survival, resistance to infectious diseases, and improvement in stress resistance. The growth enhancement of fish has also been observed as another beneficial effect of β-glucans (Table 3).

Recently, juvenile channel catfish (*Ictalurus punctatus*) fed diets supplemented with whole-cell *Saccharomyces cerevisiae* (Levucell SB20®) or yeast subcomponents such as commercial preparations of β-glucan (MacroGard ® and Betagard A ®) had a significantly higher survival rate after challenge with *Edwardsiella ictaluri* than did catfish fed with a control diet [104]. Atlantic cod (*Gadus morhua* L.) were fed for 5 weeks with a purified β-glucan product [148], after which the fish were bath-challenged with the bacterial pathogen, *Vibrio anguillarum.* The transcription of selected cytokines (proinflammatory: IL1-β, IL-8, IFNγ; anti-inflammatory: IL-10) in different intestinal segments was analyzed using qPCR, and the β-glucan product was found to have a differential effect on the expression of the cytokine genes. In the anterior intestine and rectum, the β-glucan significantly elevated the expression of IL-1β when challenged with *V. anguillarum*. Moreover, an effect on the anti-inflammatory cytokine IL-10 was also visible in the rectum after the pathogen challenge. The differential responses of cytokines in the intestine of fish upon exposure to *V. anguillarum* suggests that β-glucans impact the ability of Atlantic cod to respond to the pathogen [148].

In another recently study, different concentrations of the yeast β-glucan preparation MacroGard ® (0.1%, 1% or 2%) were orally administered to mirror carp (*Cyprinus carpio* L.) for 8 weeks [149]. Fish fed diets containing 1% and 2% MacroGard® showed significant improvements in weight gain, specific growth rate and feed conversion ratio compared to fish fed both the control and the 0.1% MacroGard® containing diet. At the end of the experiment, the haematocrit value was significantly elevated in fish fed the 2% MacroGard® diet, compared to the control fed fish, with the blood monocyte fraction significantly higher in fish fed the 1% and 2% MacroGard® diets [149].

Zebrafish (*Danio rerio*) have been suggested as a model aquacultured fish, especially for genetic [150], nutritional and comparative growth studies [151]. Furthermore, zebrafish have been suggested as a model for pathogen studies in finfish [152]. Yeast β-glucans have also been evaluated in the zebrafish (*Danio rerio*) model with promising results [153]. In the study, a 5 mg/ml β-glucan preparation derived from *S. cerevisiae* was injected intra-peritoneally into adult zebrafish, leading to a significant reduction in mortality after challenge with *Aeromonas hydrophila*. In zebrafish treated with β-glucan, the ability of kidney cells to kill *A. hydrophila* was enhanced. Moreover, the myelomonocytic cell population in the kidney at 6 h post-challenge with *A. hydrophila* was increased. The β-glucan also appears to modulate the expression of IFN-γ and chemokines in the kidney [153].

Recently, the effect of β-glucan (derived from yeast, Fibosel® (Lallemand) on the growth performance and antioxidant enzyme activity in red snapper (*Lutjanus peru*), before and after exposure to lipopolysaccharides (LPS), was investigated. The fish were fed commercial diets with 0.0%, 0.1% and 0.2% Fibosel® for 6 weeks, after which, LPS was injected intra-peritoneally. The results showed a significant increase in growth performance after 6 weeks of β-glucan feeding; the SOD activity was also significantly higher in diets containing 0.1% β-glucan in weeks 4 and 6 with respect to the control group. At 72 h after injection of LPS, samples showed a significant increase in CAT activity in fish fed diets supplemented with 0.2% β-glucan and SOD activity increased under diets containing 0.1% and 0.2% β-glucan compared to controls [154]. To explain the enhanced growth, the authors suggested that some bacterial populations modify the host's digestive enzyme activity through their ability to produce and liberate exogenous digestive enzymes, as was previously observed [155]. Other authors reported that polysaccharides used as prebiotics can stimulate the growth of beneficial microbiota in fish [156].

β-glucan sources	Administration route	Fish species	Biological effects	Ref.
β-glucan (Aqua-In-Tech, Inc.)	Oral	Nile tilapia (*Oreochromis niloticus*)	no effect	[157]
Betagard A®	oral	Nile tilapia (*Oreochromis niloticus*)	Immune modulation	[158]
S. cerevisiae (Hang Zhou Bio-Technology Co)	oral	Nile tilapia (*Oreochromis niloticus*)	Immune modulation	[159]
MacroGard® Betagard A®	oral	Channel catfish (*Ictalurus punctatus*)	improvement in stress resistance	[160]
MacroGard® Betagard A®	oral	Channel catfish (*Ictalurus punctatus*)	Immune modulation	[104]
Saccharomyces cerevisiae	oral	Large yellow croaker (*Pseudosciaena crocea*)	Immune modulation growth enhancement	[161]
MacroGard® Zymosan	oral	Fathead minnows (*Pimephales promelas*)	Immune modulation	[162]

β-glucan sources	Administration route	Fish species	Biological effects	Ref.
GY (Sigma)* GB (Sigma)**				
Saccharomyces cerevisiae	intra-peritoneal	Atlantic salmon (*Salmo salar* L)	Immune modulation	[163]
Saccharomyces cerevisiae	oral	Atlantic salmon (*Salmo salar* L)	Enhancement of salmon lice resistance	[164]
marine diatom *Chaetoceros müllerі*	oral	Atlantic cod (*Gadus morhua* L.)	Survival and growth enhancement	[165]
β-1,3/1,6 glucan: BG (Biorigin Europe, Oslo, Norway)	oral	Atlantic cod (*Gadus morhua* L.)	Immune modulation	[148]
Barley	intra-peritoneal	Rohu (*Labeo rohita)*	Immune modulation	[166]
β-glucan (Sigma)	oral	Rohu (*Labeo rohita)*	Immune modulation	[167]
β-1,3 glucan (Sigma)	oral	Rohu (*Labeo rohita*)	Immune modulation	[168]
β-1,3 glucan (Sigma)	oral	Rohu (*Labeo rohita*)	Immune modulation	[169]
Yeast cell wall preparation from *Saccharomyces cerevisiae* (Nutriferm ™)	oral	Rohu (*Labeo rohita*)	Immune modulation	[170]
β-1,3 glucan (Sigma)	oral	Asian catfish (*Clarias batrachus*)	Immune modulation	[171, 172]
Glucan (Taito Co.Ltd., Tokyo, Japan)	oral	Rainbow trout (*Oncorhynchus mykiss*)	Decrease in plasmatic cortisol	[173]
MacroGard®	oral	Rainbow trout (*Oncorhynchus mykiss*)	Immune modulation	[174]
MacroGard®	oral	Rainbow trout (*Oncorhynchus mykiss*)	Immune modulation	[175]
Barley	oral	Rainbow trout (*Oncorhynchus mykiss*)	Immune modulation	[176]
β(1,3)-D-glucan (laminaran) from *Laminaria hyperborea*	intra-peritoneal	Rainbow trout (*Oncorhynchus mykiss*)	Immune modulation	[177]
β(1,3)-D-glucan (laminaran) from *Laminaria hyperborea*	inmersion	Rainbow trout (*Oncorhynchus mykiss*)	Immune modulation	[178]

β-glucan sources	Administration route	Fish species	Biological effects	Ref.
Saccharomyces cerevisiae	intra-peritoneal oral	Rainbow trout (*Oncorhynchus mykiss*)	Immune modulation	[179]
Saccharomyces cerevisiae	intra-peritoneal	Carp (*Cyprinus carpio*)	Immune modulation	[180, 181]
Saccharomyces cerevisiae	intra-peritoneal oral	Carp (*Cyprinus carpio*)	Immune modulation	[182]
Saccharomyces cerevisiae	oral	Carp (*Cyprinus carpio*)	Immune modulation	[183]
MacroGard®	oral	Carp (*Cyprinus carpio*)	Immune modulation	[184]
MacroGard®	oral	Carp (*Cyprinus carpio*)	apoptosis modulation	[185]
MacroGard®	oral	Carp (*Cyprinus carpio*)	Growth enhancement	[186]
Saccharomyces cerevisiae	intra-peritoneal	Zebrafish (*Danio rerio*)	Immune modulation	[153]
MacroGard®	oral	Sea bass (*Dicentrarchus labrax*)	Immune modulation	[187]
MacroGard®	oral	Sea bass (*Dicentrarchus labrax*)	Immune modulation	[188]
MacroGard®	oral	European sea bass (*Dicentrarchus labrax*)	Immune modulation	[189]
EcoActiva™	oral	Pink snapper (*Pagrus auratus*)	growth enhancement Immune modulation	[190]
oyster mushroom (*Pleurotus florida*)	intra-peritoneal	Catla (*Catla catla*)	Immune modulation	[191]
Poria cocos	oral	*Ctenopharyngodon idella*	Immune modulation	[192]
Fibosel® (Lallemand) *Saccharomyces cerevisiae*	oral	*Lutjanus peru*	growth enhancement Immune modulation	

*GY: β-1,3-glucan from baker's yeast
**GB: β -1,3-glucan from barley

Table 3. Biological effect of different β-glucans in fish

8. Conclusions

It is interesting to note that even after several decades of investigation, the potential of yeast, especially those of marine origin, has not yet been fully exploited. Yeasts can be part of the gut microbiota of wild and cultivated fish; however, more information derived using molecular approaches, is needed regarding the yeast composition in fish. Although much has been reported on the molecular aspects of yeasts, the exploration of the complete yeast community through the analysis of yeast DNA or RNA is lacking. The application of such methodologies will provide us with an overview of the non-cultivated yeasts, which could play a major role in the fish host. Different enzymes can be synthesized by yeast that have biotechnological potential, but the direct contribution of this potential to fish nutrition must be explored. In contrast, several publications confirm the beneficial probiotic effects of yeast in aquaculture, but the majority of these studies are focused on two species: *S. cerevisiae* and *D. hansenii*. The identification of new yeast species/strains from other cultured fish species is required to explore new beneficial properties to improve fish health and nutrition for a more sustainable aquaculture.

Acknowledgements

This work has been supported by FONDECYT grant 11110414 from CONICYT-Chile. D. Tovar, thanks to the Mexican Council of Science and Technology (CONACYT) for the grant CB-2010 157763.

Author details

Paola Navarrete[1*] and Dariel Tovar-Ramírez[2]

*Address all correspondence to: paolanavarretew@gmail.com

1 Biotechnology Laboratory, Institute of Nutrition and Food Technology (INTA), University of Chile, Santiago, Chile

2 Physiology and Functional Genomics Laboratory, The Center for Biological Research of NW (CIBNOR), La Paz, B.C.S., México

References

[1] FAO. http://www.fao.org/fishery/aquaculture/en (accessed 5 April 2013)

[2] Rumsey GL, Siwicki AK, Anderson DP, Bowser PR. Effect of soybean protein on serological response, non-specific defense mechanisms, growth, and protein utilization in rainbow trout. Vet Immunol Immunopath 1994;41(3–4) 323-39.

[3] Van den Ing T, Olli JJ, Krogdahl A. Effects of soybean-containing diets on the proximal and distal intestine in Atlantic salmon (*Salmo salar*): a morphological study. Aquaculture 1991;94 297-305.

[4] Sealey WM, Barrows FT, Smith CE, Overturf K, LaPatra SE. Soybean meal level and probiotics in first feeding fry diets alter the ability of rainbow trout *Oncorhynchus mykiss* to utilize high levels of soybean meal during grow-out. Aquaculture 2009;293(3-4) 195-203.

[5] Rawls JF, Samuel BS, Gordon JI. Gnotobiotic zebrafish reveal evolutionarily conserved responses to the gut microbiota. P Natl Acad Sci USA 2004;101(13) 4596-601.

[6] Gatesoupe FJ. Live yeasts in the gut: Natural occurrence, dietary introduction, and their effects on fish health and development. Aquaculture 2007;267(1-4) 20-30.

[7] Kurtzman CP. Molecular taxonomy of the yeasts. Yeast 1994;10(13) 1727-40.

[8] Suh SO, Blackwell M, Kurtzman CP, Lachance MA. Phylogenetics of Saccharomycetales, the ascomycete yeasts. Mycologia 2006;98(6) 1006-17.

[9] Scorzetti G, Fell JW, Fonseca A, Statzell-Tallman A. Systematics of basidiomycetous yeasts: a comparison of large subunit D1/D2 and internal transcribed spacer rDNA regions. FEMS yeast research 2002;2(4) 495-517.

[10] Lachance MA, Starmer WT. Ecology and yeasts. In: Kurtzman CP, Fell JW, editors. The Yeasts, a Taxonomic Study. Amsterdam: Elsevier; 1998. p. 21-30.

[11] Raghukumar S. The Role of Fungi in Marine Detrital Processes. In: Ramaiah N, editor. Marine Microbiology: Facets & Opportunities: National Institute of Oceanography, Goa; 2004. p. 91-101.

[12] Rhishipal R, Philip R. Selection of marine yeasts for the generation of single cell protein from prawn-shell waste. Bioresource Technol 1998;65(3) 255-6.

[13] Rocha LL, Cordeiro RDA, Cavalcante RM, do Nascimento RF, Martins SCS, Santaella ST, et al. Isolation and characterization of phenol-degrading yeasts from an oil refinery wastewater in Brazil. Mycopathologia 2007;164(4) 183-8.

[14] Burgaud G, Arzur D, Durand L, Cambon-Bonavita MA, Barbier G. Marine culturable yeasts in deep-sea hydrothermal vents: species richness and association with fauna. Fems Microbiol Ecol 2010;73(1) 121-33.

[15] Chi ZM, Chi Z, Zhang T, Liu GL, Li J, Wang XH. Production, characterization and gene cloning of the extracellular enzymes from the marine-derived yeasts and their potential applications. Biotechnol Adv 2009;27(3) 236-55.

[16] Lee JK, Park KS, Park S, Park H, Song YH, Kang SH, et al. An extracellular ice-binding glycoprotein from an Arctic psychrophilic yeast. Cryobiology 2010;60(2) 222-8.

[17] Li J, Chi ZM, Wang XH, Peng Y, Chi Z. The selection of alkaline protease-producing yeasts from marine environments and evaluation of their bioactive peptide production. Chin J Oceanol Limn 2009;27(4) 753-61.

[18] Peng Y, Chi ZM, Wang XH, Li J. Purification and molecular characterization of exo-beta-1,3-glucanases from the marine yeast *Williopsis saturnus* WC91-2. Appl Microbiol Biot 2009;85(1) 85-94.

[19] Song CL, Chi ZM, Li J, Wang XH. beta-Galactosidase production by the psychrotolerant yeast *Guehomyces pullulans* 17-1 isolated from sea sediment in Antarctica and lactose hydrolysis. Bioproc Biosyst Eng 2010;33(9) 1025-31.

[20] Song CL, Liu GL, Xu JL, Chi ZM. Purification and characterization of extracellular beta-galactosidase from the psychrotolerant yeast *Guehomyces pullulans* 17-1 isolated from sea sediment in Antarctica. Process Biochem 2010;45(6) 954-60.

[21] Kutty SN, Philp R. Marine yeasts - a review. Yeast 2008;25(7) 465-83.

[22] Kandasamy K, Alikunhi N, Subramanian M. Yeasts in marine and estuarine environments. Journal of Yeast and Fungal Research 2012;3(6) 74-82.

[23] Lodder J, Kreger-van rij JW. The Yeasts. A taxonomic study. Amsterdam, Holland.: North Holland Publ. Co.,; 1952.

[24] Wickerham L. Taxonomy of yeasts. U S Dept Agr Tech Bull 1951;1029 1-55.

[25] Roth F, Ahean D, Fell JW, Meyers S, Meyer S. Ecology and taxonomy of yeasts isolated from various marine substrates. Limnology and Oceanography 1962;7(2) 178-85.

[26] Lodder J. The yeasts. A taxonomic study. Amsterdam: North- Holland Publishing Co.; 1970.

[27] Lazarus C, Koburger J. Identification of Yeasts From the Suwannee River Florida Estuary. Applied Microbiology 1974;27(6) 1108-11.

[28] Hernandez-Saavedra NY, Hernandez-Saavedra D, Ochoa JL. Distribution of *Sporobolomyces* (Kluyver-Et-Vanniel) Genus in the Western Coast of Baja-California-Sur, Mexico. Syst Appl Microbiol 1992;15(2) 319-22.

[29] Nagahama T, Hamamoto M, Nakase T, Horikoshi K. *Kluyveromyces nonfermentans* sp nov., a new yeast species isolated from the deep sea. International journal of systematic bacteriology 1999;49 1899-905.

[30] Nagahama T, Hamamoto M, Nakase T, Takami H, Horikoshi K. Distribution and identification of red yeasts in deep-sea environments around the northwest Pacific Ocean. Anton Leeuw Int J G 2001;80(2) 101-10.

[31] Nagahama T, Hamamoto M, Nakase T, Horikoshi K. *Rhodotorula lamellibrachii* sp nov., a new yeast species from a tubeworm collected at the deep-sea floor in Sagami Bay and its phylogenetic analysis. Anton Leeuw Int J G 2001;80(3-4) 317-23.

[32] Nagahama T, Hamamoto M, Nakase T, Horikoshi K. *Rhodotorula benthica* sp nov and *Rhodotorula calyptogenae* sp nov., novel yeast species from animals collected from the deep-sea floor, and *Rhodotorula lysiniphila* sp nov., which is related phylogenetically. International journal of systematic and evolutionary microbiology 2003;53 897-903.

[33] Gadanho M, Sampaio JP. Occurrence and diversity of yeasts in the Mid-Atlantic ridge hydrothermal fields near the Azores Archipelago. Microb Ecol 2005;50(3) 408-17.

[34] Nagahama T, Hamamoto M, Horikoshi K. *Rhodotorula pacifica* sp nov., a novel yeast species from sediment collected on the deep-sea floor of the north-west Pacific Ocean. International journal of systematic and evolutionary microbiology 2006;56 295-9.

[35] Bass D, Howe A, Brown N, Barton H, Demidova M, Michelle H, et al. Yeast forms dominate fungal diversity in the deep oceans. P R Soc B 2007;274(1629) 3069-77.

[36] Nagahama T, Abdel-Wahab MA, Nogi Y, Miyazaki M, Uematsu K, Hamamoto M, et al. *Dipodascus tetrasporeus* sp nov., an ascosporogenous yeast isolated from deep-sea sediments in the Japan Trench. International journal of systematic and evolutionary microbiology 2008;58 1040-6.

[37] Chen YS, Yanagida F, Chen LY. Isolation of marine yeasts from coastal waters of northeastern Taiwan. Aquat Biol 2009;8(1) 55-60.

[38] Kutty SN, Philip R, Damodaran R. Marine yeasts in the slope sediments of Arabian sea and bay of Bengal. European Journal of Experimental Biology 2013;3(3) 311- 27.

[39] van Uden N, Branco C. Distribution and population densities of yeast species in Pacific water, air, animals and kelp off southern California. Limnology and Oceanography 1963;8 323-9.

[40] Slavikova E, Vadkertiova R. Yeasts and yeast-like organisms isolated from fish-pond waters. Acta microbiologica Polonica 1995;44(2) 181-9.

[41] Andlid T, Vazquez-Juarez R, Gustafsson L. Yeasts isolated from the intestine of rainbow trout adhere to and grow in intestinal mucus. Mol Mar Biol Biotech 1998;7(2) 115-26.

[42] Wache Y, Auffray F, Gatesoupe FJ, Zambonino J, Gayet V, Labbe L, et al. Cross effects of the strain of dietary *Saccharomyces cerevisiae* and rearing conditions on the onset of intestinal microbiota and digestive enzymes in rainbow trout, *Onchorhynchus mykiss,* fry. Aquaculture 2006;258(1-4) 470-8.

[43] Vázquez-Juárez R, Andlid T, Gustafsson L. Cell surface hydrophobicity and its relation to adhesion of yeasts isolated from fish gut. Colloids and Surfaces B: Biointerfaces 1994;2(1-3) 199-209.

[44] Vazquez-Juarez R, Andlid T, Gustafsson L. Adhesion of yeast isolated from fish gut to crude intestinal mucus of rainbow trout, *Salmo gairdneri*. Mol Mar Biol Biotechnol 1997;6(1) 64-71.

[45] Andlid T, Juarez RV, Gustafsson L. Yeast Colonizing the Intestine of Rainbow-Trout (*Salmo-Gairdneri*) and Turbot (*Scophtalmus-Maximus*). Microb Ecol 1995;30(3) 321-34.

[46] van Uden N, Farinha M. On the significance of temperature relations and vitamin deficiency patterns in the delimitation of yeast species. Port Acta Biol, Sér B 1958;6 161-78.

[47] Kreger-van rij JW. A Taxonomic study of the Yeast genera *Endomycopsis, Pichia* and *Debaryomyces.* . Kerkwerve, Netherlands: Backhuys Biological Books 1964.

[48] Bruce J, Morris EO. Psychrophilic yeasts isolated from marine fish. Antonie van Leeuwenhoek 1973;39(1) 331-9.

[49] Aubin J, Gatesoupe FJ, Labbe L, Lebrun L. Trial of probiotics to prevent the vertebral column compression syndrome in rainbow trout (*Oncorhynchus mykiss* Walbaum). Aquaculture Research 2005;36(8) 758-67.

[50] Barnett JA, Payne RW, Yarrow D. Yeast: Characteristics and Identification. 2nd ed. Cambridge: Cambridge Univ. Press.; 1990.

[51] Laconi S, Pompei R. Study and characterization of intestinal yeasts of mullet (*Mugil* spp.) for potential probiotic use. J Food Agric Environ 2007;5(3-4) 475-80.

[52] Breuer U, Harms H. *Debaryomyces hansenii* - an extremophilic yeast with biotechnological potential. Yeast 2006;23(6) 415-37.

[53] Gao LM, Chi ZM, Sheng J, Wang L, Li J, Gong F. Inulinase-producing marine yeasts: Evaluation of their diversity and inulin hydrolysis by their crude enzymes. Microb Ecol 2007;54(4) 722-9.

[54] Bolumar T, Sanz Y, Aristoy MC, Toldra F. Purification and characterisation of proteases a and D from *Debaryomyces hansenii.* Int J Food Microbiol 2008;124(2) 135-41.

[55] Garcia-Gonzalez A, Lotz M, Ochoa JL. Anti-inflammatory activity of superoxide dismutase obtained from *Debaryomyces hansenii* on type II collagen induced arthritis in rats. Rev Invest Clin 2009;61(3) 212-20.

[56] Takac S, Sengel BS. Extracellular lipolytic enzyme activity of a newly isolated *Debaryomyces hansenii.* Preparative biochemistry & biotechnology 2010;40(1) 28-37.

[57] Segal-Kischinevzky C, Rodarte-Murguia B, Valdes-Lopez V, Mendoza-Hernandez G, Gonzalez A, Alba-Lois L. The Euryhaline Yeast *Debaryomyces hansenii* has Two Cata-

lase Genes Encoding Enzymes with Differential Activity Profile. Curr Microbiol 2011;62(3) 933-43.

[58] Viana PA, de Rezende ST, Alves AD, Manfrini RM, Alves RJ, Bemquerer MP, et al. Activity of *Debaryomyces hansenii* UFV-1 alpha-galactosidases against alpha-D-galactopyranoside derivatives. Carbohyd Res 2011;346(5) 602-5.

[59] Viana PA, de Rezende ST, Falkoski DL, Leite TD, Jose IC, Moreira MA, et al. Hydrolysis of oligosaccharides in soybean products by *Debaryomyces hansenii* UFV-1 alpha-galactosidases. Food Chem 2007;103(2) 331-7.

[60] Viana PA, de Rezende ST, Marques VM, Trevizano LM, Passos FML, Oliveira MGA, et al. Extracellular alpha-galactosidase from *Debaryomyces hansenii* UFV-1 and its use in the hydrolysis of raffinose oligosaccharides. Journal of Agricultural and Food Chemistry 2006;54(6) 2385-91.

[61] Viana PA, de Rezende ST, Passos FML, Machado SG, Maitan GP, Coelho VTD, et al. alpha-Galactosidases production by *Debaryomyces hansenii* UFV-1. Food Sci Biotechnol 2011;20(3) 601-6.

[62] Viana PA, de Rezende ST, Passos FML, Oliveira JS, Teixeira KN, Santos AMC, et al. *Debaryomyces hansenii* UFV-1 Intracellular alpha-Galactosidase Characterization and Comparative Studies with the Extracellular Enzyme. Journal of Agricultural and Food Chemistry 2009;57(6) 2515-22.

[63] Barbosa AM, Giese EC, Dekker RFH, Borsato D, Perez AIB, Iranzo JFU. Extracellular beta-glucosidase production by the yeast *Debaryomyces pseudopolymorphus* UCLM-NS7A: optimization using response surface methodology. New Biotechnol 2010;27(4) 374-81.

[64] Ragon M, Aumelas A, Chemardin P, Galvez S, Moulin G, Boze H. Complete hydrolysis of myo-inositol hexakisphosphate by a novel phytase from *Debaryomyces castellii* CBS 2923. Appl Microbiol Biot 2008;78(1) 47-53.

[65] Dujon B, Sherman D, Fischer G, Durrens P, Casaregola S, Lafontaine I, et al. Genome evolution in yeasts. Nature 2004;430(6995) 35-44.

[66] White TJ, Bruns T, Lee S, Taylor S. Amplification and direct sequencing of fungal ribosomal RNA genes for phylogenetics. In: Innis MAG, P.H.; Sninsky, J.J.; White, T.J. , editor. PCR Protocols. San Diego, USA: Academic Press Inc; 1990. p. 315-22.

[67] The Centraal bureau voor Schimmelcultures (CBS) database. http://www.cbs.knaw.nl/collections/ BioloMICSSequences.aspx (accessed 5 April 2013)

[68] The Centraal bureau voor Schimmelcultures (CBS) database. http://www.cbs.knaw.nl/collections/BiolomicsID.aspx?IdentScenario=Yeasts%20species%20online%20ID (accessed 5 April 2013)

[69] Amann RI, Ludwig W, Schleifer KH. Phylogenetic Identification and in-Situ Detection of Individual Microbial-Cells without Cultivation. Microbiol Rev 1995;59(1) 143-69.

[70] Urubschurov V, Janczyk P. Biodiversity of Yeasts in the Gastrointestinal Ecosystem with Emphasis on Its Importance for the Host. In: Grillo O, Venora G, editors. The Dynamical Processes of Biodiversity - Case Studies of Evolution and Spatial Distribution: INTECH; 2011.

[71] Hovda MB, Lunestad BT, Fontanillas R, Rosnes JT. Molecular characterisation of the intestinal microbiota of farmed Atlantic salmon (*Salmo salar* L.). Aquaculture 2007;272 581-8.

[72] Kim DH, Brunt J, Austin B. Microbial diversity of intestinal contents and mucus in rainbow trout (*Oncorhynchus mykiss*). J Appl Microbiol 2007;102(6) 1654-64.

[73] Navarrete P, Espejo RT, Romero J. Molecular analysis of microbiota along the digestive tract of juvenile Atlantic salmon (*Salmo salar* L.). Microb Ecol 2009;57(3) 550-61.

[74] Navarrete P, Magne F, Mardones P, Riveros M, Opazo R, Suau A, et al. Molecular analysis of intestinal microbiota of rainbow trout (*Oncorhynchus mykiss*). Fems Microbiol Ecol 2010;71(1) 148-56.

[75] Pond MJ, Stone DM, Alderman DJ. Comparison of conventional and molecular techniques to investigate the intestinal microflora of rainbow trout (*Oncorhynchus mykiss*). Aquaculture 2006;261(1) 194-203.

[76] Romero J, Navarrete P. 16S rDNA-based analysis of dominant bacterial populations associated with early life stages of coho salmon (*Oncorhynchus kisutch*). Microb Ecol 2006;51(4) 422-30.

[77] Navarrete P, Magne F, Araneda C, Fuentes P, Barros L, Opazo R, et al. PCR-TTGE analysis of 16S rRNA from rainbow trout (*Oncorhynchus mykiss*) gut microbiota reveals host-specific communities of active bacteria. Plos One 2012;7(2).

[78] Muyzer G, Dewaal EC, Uitterlinden AG. Profiling of Complex Microbial-Populations by Denaturing Gradient Gel-Electrophoresis Analysis of Polymerase Chain Reaction-Amplified Genes-Coding for 16s Ribosomal-Rna. Appl Environ Microb 1993;59(3) 695-700.

[79] Lachance MA, Daniel HM, Meyer W, Prasad GS, Gautam SP, Boundy-Mills K. The D1/D2 domain of the large-subunit rDNA of the yeast species *Clavispora lusitaniae* is unusually polymorphic. FEMS yeast research 2003;4(3) 253-8.

[80] Cocolin L, Aggio D, Manzano M, Cantoni C, Comi G. An application of PCR-DGGE analysis to profile the yeast populations in raw milk. Int Dairy J 2002;12(5) 407-11.

[81] Cocolin L, Manzano M, Rebecca S, Comi G. Monitoring of yeast population changes during a continuous wine fermentation by molecular methods. Am J Enol Viticult 2002;53(1) 24-7.

[82] Gatto V, Torriani S. Microbial population changes during sourdough fermentation monitored by DGGE analysis of 16S and 26S rRNA gene fragments. Ann Microbiol 2004;54(1) 31-42.

[83] Molnar O, Wuczkowski M, Prillinger H. Yeast biodiversity in the guts of several pests on maize; comparison of three methods: classical isolation, cloning and DGGE. Mycol Prog 2008;7(2) 111-23.

[84] Prakitchaiwattana CJ, Fleet GH, Heard GM. Application and evaluation of denaturing gradient gel electrophoresis to analyse the yeast ecology of wine grapes. FEMS yeast research 2004;4(8) 865-77.

[85] Valmorri S, Tofalo R, Settanni L, Corsetti A, Suzzi G. Yeast microbiota associated with spontaneous sourdough fermentations in the production of traditional wheat sourdough breads of the Abruzzo region (Italy). Anton Leeuw Int J G 2010;97(2) 119-29.

[86] Ubeda JF, Fernandez-Gonzales M, Briones AI. Application of PCR-TTGE and PCR-RFLP for Intraspecific and Interspecific Characterization of the Genus *Saccharomyces* Using Actin Gene (ACT1) Primers. Curr Microbiol 2009;58(1) 58-63.

[87] Hamady M, Knight R. Microbial community profiling for human microbiome projects: Tools, techniques, and challenges. Genome research 2009;19(7) 1141-52.

[88] Beloshapka AN, Dowd SE, Suchodolski JS, Steiner JM, Duclos L, Swanson KS. Fecal microbial communities of healthy adult dogs fed raw meat-based diets with or without inulin or yeast cell wall extracts as assessed by 454 pyrosequencing. Fems Microbiol Ecol 2013;84(3) 532-41.

[89] Delgado S, Cabrera-Rubio R, Mira A, Suarez A, Mayo B. Microbiological Survey of the Human Gastric Ecosystem Using Culturing and Pyrosequencing Methods. Microb Ecol 2013;65(3) 763-72.

[90] Marzorati M, Maignien L, Verhelst A, Luta G, Sinnott R, Kerckhof FM, et al. Barcoded pyrosequencing analysis of the microbial community in a simulator of the human gastrointestinal tract showed a colon region-specific microbiota modulation for two plant-derived polysaccharide blends. Anton Leeuw Int J G 2013;103(2) 409-20.

[91] Oikonomou G, Teixeira AGV, Foditsch C, Bicalho ML, Machado VS, Bicalho RC. Fecal Microbial Diversity in Pre-Weaned Dairy Calves as Described by Pyrosequencing of Metagenomic 16S rDNA. Associations of *Faecalibacterium* Species with Health and Growth. Plos One 2013;8(4).

[92] Roos S, Dicksved J, Tarasco V, Locatelli E, Ricceri F, Grandin U, et al. 454 Pyrosequencing Analysis on Faecal Samples from a Randomized DBPC Trial of Colicky Infants Treated with *Lactobacillus reuteri* DSM 17938. Plos One 2013;8(2).

[93] Zened A, Combes S, Cauquil L, Mariette J, Klopp C, Bouchez O, et al. Microbial ecology of the rumen evaluated by 454 GS FLX pyrosequencing is affected by starch and oil supplementation of diets. Fems Microbiol Ecol 2013;83(2) 504-14.

[94] Heinrichs G, Hubner I, Schmidt CK, de Hoog GS, Haase G. Analysis of Black Fungal Biofilms Occurring at Domestic Water Taps (I): Compositional Analysis Using Tag-Encoded FLX Amplicon Pyrosequencing. Mycopathologia 2013;175(5-6) 387-97.

[95] Monard C, Gantner S, Stenlid J. Utilizing ITS1 and ITS2 to study environmental fungal diversity using pyrosequencing. Fems Microbiol Ecol 2013;84(1) 165-75.

[96] Pannanusorn S, Elings MA, Romling U, Fernandez V. Pyrosequencing of a hypervariable region in the internal transcribed spacer 2 to identify clinical yeast isolates. Mycoses 2012;55(2) 172-80.

[97] Heinrichs G, de Hoog GS, Haase G. Barcode Identifiers as a Practical Tool for Reliable Species Assignment of Medically Important Black Yeast Species. J Clin Microbiol 2012;50(9) 3023-30.

[98] Kim JY, Hahn HJ, Choe YB, Lee YW, Ahn KJ, Moon KC. Molecular Biological Identification of *Malassezia* Yeasts Using Pyrosequencing. Ann Dermatol 2013;25(1) 73-9.

[99] Ortuno J, Cuesta A, Rodriguez A, Esteban MA, Meseguer J. Oral administration of yeast, *Saccharomyces cerevisiae,* enhances the cellular innate immune response of gilthead seabream (*Sparus aurata* L.). Vet Immunol Immunopathol 2002;85(1-2) 41-50.

[100] Harikrishnan R, Kim MC, Kim JS, Balasundaram C, Heo MS. Immunomodulatory effect of probiotics enriched diets on Uronema marinum infected olive flounder. Fish Shellfish Immunol 2011;30(3) 964-71.

[101] Tukmechi A, Rahmati Andani HR, Manaffar R, Sheikhzadeh N. Dietary administration of beta-mercapto-ethanol treated *Saccharomyces cerevisiae* enhanced the growth, innate immune response and disease resistance of the rainbow trout, *Oncorhynchus mykiss*. Fish Shellfish Immunol 2011;30(3) 923-8.

[102] Chiu CH, Cheng CH, Gua WR, Guu YK, Cheng W. Dietary administration of the probiotic, *Saccharomyces cerevisiae* P13, enhanced the growth, innate immune responses, and disease resistance of the grouper, *Epinephelus coioides*. Fish Shellfish Immunol 2010;29(6) 1053-9.

[103] Torrecillas S, Makol A, Benitez-Santana T, Caballero MJ, Montero D, Sweetman J, et al. Reduced gut bacterial translocation in European sea bass (*Dicentrarchus labrax*) fed mannan oligosaccharides (MOS). Fish Shellfish Immunol 2011;30(2) 674-81.

[104] Welker TL, Lim C, Yildirim-Aksoy M, Klesius PH. Effect of short-term feeding duration of diets containing commercial whole-cell yeast or yeast subcomponents on im-

mune function and disease resistance in channel catfish, *Ictalurus punctatus*. Journal of animal physiology and animal nutrition 2012;96(2) 159-71.

[105] Reyes-Becerril M, Tovar-Ramirez D, Ascencio-Valle F, Civera-Cerecedo R, Gracia-Lopez V, Barbosa-Solomieu V. Effects of dietary live yeast *Debaryomyces hansenii* on the immune and antioxidant system in juvenile leopard grouper *Mycteroperca rosacea* exposed to stress. Aquaculture 2008;280(1-4) 39-44.

[106] Reyes-Becerril M, Tovar-Ramirez D, Ascencio-Valle F, Civera-Cerecedo R, Gracia-Lopez V, Barbosa-Solomieu V, et al. Effects of dietary supplementation with probiotic live yeast *Debaryomyces hansenii* on the immune and antioxidant systems of leopard grouper *Mycteroperca rosacea* infected with *Aeromonas hydrophila*. Aquaculture Research 2011;42(11) 1676-86.

[107] Reyes-Becerril M, Salinas I, Cuesta A, Meseguer J, Tovar-Ramirez D, Ascencio-Valle F, et al. Oral delivery of live yeast *Debaryomyces hansenii* modulates the main innate immune parameters and the expression of immune-relevant genes in the gilthead seabream (*Sparus aurata* L.). Fish Shellfish Immunol 2008;25(6) 731-9.

[108] Tovar-Ramirez D, Mazurais D, Gatesoupe JF, Quazuguel P, Cahu CL, Zambonino-Infante JL. Dietary probiotic live yeast modulates antioxidant enzyme activities and gene expression of sea bass (*Dicentrarchus labrax*) larvae. Aquaculture 2010;300(1-4) 142-7.

[109] Buetler TM, Krauskopf A, Ruegg UT. Role of superoxide as a signaling molecule. News Physiol Sci 2004;19 120-3.

[110] Tovar D, Zambonino J, Cahu C, Gatesoupe FJ, Vazquez-Juarez R, Lesel R. Effect of live yeast incorporation in compound diet on digestive enzyme activity in sea bass (*Dicentrarchus labrax*) larvae. Aquaculture 2002;204(1-2) 113-23.

[111] Tovar-Ramirez D, Infante JZ, Cahu C, Gatesoupe FJ, Vazquez-Juarez R. Influence of dietary live yeast on European sea bass (*Dicentrarchus labrax*) larval development. Aquaculture 2004;234(1-4) 415-27.

[112] Tovar-Ramirez D. El efecto probiótico de la levadura *Debaryomyces hansenii* CBS 8339 sobre el desarrollo larvario del pargo Lunarejo *Lutjanus guttatus*. La Paz, BCS, México: Personal Communication; 2013.

[113] Zacarias-Soto M, Lazo JP, Viana MT. Effect of three probiotics administered through live feed on digestive enzyme activity in California halibut, *Paralichthys californicus*, larvae. J World Aquacult Soc 2011;42(3) 321-31.

[114] Klaenhammer TR, Kullen MJ. Selection and design of probiotics. Int J Food Microbiol 1999;50(1-2) 45-57.

[115] Gatesoupe FJ. Probiotic and formaldehyde treatments of *Artemia nauplii* as food for larval pollack, *Pollachius pollachius*. Aquaculture 2002;212(1-4) 347-60.

[116] Lara-Flores M, Olvera-Novoa MA, Guzman-Mendez BE, Lopez-Madrid W. Use of the bacteria *Streptococcus faecium* and *Lactobacillus acidophilus*, and the yeast *Saccharomyces cerevisiae* as growth promoters in Nile tilapia (*Oreochromis niloticus*). Aquaculture 2003;216(1-4) 193-201.

[117] Noh SH, Han K, Won TH, Choi YJ. Effect of antibiotics, enzyme, yeast culture and probiotics on the growth performance of Israeli carp. Korean Journal of Animal Science 1994;36(480-486).

[118] Mohanty SN, Swain SK, Tripathi SD. Growth and survival of rohu spawn fed on liver based diet. Journal of Inland Fisheries Society of India 1993;25(2) 41-5.

[119] Guzmán-Villanueva L, Tovar-Ramírez D, Civera-Cerecedo R, editors. Effect of wild and ornithine decarboxylase deficient *Debaryomyces hansenii,* on *Paralabrax maculatofasciatus* larvae development. Caribbean and Latin American Aquaculture; 2007 6–9 November; San Juan Puerto Rico.

[120] Dufour C, Dandrifosse G, Forget P, Vermesse F, Romain N, Lepoint P. Spermine and spermidine induce intestinal maturation in the rat. Gastroenterology 1988;95(1) 112-6.

[121] Reyes-Becerril M, Ángeles Esteban MA, Tovar-Ramírez D, Ascencio-Valle F. Polyamine determination in different strains of the yeast *Debaryomyces hansenii* by high pressure liquid chromatography. . Food Chem 2011;127 1862-5.

[122] Reyes-Becerril M, Ascencio-Valle F, Meseguer J, Tapia-Paniagua ST, Moriñigo MA, Esteban MÁ. *Debaryomyces hansenii* L2-enriched diet enhances the immunity status, gene expression and intestine functionality in gilthead seabream (*Sparus aurata* L.). Aquaculture Research 2012;43 1107-18.

[123] Aleem E. beta -Glucans and their Applications in Cancer Therapy: Focus on human studies. Anti-cancer agents in medicinal chemistry 2013;13(5) 709-19.

[124] Cloetens L, Ulmius M, Johansson-Persson A, Akesson B, Onning G. Role of dietary beta-glucans in the prevention of the metabolic syndrome. Nutrition reviews 2012;70(8) 444-58.

[125] Tosh SM. Review of human studies investigating the post-prandial blood-glucose lowering ability of oat and barley food products. Eur J Clin Nutr 2013;67(4) 310-7.

[126] Othman RA, Moghadasian MH, Jones PJH. Cholesterol-lowering effects of oat beta-glucan. Nutrition reviews 2011;69(6) 299-309.

[127] Andersson KE, Hellstrand P. Dietary oats and modulation of atherogenic pathways. Molecular nutrition & food research 2012;56(7) 1003-13.

[128] Du B, Bian Z, Xu B. Skin Health Promotion Effects of Natural Beta-Glucan Derived from Cereals and Microorganisms: A Review. Phytotherapy research : PTR 2013.

[129] Tanioka A, Tanabe K, Hosono A, Kawakami H, Kaminogawa S, Tsubaki K, et al. Enhancement of intestinal immune function in mice by β-D-glucan from *Aureobasidium pullulans* ADK-34. Scand J Immunol 2013;78(1) 61-8.

[130] Ganner A, Schatzmayr G. Capability of yeast derivatives to adhere enteropathogenic bacteria and to modulate cells of the innate immune system. Appl Microbiol Biot 2012;95(2) 289-97.

[131] Muramatsu D, Iwai A, Aoki S, Uchiyama H, Kawata K, Nakayama Y, et al. beta-glucan derived from *Aureobasidium pullulans* is effective for the prevention of influenza in mice. Plos One 2012;7(7).

[132] Volman JJ, Ramakers JD, Plat J. Dietary modulation of immune function by beta-glucans. Physiol Behav 2008;94(2) 276-84.

[133] Novak M, Vetvicka V. beta-glucans, history, and the present: Immunomodulatory aspects and mechanisms of action. J Immunotoxicol 2008;5(1) 47-57.

[134] Goodridge HS, Wolf AJ, Underhill DM. beta-glucan recognition by the innate immune system. Immunol Rev 2009;230 38-50.

[135] Brown GD, Gordon S. Immune recognition of fungal beta-glucans. Cell Microbiol 2005;7(4) 471-9.

[136] Dalmo RA, Bogwald J. beta-glucans as conductors of immune symphonies. Fish Shellfish Immun 2008;25(4) 384-96.

[137] Zimmerman JW, Lindermuth J, Fish PA, Palace GP, Stevenson TT, DeMong DE. A novel carbohydrate-glycosphingolipid interaction between a beta-(1-3)-glucan immunomodulator, PGG-glucan, and lactosylceramide of human leukocytes. J Biol Chem 1998;273(34) 22014-20.

[138] Palma AS, Feizi T, Zhang YB, Stoll MS. Ligands for the beta-glucan receptor, Dectin-1, assigned using "designer" microarrays of oligosaccharide probes (neoglycolipids) generated from glucan polysaccharides. (vol 281, pg 5771, 2006). J Biol Chem 2006;281(34) 24999-.

[139] Brown GD, Taylor PR, Reid DM, Willment JA, Williams DL, Martinez-Pomares L, et al. Dectin-1 is a major beta-glucan receptor on macrophages. J Exp Med 2002;196(3) 407-12.

[140] Grunebach F, Weck MM, Reichert J, Brossart P. Molecular and functional characterization of human Dectin-1. Exp Hematol 2002;30(11) 1309-15.

[141] Xu S, Huo J, Lee KG, Kurosaki T, Lam KP. Phospholipase C gamma 2 Is Critical for Dectin-1-mediated Ca(2+) Flux and Cytokine Production in Dendritic Cells. J Biol Chem 2009;284(11) 7038-46.

[142] Shah VB, Ozment-Skelton TR, Williams DL, Keshvaraa L. Vav1 and PI3K are required for phagocytosis of beta-glucan and subsequent superoxide generation by microglia. Mol Immunol 2009;46(8-9) 1845-53.

[143] Gantner BN, Simmons RM, Canavera SJ, Akira S, Underhill DM. Collaborative induction of inflammatory responses by dectin-1 and toll-like receptor 2. J Exp Med 2003;197(9) 1107-17.

[144] Bricknell I, Dalmo RA. The use of immunostimulants in fish larval aquaculture. Fish Shellfish Immun 2005;19(5) 457-72.

[145] Meena DK, Das P, Kumar S, Mandal SC, Prusty AK, Singh SK, et al. Beta-glucan: an ideal immunostimulant in aquaculture (a review). Fish Physiol Biochem 2013;39(3) 431-57.

[146] Jorgensen JB, Sharp GJE, Secombes CJ, Robertsen B. Effect of a Yeast-Cell-Wall Glucan on the Bactericidal Activity of Rainbow-Trout Macrophages. Fish Shellfish Immun 1993;3(4) 267-77.

[147] Paulsen SM, Engstad RE, Robertsen B. Enhanced lysozyme production in Atlantic salmon (*Salmo salar* L.) macrophages treated with yeast beta-glucan and bacterial lipopolysaccharide. Fish Shellfish Immun 2001;11(1) 23-37.

[148] Lokesh J, Fernandes JMO, Korsnes K, Bergh O, Brinchmann MF, Kiron V. Transcriptional regulation of cytokines in the intestine of Atlantic cod fed yeast derived mannan oligosaccharide or beta-Glucan and challenged with *Vibrio anguillarum*. Fish Shellfish Immun 2012;33(3) 626-31.

[149] Kühlwein H, Merrifield D, Rawling M, Foey A, Davies S. Effects of dietary β-(1,3)(1,6)-D-glucan supplementation on growth performance, intestinal morphology and haemato-immunological profile of mirror carp (*Cyprinus carpio* L.). Journal of Animal Physiology and Animal Nutrition (Berl) 2013;doi: 10.1111/jpn.12078.

[150] Dahm R, Geisler R. Learning from small fry: The zebrafish as a genetic model organism for aquaculture fish species. Mar Biotechnol 2006;8(4) 329-45.

[151] Ulloa PE, Iturra P, Neira R, Araneda C. Zebrafish as a model organism for nutrition and growth: towards comparative studies of nutritional genomics applied to aquacultured fishes. Rev Fish Biol Fisher 2011;21(4) 649-66.

[152] Sullivan C, Kim CH. Zebrafish as a model for infectious disease and immune function. Fish Shellfish Immun 2008;25(4) 341-50.

[153] Rodriguez I, Chamorro R, Novoa B, Figueras A. beta-Glucan administration enhances disease resistance and some innate immune responses in zebrafish (*Danio rerio*). Fish Shellfish Immun 2009;27(2) 369-73.

[154] Guzmán-Villanueva LT, Tovar-Ramírez D, Ascencio-Valle F, Macías-Rodríguez ME, Esteban MA, Gisbert E, editors. Effects of β-Glucan 1,3/1,6 on antioxidant enzyme ac-

tivity in *Lutjanus peru* exposed to lipopolysaccharides. . Aquaculture 2012; 2012 1-5 September; Prague, Czech Republic: World Aquaculture Society.

[155] Dimitroglou A, Merrifield DL, Carnevali O, Picchietti S, Avella M, Daniels C, et al. Microbial manipulations to improve fish health and production - A Mediterranean perspective. Fish Shellfish Immun 2011;30(1) 1-16.

[156] Ringo E, Olsen RE. The effect of diet on aerobic bacterial flora associated with intestine of Arctic charr (Salvelinus alpinus L.). J Appl Microbiol 1999;86(1) 22-8.

[157] Whittington R, Lim C, Klesius PH. Effect of dietary beta-glucan levels on the growth response and efficacy of *Streptococcus iniae* vaccine in Nile tilapia, *Oreochromis niloticus*. Aquaculture 2005;248(1-4) 217-25.

[158] R. S, C. L, M. Y-A, Welker TL, Klesius PH. Effect of yeast oligosaccharide diet supplements on growth and disease resistance in juvenile Nile tilpia (*Oreochromis niloticus*). Journal of applied aquaculture 2009;21 61-71.

[159] El-Boshy ME, El-Ashram AM, AbdelHamid FM, Gadalla HA. Immunomodulatory effect of dietary *Saccharomyces cerevisiae*, beta-glucan and laminaran in mercuric chloride treated Nile tilapia (*Oreochromis niloticus*) and experimentally infected with *Aeromonas hydrophila*. Fish Shellfish Immun 2010;28(5-6) 802-8.

[160] Welker TL, Lim C, Yildirim-Aksoy M, Shelby R, Klesius PH. Immune response and resistance to stress and *Edwardsiella ictaluri* challenge in channel catfish, *Ictalurus punctatus*, fed diets containing commercial whole-cell yeast or yeast subcomponents. J World Aquacult Soc 2007;38(1) 24-35.

[161] Ai QH, Mai KS, Zhang L, Tan BP, Zhang WB, Xu W, et al. Effects of dietary beta-1,3 glucan on innate immune response of large yellow croaker, *Pseudosciaena crocea*. Fish Shellfish Immun 2007;22(4) 394-402.

[162] Palic D, Andreasen C, Herolt D, Menzel B, Roth J. Immunomodulatory effects of b-glucan on neutrophil function in fathead minnows (*Pimephales promelas* Rafinesque, 1820). Dev Comp Immunol 2006;30 817-30.

[163] Rorstad G, Aasjord PM, Robertsen B. Adjuvant Effect of a Yeast Glucan in Vaccines against Furunculosis in Atlantic Salmon (*Salmo salar* L). Fish Shellfish Immun 1993;3(3) 179-90.

[164] Refstie S, Baeverfjord G, Seim RR, Elvebo O. Effects of dietary yeast cell wall beta-glucans and MOS on performance, gut health, and salmon lice resistance in Atlantic salmon (*Salmo salar*) fed sunflower and soybean meal. Aquaculture 2010;305(1-4) 109-16.

[165] Skjermo J, Storseth TR, Hansen K, Handa A, Oie G. Evaluation of beta-(1 -> 3, 1 -> 6)-glucans and high-M alginate used as immunostimulatory dietary supplement during first feeding and weaning of Atlantic cod (*Gadus morhua* L.). Aquaculture 2006;261(3) 1088-101.

[166] Misra CK, Das BK, Mukherjee SC, Pattnaik P. Effect of multiple injections of beta-glucan on non-specific immune response and disease resistance in Labeo rohita fingerlings. Fish Shellfish Immunol 2006;20(3) 305-19.

[167] Misra CK, Das BK, Mukherjee SC, Pattnaik P. Effect of long term administration of dietary beta-glucan on immunity, growth and survival of Labeo rohita fingerlings. Aquaculture 2006;255(1-4) 82-94.

[168] Sahoo PK, Mukherjee SC. Effect of dietary beta-1,3 glucan on immune responses and disease resistance of healthy and aflatoxin B-1-induced immunocompromised rohu (*Labeo rohita* Hamilton). Fish Shellfish Immun 2001;11(8) 683-95.

[169] Sahoo PK, Mukherjee SC. The effect of dietary immunomodulation upon *Edwardsiella tarda* vaccination in healthy and immunocompromised Indian major carp (*Labeo rohita*). Fish Shellfish Immun 2002;12(1) 1-16.

[170] Pal D, Joardar SN, Roy B. Immunostimulatory effects of a yeast (*Saccharomyces cerevisiae*) cell wall feed supplement on rohu (*Labeo rohita*), an Indian major carp. Isr J Aquacult-Bamid 2007;59(3) 175-81.

[171] Kumari J, Sahoo PK. Dietary immunostimulants influence specific immune response and resistance of healthy and immunocompromised Asian catfish *Clarias batrachus* to *Aeromonas hydrophila* infection. Dis Aquat Organ 2006;70(1-2) 63-70.

[172] Kumari J, Sahoo PK. Dietary beta-1,3 glucan potentiates innate immunity and disease resistance of Asian catfish, *Clarias batrachus* (L.). J Fish Dis 2006;29(2) 95-101.

[173] Jeney G, Galeotti M, Volpatti D, Jeney Z, Anderson DP. Prevention of stress in rainbow trout (*Oncorhynchus mykiss*) fed diets containing different doses of glucan. Aquaculture 1997;154(1) 1-15.

[174] Verlhac V, Obach A, Gabaudan J, Schuep W, Hole R. Immunomodulation by dietary vitamin C and glucan in rainbow trout (*Oncorhynchus mykiss*). Fish Shellfish Immun 1998;8(6) 409-24.

[175] Siwicki A, Kazuñ K, Glabski E, Terech-Majewska E, Baranowski P, Trapkowska S. The effect of beta-1.3/1.6 – glucan in diets on the effectiveness of anti-*Yersinia ruckeri* vaccine – an experimental study in rainbow trout (*Oncorhynchus mykiss*). Polish journal of food and nutrition sciences 2004;13(54) 59-61.

[176] Sealey WM, Barrows FT, Hang A, Johansen KA, Overturf K, LaPatra SE, et al. Evaluation of the ability of barley genotypes containing different amounts of beta-glucan to alter growth and disease resistance of rainbow trout *Oncorhynchus mykiss*. Anim Feed Sci Tech 2008;141(1-2) 115-28.

[177] Lovoll M, Fischer U, Mathisen GS, Bogwald J, Ototake M, Dalmo RA. The C3 subtypes are differentially regulated after immunostimulation in rainbow trout, but head kidney macrophages do not contribute to C3 transcription. Vet Immunol Immunopathol 2007;117(3-4) 284-95.

[178] Zhang Z, Swain T, Bogwald J, Dalmo RA, Kumari J. Bath immunostimulation of rainbow trout (*Oncorhynchus mykiss*) fry induces enhancement of inflammatory cytokine transcripts, while repeated bath induce no changes. Fish Shellfish Immunol 2009;26(5) 677-84.

[179] Kunttu HMT, Valtonen ET, Suomalainen LR, Vielma J, Jokinen IE. The efficacy of two immunostimulants against *Flavobacterium columnare* infection in juvenile rainbow trout (*Oncorhynchus mykiss*). Fish Shellfish Immun 2009;26(6) 850-7.

[180] Selvaraj V, Sampath K, Sekar V. Administration of yeast glucan enhances survival and some non-specific and specific immune parameters in carp (*Cyprinus carpio*) infected with *Aeromonas hydrophila*. Fish Shellfish Immun 2005;19(4) 293-306.

[181] Selvaraj V, Sampath K, Sekar V. Use of glucan from *Saccharomyces cerevisiae* as an immunostimulant in carp: Impact on hematology, phagocyte function, and infection with *Aeromonas hydrophila*. Isr J Aquacult-Bamid 2005;57(1) 39-48.

[182] Selvaraj V, Sampath K, Vaithilingam S. Adjuvant and immunostimulatory effects of b-glucan administration in combination with lipopolysaccharide enhances survival and some immune parameters in carp challenged with *Aeromonas hydrophila*. Veterinary Immunology and Immunopathology 2006;114 15-24.

[183] Lin S, Pan Y, Luo L, Luo L. Effects of dietary beta-1,3-glucan, chitosan or raffinose on the growth, innate immunity and resistance of koi (*Cyprinus carpio* koi). Fish Shellfish Immunol 2011;31(6) 788-94.

[184] Falco A, Frost P, Miest J, Pionnier N, Irnazarow I, Hoole D. Reduced inflammatory response to *Aeromonas salmonicida* infection in common carp (*Cyprinus carpio* L.) fed with beta-glucan supplements. Fish Shellfish Immunol 2012;32(6) 1051-7.

[185] Miest JJ, Falco A, Pionnier NP, Frost P, Irnazarow I, Williams GT, et al. The influence of dietary beta-glucan, PAMP exposure and Aeromonas salmonicida on apoptosis modulation in common carp (*Cyprinus carpio*). Fish Shellfish Immunol 2012;33(4) 846-56.

[186] Kuhlwein H, Merrifield DL, Rawling MD, Foey AD, Davies SJ. Effects of dietary beta-(1,3)(1,6)-D-glucan supplementation on growth performance, intestinal morphology and haemato-immunological profile of mirror carp (*Cyprinus carpio* L.). Journal of animal physiology and animal nutrition 2013.

[187] Bagni M, Archetti L, Amadori M, Marino G. Effect of long-term oral administration of an immunostimulant diet on innate immunity in sea bass (*Dicentrarchus labrax*). J Vet Med B 2000;47(10) 745-51.

[188] Bagni M, Romano N, Finoia MG, Abelli L, Scapigliati G, Tiscar PG, et al. Short- and long-term effects of a dietary yeast beta-glucan (Macrogard) and alginic acid (Ergosan) preparation on immune response in sea bass (*Dicentrarchus labrax*). Fish Shellfish Immunol 2005;18(4) 311-25.

[189] Bonaldo A, Thompson KD, Manfrin A, Adams A, Murano E, Mordenti AL, et al. The influence of dietary beta-glucans on the adaptive and innate immune responses of European sea bass (*Dicentrarchus labrax)* vaccinated against vibriosis. Ital J Anim Sci 2007;6(2) 151-64.

[190] CooK MT, Hayball PJ, Hutchinson W, Nowak BF, Hayball JD. Administration of a commercial immunostimulant preparation, EcoActiva (TM) as a feed supplement enhances macrophage respiratory burst and the growth rate of snapper (*Pagrus auratus*, Sparidae (Bloch and Schneider)) in winter. Fish Shellfish Immun 2003;14(4) 333-45.

[191] Kamilya D, Maiti TK, Joardar SN, Mal BC. Adjuvant effect of mushroom glucan and bovine lactoferrin upon Aeromonas hydrophila vaccination in catla, *Catla catla* (Hamilton). J Fish Dis 2006;29(6) 331-7.

[192] Y. K, Q. Z. Effect of b-glucan on activity of antioxidant enzymes and Mx gene expression in virus infected grass carp. Fish & Shellfish Immunology 2009;27 336-40.

Chapter 6

The Use and Benefits of *Bacillus* Based Biological Agents in Aquaculture

Mulalo Edna Nemutanzhela, Yrielle Roets,
Neil Gardiner and Rajesh Lalloo

Additional information is available at the end of the chapter

http://dx.doi.org/10.5772/57375

1. Introduction

Global shortages in seafood resources have driven the growth of aquaculture as an economic activity, predominantly in developing countries [1-2]. As a consequence of space and resource constraints, traditional aquaculture has been intensified into reticulated systems with high stocking densities of the cultured species [3-4]. This results in an artificial environment that has a propensity for supporting the growth of pathogenic bacteria and the accumulation of waste metabolites in aquaculture systems [5]. The indiscriminate release of spent aquaculture wastes into surrounding environments is also problematic [6-7].

The outbreak of disease in aquaculture systems, caused by bacterial pathogens, is a complex phenomenon associated with stressful environmental conditions such as poor water quality and can ultimately result in mass mortality and significant loss to the industry [8-9]. The main cause of poor water quality is waste accumulation through hyper-nutrification resulting from excessive feeding rates and high nutrient dietary composition, both of which are common phenomena in intensive aquaculture systems [13-15]. High levels of nitrogenous and phosphorous waste accumulation predispose fish to infestation by parasites and pathogens and also pose a threat to the environment [13,16-17]. Selection for certain characteristics by breeders has also in some cases reduced the vigour in breeding lines, making fish less hardy and more susceptible to disease [10]. Of particular importance is the prevalence of bacterial disease, which results in damage and often leads to death of fish [11]. Gram-negative bacteria such as *Aeromonas hydrophila* are amongst the main pathogenic micro-organisms responsible for bacterial disease [8,12]. Conventional methods of dealing with disease include the use of chemicals and antibiotics, which alter natural microbial populations, damage the environment and increase resistance and virulence of pathogenic micro-organisms [5,17-21].

Useful micro-organisms play a number of roles in pond culture, particularly with respect to productivity, nutrient cycling, nutrition of the cultured animals, water quality, disease control and environmental impact of effluents [22-24]. Bacterial additives demonstrate the potential to improve water quality and reduce pathogen load and mortality, and have thus emerged in modern day aquaculture as alternatives to chemicals and antibiotics [17,24]. Many bacterial strains have also demonstrated a significant algaecidal effect, which is advantageous in aquaculture systems through reduction of algal growth and hence algal blooms which can destabilise these systems [25-26]. Biological agents such as Gram-positive *Bacillus spp.* offer an attractive solution to the challenges facing modern aquaculture. Advantages of this genus include the ability to grow rapidly, tolerate a wide range of physiological conditions and the ability to sporulate. The robust spores of *Bacillus spp.* are also amenable to simple and cost effective production processes and the end products are stable for long periods [24, 27].

2. Aquaculture as an economic activity

The Food and Agriculture Organization of the United Nations [28] reported that capture fisheries and aquaculture supplied the world with about 154 million tonnes of fish in 2011, of which 131 million tonnes were used for human consumption [28]. Aquaculture contributed 79 million tonnes to the global fisheries market in 2010 at a value of $125 billion. Aquaculture farming used for food consumption comprised 60 million tonnes ($119 billion), 15 million tonnes was used for fish meal and fish oil production, while the remainder was used for ornamental fish production. With sustained growth in fish production and improved distribution channels, world supply of fish for human consumption has grown dramatically in the last five decades. An average growth rate of 3.2% per year in the period 1961–2009, has outpaced the increase of 1.7% per year in the world's population. The global aquaculture market comprises both marine and inland (freshwater) farming. The majority (90%) of fresh water ornamental fish are captive bred, compared to only 25 of the 8000 species of marine fish. In 2010, 75% of the quantity of fish and fishery products produced consisted of products destined for human consumption, with ornamental aquaculture contributing a smaller volume.

Aquaculture production is dominated by developing countries, and predominates in Asian countries. The methods of practice of aquaculture have evolved into intensive reticulated systems, in contrast to traditional extensive systems, due to restrictions in availability of land and as a consequence of increased environmental awareness. Aquaculture is probably the fastest growing food-producing sector globally, and the most recent estimates for worldwide aquaculture show that it contributes just over 50% of total fish production. This has been an astonishingly fast growth rate from only 16% of total consumption 15 years ago. The key impetus for growth of the market is global food security and a resistance towards resource exploitation through over-harvesting of natural waters [29]. The consumer drives the aquaculture practice, product quality and branding. End products must thus address consumer food concerns and must at least be as desirable as naturally harvested products.

3. Current challenges of the aquaculture industry

Key challenges to the development and growth of aquaculture as an economic activity are limited water resources, energy requirements and the environmental impact of aqua-farming methods. To address these challenges water is re-cycled and farming activities are intensified, resulting in an increase in stocking density, deterioration in water quality, increased incidence of disease, poor feed to body mass conversion efficiencies and higher mortality rates. The net result is reduced yield. Annual losses to the market due to disease, water quality and nutrition are estimated at 40% [30].

3.1. Disease in aquaculture

Definitions of disease include an unhealthy condition or infection with a pathogen. Disease is a complex phenomenon, leading to some form of measurable damage to the host [12]. Outbreaks of disease either begin suddenly and progress rapidly, often with high mortalities, and disappear with equal rapidity (acute disease) or develop more slowly with less severity, but persist for greater periods (chronic disease). Fish disease is the outcome of aberrations to the delicate interaction between the hosts, the disease-causing agent, and external conditions such as unsuitable changes in the environment, poor hygiene and overcrowding. Disease outbreak is generally associated with a primary invasion by parasites or mechanical injury, coupled to stressful environmental conditions such as changing temperature and poor water quality [8]. The prevalence of infectious agents can result in mass mortality causing significant losses to aquaculture operations [9]. Fish diseases such as rotting fins and ulceration of the skin are more prevalent when fluctuation in temperature causes immuno-modulation, resulting in inferior disease resistance and increased mortality [31-32]. An array of stress factors such as poor water quality, parasite load or a natural physiological state (e.g. during the reproductive phase) in the life cycle of the fish are also often associated with outbreaks of disease [12]. Strict selection for desirable characteristics by breeders has also reduced the vigour in breeding lines, making fish less hardy and more susceptible to disease [35]. Disease is not necessarily caused by the action of a single bacterial taxon, as representatives of many bacterial taxa have at one time or another been associated with disease outbreaks. *Aer. hydrophila* and *Pseudomonas spp.* are among the predominant species responsible for causing fish diseases [33]. Many bacterial pathogens are members of the normal microflora of water and/or fish. However not all of these bacteria are primary pathogens as many can be categorized as opportunistic pathogens, which colonize and cause disease in already damaged hosts.

Environmental factors play a key role in the onset of disease which is reported as being a consequence of the interaction between the host, environmental stress and prevalence of disease causing agents [8,12,34]. Some diseases are prevalent in spring and associated with environmental change to warmer temperatures, a period which is also characterised by an increase in the activity of pathogenic bacteria and parasites. Temperature fluctuation causes transient immuno-modulation of fish, which can result in reduced disease resistance [31-32]. Haemorrhagic septicaemia is an example of this phenomenon, with the disease resulting from infection by a wide range of pathogens that cause open ulcerated lesions and haemorrhages

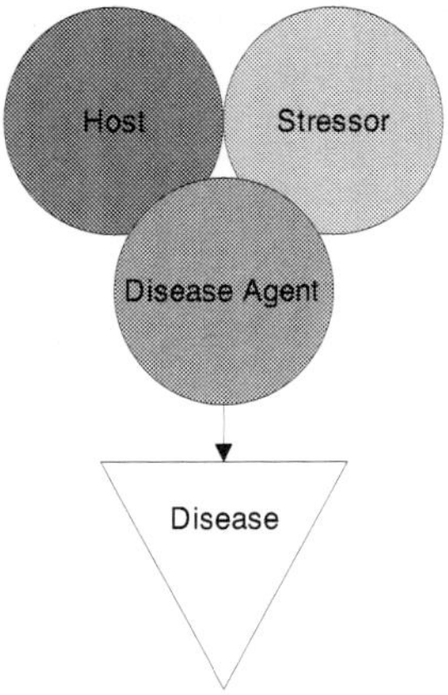

Figure 1. Interactive factors influencing disease in aquaculture (modified from [12]).

on the infected fish [12,36-37]. Additional clinical symptoms can include fin and tail rot, the loss of scales, localized haemorrhages, particularly in the gills and vent, exophthalmia and abdominal distension [12]. The acute form of this disease is of sudden onset, and the fish usually die within 2-3 days [38-40]. The main pathogenic micro-organisms involved in septicaemia are *Aer. hydrophila, Aer. salmonicida,* and to a lesser extent *Pseudomonas floures-cens* [8,12]. *Aer. hydrophila* is known to produce haemolysin, cytotoxins and enterotoxins which cause tissue necrosis resulting in ulcers, dropsy and abdominal oedema associated with haemorrhagic septicaemia [8]. *Aer. salmonicida* has been specifically associated with ulcerative erythrodermatitis and furunculosis [8,12,41]. *P. flourescens,* which is ubiquitous in fresh water and is generally regarded as a secondary invader of damaged tissue, has also been associated with outbreaks of septicaemia [42-45]. There is therefore merit in reducing the prevalence of these bacteria in aquaculture systems.

Pathogen	Disease
Aeromonas hydrophila	Haemorrhagic septicaemia, motile *Aeromonas* septicaemia, redsore disease, fin rot
Aeromonas salmonicida	Furunculosis, carp erythrodermatitis, ulcer disease.
Pseudomonas fluorescens	Generalized septicaemia
Pseudomonas pseudoalcaligenes	Skin ulceration

Table 1. Predominant bacterial pathogens causing disease of *Cyprinus carpio* (modified from [12]).

3.2. Water quality

Use of reticulated systems for intensive culture results in substantial amounts of particulate organic and soluble inorganic excretory waste, due mainly to increased stocking density [17]. The main source of this waste is hyper-nutrification, resulting from excessive feeding rates and high nutrient dietary composition, which has a significant influence on the survival, growth and reproduction of fish [13-15,17,46]. Nitrogen and phosphorous waste accumulation pose a threat to the environment and can predispose fish to infestation by parasites and pathogens due to a reduction in immunity [13,17].

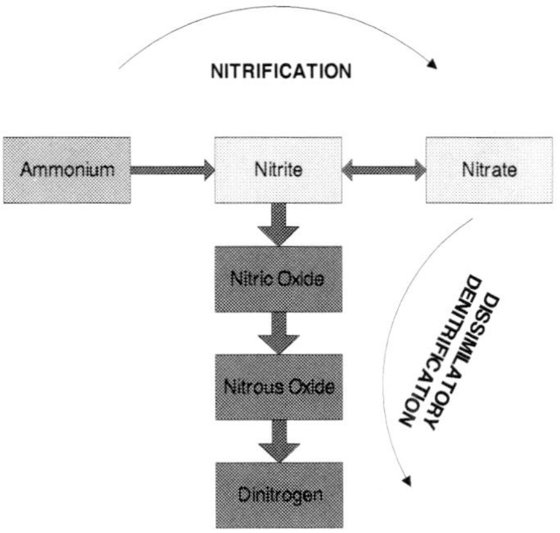

Figure 2. Nitrification and denitrification cycle [30].

Ammonia is a primary metabolic waste of fish and is excreted through the gills by bronchial diffusion. It is also produced by bacterial ammoniafication of uneaten food and faeces and is released from the mineralization of sediment [47-50]. Ammonia is oxidised to nitrite and finally to nitrate through the process of nitrification, with ammonia and nitrite being the most toxic of these metabolites to fish. Nitrite can also be produced through the process of denitrification [48]. Ammonia concentrations above 0.3 mg/l have been reported as toxic to fish, with hyperplasia of gill tissue, gill necrosis, pathological evidence of kidney and liver damage and reduction in growth rate occurring at this and higher concentrations [51-53]. Exposure to high ammonia concentration also causes epithelial lifting on gill filaments resulting in respiratory impairment and mortality [54]. Nitrite is usually present at low concentrations in natural systems, except when there is an imbalance, because it is a common intermediate in nitrification and denitrification, catabolic ammoniafication and nitrate assimilation [55]. Through denitrification, nitrite can be produced as an intermediate in the conversion of nitrate to nitric oxide, nitrous oxide and nitrogen gas [56]. Nitrite is considered harmful to fish at levels of 0.15 mg/l and above, causing conversion of haemoglobin to methaemoglobin in blood, which results in inhibition of oxygen transport and mortality due to brown blood disease [13]. Increased concentrations of nitrite also significantly affect weight gain, specific growth rate and food conversion efficiency [57].

Dietary phosphorous is an essential component of fish feeds as it improves weight gain and feed conversion ratio. It is however poorly utilized due to the absence of an acidic stomach in some species and because phosphate is often bound to phytic acid in vegetable protein [58]. Ingested phosphorous is therefore lost in faeces and results in poor water quality with increased algal growth and eutrophication [59-60].

4. Conventional approaches for addressing challenges in aquaculture

The rearing of fish in reticulated systems results in a highly artificial environment which has a propensity for the accumulation of waste metabolites and which promotes the growth of pathogenic bacteria. Management considerations for aquaculture operations include nutrition, water quality, physical parameters and pathogen and disease control [61]. Chemicals are often used to control disease and include a wide range of topical disinfectants, organophosphates, antimicrobials and parasiticides to deal with disease and water quality [18,26]. Water quality is traditionally managed through conventional reticulated filtration systems, which are sensitive to process fluctuations and can result in mass mortality when the systems crash.

4.1. Use of chemicals in aquaculture

Antimicrobial agents are extensively used for treatment during disease outbreak or at prophylactic doses to prevent outbreak of disease. This can lead to antibiotic resistance and increased virulence of pathogenic organisms, leading to a requirement for high doses of existing drugs or new drugs to control disease [5,17,20]. Antibiotic resistance can pose a risk to human health and can cause mass mortality of fish [63]. Studies have also demonstrated that chemicals used in aquaculture can be toxic to the fish themselves, with exposure to some chemicals causing a stress response and blood biochemical changes [17,21,64]. The presence of higher drug concentrations, and an ever increasing spectrum of chemical residues, can result in detrimental effects to consumers and the environment [62]. These chemicals also have a negative impact on the aquaculture filtration systems themselves, resulting in a deterioration in water quality. Chemicals are often recalcitrant, persisting for several days to months, and can cause alterations in naturally occurring bacterial populations, Regulators have recognised the risks posed by use of chemicals as substantiated by the ever increasing list of banned substances, a consequence of which is a reduction in treatment options for aquaculture [24,65-66]. Governments and organizations have recently introduced much tighter restrictions on the use of antibiotics in animal production. As an example, the European Union (EU) banned the use of avoparcin in 1997 and in 1999 included virginiamycin, spiramycin, tylosin and bacitracin as banned growth promoters in animal feeds [67-68].

4.2. Conventional biofiltration

Normally the oxidation of ammonia to the more benign nitrate ion occurs through ammonia and nitrite oxidising obligate chemoautotrophs such as *Nitrosomonas* and *Nitrobacter spp.* which are slow growing and sensitive to fluctuations in environmental conditions [55,69]. Removal of nitrate and nitrite is a challenge in intensive aquaculture operations. System fluctuations, resulting from the sensitivity of natural filter bacteria, often lead to accumulation of ammonia,

http://dx.doi.org/10.5772/57375

nitrite, nitrate and phosphate. Although the concentration of these residues can be reduced by the addition of fresh water, purges of effluent containing high concentrations of these compounds into natural river and seawaters results in a deterioration of the environment and can lead to algal blooms, which may be detrimental to natural ecosystems [60]. High capital investment is thus required for installation of larger scale filtration systems to compensate for these inefficiencies of conventional filtration.

5. Biological solutions as alternatives for addressing challenges in aquaculture

Given the challenges in conventional aquaculture practise, alternative methods for disease control and enhancement of water quality are desperately required. Micro-organisms play important roles in aquaculture, particularly with respect to nutrient cycling and the nutrition of the cultured animals, water quality, disease control and the environmental impact of effluent [22]. Beneficial microbes can be used to alter or regulate the composition of bacterial flora in a water system to optimise fish production by reducing pathogen concentration, by improving water quality through reduction of waste ions and through accelerated mineralization and nitrification, by reducing algal growth and by accelerating sediment decomposition [17,20,70-71]. These biological agents also confer the added advantage of natural integration into existing ecosystems and present opportunities for development of multi-effect products which are attractive to end users. The marketing of biological and "organic certified" solutions for enhancement of fish health has also gained consumer acceptance. The use of beneficial microbes is a more appropriate remedy than the use of chemicals but successful application requires an understanding of the ecological processes occurring in aquaculture systems, of the agents responsible for disease and knowledge of the beneficial characteristics of bacteria to be used as biological agents [5,72].

5.1. Biological agents

Microbial webs are an integral part of all aquaculture systems and have a direct impact on productivity, especially in intensive culture operations. The quality of water and health of the cultured species is governed by the activities of a diversity of microbes with different roles and interactions in the ecosystem [61]. There are distinct uses of bacterial supplements in aquaculture for bio-augmentation as probiotics and as biocontrol and bioremediation agents [19]. Bio-augmentation refers to the augmentation of the environment and/or the microbes to result in enhanced fish health while probiotics are normally associated with feed and digestion. A strict definition of biocontrol agents are microorganisms that are antagonistic to pathogens. In some instances however the description of biocontrol agents transcends the boundary between bio-augmentation, and the exclusion of pathogens [73]. Bioremediation refers to the breakdown of pollutants or waste by microbes [5,61].

A probiotic can be defined as a cultured product or live microbial feed supplement which beneficially affects the host by improving its intestinal balance [74]. The important components of this definition reflect the need for a living microorganism and application to the host as a feed supplement. A broader definition is that of a live microbial supplement, which beneficially affects the host animal by improving its microbial balance [75]. In a third proposed definition, a probiotic is any microbial preparation, or the components of microbial cells, with a beneficial effect on the health of the host [76]. It is thus apparent that there are variations in the actual application of the terminology associated with biological agents [77]. Based on the observation that organisms are capable of temporarily modifying the bacterial composition of water and sediment, it was suggested that the definition should include the addition of live naturally occurring bacteria to tanks and ponds [73]. Verschuere *et al.* [26] presented a wider and useful description, given the broad spectrum effects of microbial consortia used in aquaculture. They described a biological agent as a live microbial adjunct, which has a beneficial effect on the host by modifying the host-associated or ambient microbial community, by ensuring improved use of the feed or enhancing its nutritional value, by enhancing the host response towards disease, or by improving the quality of its ambient environment.

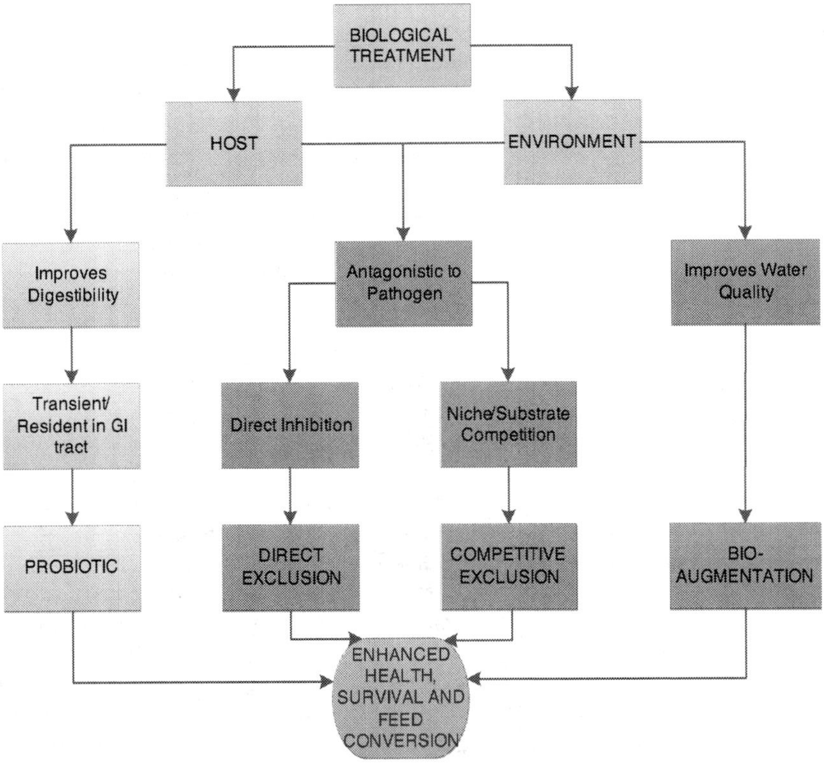

Figure 3. Schematic representation of the effects of biological agents in addressing aquaculture challenges.

The range of biological treatments examined for use in aquaculture has encompassed both Gram-negative and Gram-positive bacteria, bacteriophages, yeasts, unicellular algae, enzyme preparations and plant extracts. Microbes have been successfully applied to aquaculture systems via inclusion in artificial or live feed, by addition to biofiltration systems and by direct addition to water [77]. Most biological treatments used in aquaculture belong to the genera *Lactobacillus, Vibrio, Bacillus,* or *Pseudomonas,* although other genera have been applied to a lesser extent [26].

5.2. Modes of action of biological agents

Mechanisms of probiosis include competition with pathogens for adhesion sites, immune stimulation, synthesis of antimicrobials, competitive exclusion, bioaugmentation and bioremediation [23-24,26,78]. Although many biological treatments have been developed over the last decade, the approach used has generally been empirical and the exact modes of action were rarely elucidated, negatively affecting technology adoption and implementation in aquaculture [26].

One possible mechanism for preventing colonization by pathogens is competition for adhesion sites on gut or other tissue surfaces [78]. It is known that the ability to adhere to enteric mucus and cell wall surfaces is necessary for bacteria to become established in fish intestines [79-80]. The ability to adhere and grow on or in intestinal or external mucus has been demonstrated for fish pathogens in *in vitro* environments [81-82]. Since bacterial adhesion is important during the initial stage of pathogenic infection, competition with pathogens for adhesion receptors might be the first probiotic effect of a biological agent [81,83].

Immuno-stimulants are chemical compounds that activate the immune system of animals and render them more resistant to infections [84]. Fish larvae, shrimps, and other invertebrates have immune systems that are less well developed than their adult counterparts and are dependent primarily on non-specific immune responses for their resistance to infection [85]. Bacteria may act as immuno-stimulants in fish and shrimp, but it has not yet been conclusively demonstrated that they have a beneficial effect on the immune response of cultured aquatic species [26,86].

Microbial populations may release chemical substances that have a bacteriocidal or bacteriostatic effect on other microbial populations, which can alter inter-population relationships. The presence of bacteria producing inhibitory substances is thought to constitute a barrier against the proliferation of opportunistic pathogens. In general, the antibacterial effect of bacteria is due to the production of antibiotics, bacteriocins, siderophores, enzymes, hydrogen peroxide or alteration of pH by the production of organic acids, ammonia or diacetyl [26]. Many authors assign the inhibitory effects detected in *in vitro* antagonism tests to bacteriocins or antibiotics without investigating other possible mechanisms. It has been argued that growth inhibition could, in many cases, be accounted for by primary metabolites or simply by a decrease of the pH [26]. At this stage however, the association between amensalistic activity and *in vivo* probiotic activity is very weak and circumstantial. Typically, a correlation is made between the *in vitro* ability of the probiotics to inhibit pathogens and the *in vivo* protection of the cultured aquatic species, but in none of the studies investigated has it been shown unequivocally that the production of inhibitory compounds is the cause of the *in vivo* probiotic activity of the strains [26]. Further research is thus required in this field.

Competition for nutrients or available energy may determine how different microbial populations coexist in the same ecosystem, but to date there have been no comprehensive studies on this subject [87]. Competitive exclusion is an ecological process that allows manipulation of the bacterial species composition in water, sediment or the host itself, by competitive assimilation of nutrients and/or an intrinsically higher growth rate [5,23-24]. The microbial ecosystem in aquaculture environments is generally dominated by heterotrophs competing for organic substrates as both carbon and energy sources. Competitive utilization of these substrates can thus attenuate target pathogenic microorganisms as demonstrated by several studies. A bacterial strain selected for its active growth in organic-poor medium, was reported to prevent the establishment of a *Vibrio alginolyticus* infection *in vivo.* Since the strain had demonstrated no *in vitro* inhibitory effect on the pathogen it was thought to be a consequence of competitive exclusion [25]. In another example, *in vitro* antagonism tests did not show production of extracellular inhibitory compounds, yet living cells were required to protect *Artemia* against pathogenic *V. alginolyticus.* It was suggested that the selected bacteria exerted their protective action by competing with the pathogen for chemicals and available energy [26].

Virtually all microorganisms require iron for growth [88]. Siderophores are low molecular weight (< 1,500), ferric ion-specific chelating agents that can dissolve precipitated iron thus making it available for microbial growth [89]. The ecological significance of siderophores resides in their capacity to scavenge an essential nutrient from the environment and deprive competitors from accessing it. The requirement for iron is high for many pathogens in highly iron limited environments [88,90] and several studies have reported a correlation between iron availability and pathogen growth. In a challenge test with pathogenic *V. anguillarum,* salmon mortality was reported to increase linearly with dietary iron content [91]. Siderophore-producing *P. fluorescens* AH2 was demonstrated to inhibit several Gram-positive and Gram-negative bacteria, particularly when iron availability was limited [75]. *In vitro* co-culture tests revealed that the growth of *V. anguillarum* was inhibited by the filter-sterilized supernatants from iron-limited cultures of *P. fluorescens* AH2 but not from iron-replete cultures. *In vivo* studies using rainbow trout juveniles demonstrated a 46% reduction in mortality due to *V. anguillarum* infection when the culture was treated by *P. fluorescens* AH2 *in vivo.* Non-pathogenic bacteria which produce siderophores could thus be used to compete with pathogens whose pathogenicity is known to be dependent on the availability of iron [26]. It must be noted however that the body of evidence supporting the competition for free iron as a mode of action of biological agents is still scant and at present still circumstantial [26]. More recently Lalloo *et al.* [92] were able to demonstrate siderophore production as the mode of action responsible for attenuation of pathogen growth in both *in vitro* and *in vivo* studies.

Improvement in water quality has been recorded in studies involving the addition of biological agents. These improvements include the reduction of total and dissolved solids concentrations, lower concentrations of waste ions and a reduction in algal populations. Gram-positive bacteria are generally more efficient in converting organic matter to CO_2 than Gram-negative bacteria, which convert a greater percentage of organic carbon to bacterial biomass or slime.

By maintaining higher levels of these Gram-positive bacteria in production systems, farmers can reduce the build-up of dissolved and particulate organic carbon during the culture cycle [26]. Nitrite accumulation may be caused by imbalances in the activities of nitrate and nitrite reductase and inhibition of nitrite reductase by oxygen. Bio-communities however usually contain bacteria with different nitrate and nitrite reductase activities enhancing the denitrification efficiency of the overall bio-community [93]. Although the specific nitrification activity of heterotrophic bacteria is generally lower than that of chemoautotrophs, the overall impact on denitrification could be greater due to the higher cell numbers of heterotrophic bacteria and their robustness to process fluctuations. There is therefore merit in utilizing biological agents for nitrification and phosphate bioremediation to improve water quality in aquaculture [26,86]. Many bacterial strains have been shown to have a significant algaecidal effect on various species of micro algae [26,94-95]. This effect is valuable where algal blooms may be problematic, causing blockages to flow systems and changes in oxygen concentration due to algal cellular respiration.

Formulation of bacterial consortia with interactive effects, including pathogen inhibition, high growth rate and improvement in water quality, provides broad spectrum effects in a single product [72]. Lalloo *et al.* [72] obtained natural isolates from mud sediment and *Cyprinus carpio* tissue samples, which were purified and assessed in *in vitro* studies for growth inhibition of pathogenic *Aer. hydrophila* and for their ability to reduce the concentrations of ammonium, nitrate, nitrite and phosphate ions. A consortium of *Bacillus* isolates was formulated for *in vivo* trials using *C. carpio,* and demonstrated positive results for pathogen inhibition and waste ion reduction.

5.3. *Bacillus spp.* as attractive biological agents

The application of *Bacillus* species in aquaculture is growing rapidly, especially in countries where intensive systems for farming of fish and shellfish are utilised [23-24,72]. *Bacilli* are used as components of biocontrol products which are often composed of mixtures of species, which are able to exert a range of beneficial effects on aquaculture systems [24,72]. They are ubiquitous in sediments and are naturally ingested by animals [5]. An advantage of using *Bacillus spp.* is that they are not generally involved in horizontal gene transfer processes with Gram-negative organisms such as *Vibrio* and *Aeromonas spp.* and are thus unlikely to acquire genes for antibiotic resistance or virulence from these species [5]. Other key positive characteristics of this genus are the ability to replicate rapidly, tolerate a multitude of environmental conditions and provide a broad range of beneficial effects that can improve aquaculture productivity [24,27]. Additionally, the ability of *Bacilli* to sporulate enables downstream processing and formulation of shelf stable spore based products [88]. Many spore forming *Bacilli* are sold worldwide as components of products for human and animal use, including *B. coagulans, B. subtilis, B. clausii, B. cereus* and *B. toyoi* [23].

Several studies have demonstrated the application of *Bacillus* based products in aquaculture. *Bacillus* strain IP5832 spores fed to turbot larvae resulted in a decrease in the *Vibrionaceae* population with significant improvement in weight gain and survival of the larvae [19]. In a further study it was reported that a *Bacillus spp.* improved food absorption by enhancing protease levels, decreased the number of pathogenic bacteria in the system and improved turbot larval growth [77]. The survival and net production of channel catfish was improved in a farm trial using a mixed culture of *Bacillus*, but the mode of action was not specified [96]. It was reported that *Penaeus monodon* larvae fed with *Bacillus* S11 fortified *Artemia* had significantly reduced development times and fewer disease problems than larvae reared in the absence of the *Bacillus* strain. When challenged with a pathogenic *V. harveyi* strain D331, survival was also significantly improved in treated groups compared to untreated controls [97]. It was also concluded, based on studies on several farms in Indonesia that the use of *Bacillus* in penaeid culture ponds enhanced the production of shrimps by preventing mortality normally caused by luminescent *Vibrio spp.* [61].

Bacillus spp. also contribute to nitrogen removal in spite of the classical belief that this process is predominated by autotrophic bacteria [55,72,93,98-102]. Some members of this group, such as *B. subtilis* and *B. cereus*, are able to grow under aerobic, facultative aerobic and anaerobic conditions, allowing for switches in nitrogen metabolism that facilitate both nitrification and denitrification [86,93,103]. The pattern of nitrite metabolism by *B. subtilis* I-41 was demonstrated as exceptional among strains which showed switching of nitrite and nitrate metabolism [55]. Nitrite oxidation might thus be common, rather than the exception, in heterotrophic bacteria such as *Bacillus spp.* [86]. The reduction of phosphate concentration in *C. carpio* culture systems has also been demonstrated through addition of *Bacillus* species [72]. The improvement in bio-availability of bound phosphate, through solubilisation, is also thought to facilitate removal of phosphate and reduce the propensity of algal blooms [60,104].

Identity of probiotic	Used on	Method of application	Reference
Bacillus sp. S11	*Penaeus monodon*	Premixed with feed	[105]
Bacillus sp. 48	*Centropomus undecimalis*	Added to water	[106]
Bacillus sp.	Penaeids	In water	[61]
B. megaterium, B. polymyxa, B. licheniformis, B. subtilis	Channel catfish	In water	[96]
Mixed culture, mostly *Bacillus spp.*	*Brachionus plicatilis*	Mixed with water	[107]
Bacillus spp.	*C. carpio*	In water	[72]
Bacillus strain IP5832	Turbot larvae	In water	[19]

Table 2. Summary of studies investigating the application of *Bacillus* based biological treatments.

Bacillus spp. have the ability to form endospores which are rigid structures that are capable of surviving under harsh conditions. Spores are considered metabolically inert, but can be used as biological agents due to the many advantages of this form over vegetative cells. These include their higher resistance to external factors such as mechanical force, desiccation, solar radiation and high temperatures [108]. As a consequence of this resistance to environmental stress, spores are attractive for commercial application as they can endure harsh processing steps during production and are resilient to fluctuations in systems where they are applied, thus ensuring better survival and efficacy than vegetative cells [23]. Products containing spores can be stored in a stable form for long periods under challenging conditions normally prevalent on aquaculture farms [24,109]. *Bacillus* spores are found in the bottom of ponds, lakes and rivers and many aquatic species will naturally ingest these microbes. They generally exist in symbiotic relationships with their host [24]. Their ability to germinate selectively in response to external triggers is advantageous for application as biological agents in aquaculture, as they have the ability to recover the characteristics of a metabolically active cell in response to specific nutrients when these effects are required [108,110]. Lalloo *et al.* [88] showed that a *Bacillus* spore concentrate and powder blend were stable over a 42-day test period without significant loss in viability of spores, while final product formulations were stable for at least two years.

6. Isolation, screening and selection of candidate biological treatment agents

There is an elegant logic in isolating putative biological agents from the host or the environment in which the agents are likely to exert a beneficial effect, but there is no unequivocal indication that these isolates perform better than isolates completely alien to the cultured species or originating from a different habitat [26]. A combination of methods and incubation conditions need to be used to achieve pure cultures of target organisms. To an extent, the range of media to be used is governed by personal choice and experience [12]. Many bacteria that are residents of soil and aquatic habitats low in nutrients have difficulty growing in rich media. Also, many potential contaminants cannot compete in dilute media, so the limitation in nutrient availability becomes a selective factor. In order to appropriately select biological agents it is essential to understand the mechanisms of action and to define selection criteria for potential microbes. A classical screening and selection rationale may include collection of background information, acquisition of isolates, purification of isolates and evaluation based on pre-determined criteria for both *in vitro* and *in vivo* environments [71]. Good pre-selection criteria can include the viability of the potential probiotic within the host and/or within its culture environment, adherence to host surfaces, the ability to prevent infection by pathogenic bacteria and ability to utilise waste ions. Other selection criteria include biosafety considerations, methods of production and processing, the method of administering the probiotic and the robustness of the biological agent in the environment where the microorganisms are expected to be active. Possible pathogenicity to different life stages of the target species should also be considered. Verschuere *et al.* [26] tested their probiotics on *Artemia* to verify that the defence systems of the shrimps were able to cope with the presence of the putative probiotics.

6.1. Isolation of biological agents

When selecting desirable biological agents enrichment techniques that make it possible to exploit the differential characteristics of target isolates in mixed microbial populations should be applied. *Bacillus spp.* are isolated almost ubiquitously from soil, water, mud, sediment, dust, air and the surfaces and organs of aquatic animals [23]. They have been isolated from fish, crustaceans, bivalves and shrimps and have been found in the microflora of the gills, skin and intestinal tract [19,24]. One effective strategy being used in developing countries is the isolation of *Bacillus spp.* from commercial ponds and then using selected isolates as commercial products [24]. *Bacilli* are classified as endospore forming Gram-positive rods and cocci and isolation procedures must selectively enrich for these organisms while excluding other genera in the same group. In one study, methods used for isolating various *Bacillus* strains were based mainly on the resistance of their endospores to elevated temperatures [111]. The technique used involved blending of samples with an enrichment medium, which also induced vegetative cells to sporulate, followed by incubation to allow formation of mature spores in large quantities. The isolation involved heat treatment for the selection of spores from *Bacillus* species. Ethanol is also a useful disinfectant and dehydration agent to use for isolation of *Bacillus* strains as its application kills vegetative cells, whereas the more resistant endospores survive. The resistance of *Bacilli* to the antibiotic polymyxin B also enables use of this antibiotic for the selection of this group of bacteria whilst eliminating most Gram-negative bacteria. Once selected, cells can be characterised by their morphology, typically using microscopic techniques, by use of gram staining and by quantification of the activity of enzymes such as catalase [111].

6.2. *In vitro* screening and selection of aquaculture biological agents

To appropriately select biological agents it is essential to understand the mechanisms of action and to define selection criteria for potential probiotics [112]. Many bacteria have been exploited as biological agents but their selection has been based mainly on empirical observations rather than scientific data [71].

A common protocol for screening candidate biological agents is to perform *in vitro* antagonism tests, in which pathogens are exposed to antagonists in culture medium [75,80,113-116]. Assays for the production of inhibitory compounds and siderophores, or the competition for nutrients, are some common strategies that have also been used [75,80,117-119]. Results of *in vitro* antagonism tests should however be interpreted with caution as growth media and conditions can influence the effects observed which may differ from the actual activity *in vivo* [80,120]. The pre-selection of candidate probionts based on *in vitro* antagonism tests has however led to the discovery of many effective probionts and is a useful first step in selection [116]. The use of the target organism in the screening procedures provides a stronger basis for selection of antagonists [26]. The target species should be challenged under normal or stress conditions with the candidate biological agent. Growth inhibition may not always be a consequence of the production of inhibitory substances such as antibiotics, as inhibition caused by other mechanisms must also be considered during *in vitro* screening tests [121-122]. As an example Lalloo *et al.* [88] investigated the mode of action of a novel *B. cereus* isolate for the inhibition of pathogenic *Aer. hydrophila*. The production of antimicrobial compounds was excluded as the mode of action based on the absence of growth inhibition of *Aer. hydrophila* by intracellular or extracellular fractions of *B. cereus*. In contrast, actively growing *B. cereus* cells inhibited the growth of the *Aer. hydrophila*. Based on co-culture data, competitive exclusion through an intrinsically higher growth rate and competitive uptake of essential nutrients was identified as the mode of action. Co-cultivation of *B. cereus* with *Aer. hydrophila* resulted in a 70% reduction in the cell density of the pathogenic organism in a remarkably short time period. These findings confirmed previous work where a decline in pathogen levels was demonstrated in both *in vitro* and *in vivo* studies when *B. cereus* was administered as a biological agent [72]. Further studies investigated the effect of iron availability on pathogen growth and demonstrated the superior efficiency of *B. cereus* in assimilating iron, resulting in a decline in pathogen levels in iron deficient medium. [88].

In aquaculture, bioremediation or bioaugmentation is an important selection criterion, particularly under conditions that mimic the application environment [5]. While some studies have reported screening strategies to select for the bioremediation capabilities of potential aquaculture biological agents, this area has regrettably not been well reported to date [27,72]. Recent studies by Lalloo *et al.* [72] described methodology applied for the selection of *Bacillus spp.* based on their ability to utilise ammonia, nitrate, nitrite and phosphate ions.

Once candidate biological agents are selected, proper identification and safety assessment is an important requirement prior to application *in vivo*. Identification can be performed using techniques such as 16S RNA sequence homology. Where close sequence homology is found between species of potentially dangerous genera, additional assessment may be necessary. Lalloo *et al.* [72] demonstrated that their *B. cereus* isolate was negative for anthrax toxins and did not contain the anthrax virulence plasmids pXO1, pXO2 or the *B. cereus* enterotoxin. These studies were necessitated by the high sequence homology found between *Bacillus* species. Toxicity towards the cultured species can also be employed in screening strategies. As an example, Austin and Austin [12] tested their candidate biological agent by injection into Atlantic salmon followed by histopathological examination of the kidneys, spleen and muscles.

6.3. *In vivo* validation of the efficacy of putative biological agents

Once candidate biological agents have been selected, the next important step is confirmation of observed efficacy using *in vivo* tests. The use of small scale model *in vivo* systems is a cost effective method that allows more certainty in selection of candidate biological agents [26,94]. These tests may measure various effects, including antagonism, by including an experimental infection with a representative pathogen. Pathogens can be administered via the diet, through immersion, by injection or via the culture water [123]. To determine the effects of a specific bacterial strain on a cultured organism, the elimination of other microbes from the culture system may be necessary [124]. This approach can also be used to examine other effects on water quality and the impact on other trophic levels, such as algae [94-95,125]. With *in vivo* challenge tests, changes in population dynamics of the antagonist and the pathogen as well as other effects on the culture system should be studied. Of importance are the unintended negative effects on the target species and interference with filtration efficiency in reticulated culture systems [26,123]. As an example, in an oyster culture system, a decrease in the level of the pathogen *V. tubiashii* was observed when an *Aeromonas* probiotic strain was added together with the growth media of the probiotic strain. The putative antagonist itself could not however be detected in the culture after four days. This example shows the importance of measuring interactions, including mortality or disease, after experimental infection and to include appropriate controls in study designs [116]. In another study, the efficacy of a *B. cereus* isolate was demonstrated *in vivo* based on predefined criteria. In addition to this, the tolerance to, and functionality across, a range of physiological conditions in systems used to rear *C. carpio* was also proven [27,92]. Furthermore, the *in vivo* treatment did not result in a negative impact on oxygen sufficiency, growth or health of the specimens, which are all important considerations for application of biological agents [21].

6.4. Other considerations during selection of biological agents

Strains showing well-established biological effects in *in vitro* and *in vivo* studies need to be tested for their suitability to real world biological treatment applications. Additional criteria such as biosafety considerations, methods of production and processing, the method of administering the probiotic and the environmental conditions where the microorganisms are expected to be active are important considerations [112]. An isolate cannot be used as a probiotic unless it has been confirmed as non-pathogenic to the host, to humans and to the environment [26]. Relevant legislation, if any, should be taken into account before commencement of commercial application. Finally, a cost-benefit analysis will determine whether the probiotics can be applied in practice or not [26].

7. Bio-production of biological agents

Large scale production of probiotics is an essential step towards application in the aquaculture industry as production cost is an important consideration in the development of commercially relevant biological products [126]. The cultivation of microorganisms at a large scale is influenced by various factors such as the composition of the media, as well as physical and chemical variables [127]. It has been widely documented that nutrient sources influence the growth, spore production, and synthesis of commercially useful metabolites in *Bacillus spp.* [128-130]. The nutritional and physicochemical parameters of the fermentation process thus need to be optimized, with use of economical and commercially available media a key consideration to reduce costs of bio-production [131-132]. Media formulation and optimization are key considerations for the production of affordable aquaculture biological agents, yet limited progress has been made in this area to satisfy market opportunities for affordable commercial aquaculture products [126,77]. With increased cell yield, productivity and cost reduction, the fermentation production process can be made feasible and economically attractive for application of aquaculture products [128]. Another key consideration is that scale up of production must not compromise product efficacy or amenability to stabilization and formulation [133]. Genetic engineering provides an option for improvement of biological agents; however public resistance to genetically modified organisms, particularly when associated with food production is an important consideration before adopting this approach.

7.1. High cell density cultivation of *Bacillus spp.*

Although *Bacillus* biological agents are widely used in aquaculture, there are limited studies on their production and little is known about the impact of nutrient supplementation on high-density production of bacterial spores by fermentation [134-135]. Carbon and nitrogen sources generally play a dominant role in the productivity of a fermentation process as these nutrients are directly linked with the production of the biological agent [136-137]. According to current understanding, the development pathway leading from a vegetative cell to a spore is triggered by depletion of carbon, nitrogen, phosphate or essential micronutrients [138-140]. A suitable medium must thus support vegetative growth and also the production of spores [141]. It has been widely documented that nutrient sources influence the growth, spore production and synthesis of commercially useful metabolites in this species [128-130].

The type of carbon source and the carbon to nitrogen ratio play an important role in microbial growth [142]. It has been observed that *B. subtilis* uses glucose as its major carbon source and the efficiency of carbon utilisation towards biomass formation is low when the glucose concentration exceeds ~10 g/l in batch culture [143]. The production of by-products is increased in the presence of excess glucose, resulting in reduced yields of biomass, which is undesirable when producing biological agents [143]. Certain over-flow metabolites can also inhibit cell growth [143]. Monteiro *et al.* [134] also observed that an increase in glucose concentration up to 5 g/l led to an increase in the vegetative cell and spore concentration of *B. subtilis*, while higher sugar concentrations inhibited sporulation. It is therefore of great importance to regulate carbon availability to optimise growth and sporulation parameters precisely [127]. It is also noteworthy that the glucose consumption rate depends significantly on factors such as pH and oxygen sufficiency. Mass transfer parameters such as agitation and aeration are thus important in maximising vegetative cell growth, without inducing a premature onset of sporulation [144]. In most studies, glucose was found to be the best carbon substrate for the production of *Bacillus spp.* and their spores [145].

Various protein substrates have been tested for the growth and synthesis of commercially useful metabolites by *Bacillus spp.* [27,128-130,146]. It has furthermore been widely documented that protein sources influence spore production in this species [141,147-149]. Commonly used nitrogen based nutrient sources include a wide range of peptones, extracts and hydrolysates, many of which are too expensive for industrial scale manufacture of large volume products and have negative market acceptance if they contain animal by-products [143,150]. Media formulated to support high productivities are thus predominantly formulated with inexpensive complex nitrogen sources [137,151]. Although yeast extract, peptones and meat extracts have been shown to improve bacterial growth rate as they are good sources of protein, vitamins and co-factors, there have been reports suggesting that metabolite production, and particularly spore production, are often better when corn steep liquor (CSL) is used [128-129, 135,141,146,152-153]. CSL contains a wide range of macro and micro elements known to be important for spore production [154]. Lalloo *et al.* [27] demonstrated an attractive material cost of production at the optimal supplementation level for CSL, reducing the overall production cost by using this inexpensive source of nitrogen. They also demonstrated that CSL was a preferred nutrient substrate for the production of *Bacillus* spores, in comparison to conventional nutrient substrates. The use of CSL resulted in a higher spore concentration, productivity and spore yield on protein, in comparison to yeast extract and nutrient broth. Apart from the nature of protein source, the protein concentration in culture media also affects growth and spore production [155]. *B. subtilis* spore productivity increased, but spore yield decreased, with an increase in CSL concentration [149]. The yield of spores on carbohydrate increased with increasing concentration of CSL, suggesting that a higher protein to carbon ratio was preferable for production of *B. subtilis* and *B. licheniformis* spores [143,156]. High levels of CSL supplementation (~60 g/l) however resulted in slow growth, cell lysis and poor spore formation as sporulation efficiency is known to be low following poor growth [141,157-158]. Sporulation takes longer in high cell density cultivations, thus resulting in a compromise between spore concentration and productivity [138]. A major advantage of CSL is that it is available in an ultra-filtered phytase treated variant, which is cost competitive and offers processing advantages in both up-stream and down-stream process unit operations [88,156,158]. Precipitation and mass transfer issues are reduced when using this form of CSL for high cell density cultivation, due to hydrolysis of phytic acid and the removal of solids through the ultra-filtration process [134,138]. As this CSL is not spray dried, as is typical of the conventional type, degradation of vitamins and key nutrients is reduced which improves growth performance [129]. Of peripheral benefit is the use of a corn wet processing waste which improves value addition and reduces environmental pollution normally caused by such materials [128].

7.2. Production of spores

The key challenge in spore production is to maximize sporulation from a high density vegetative cell culture [134,139]. Environmental signals for sporulation include culture density dependant peptides, oxygen availability and limitation of carbon, nitrogen or phosphorous [140]. The life-cycle of a spore forming bacteria consists of four stages i.e. vegetative growth, sporulation, germination and outgrowth [139,159]. Cells enter a sporulation pathway, which involves three differentiating cell types, namely the predivisional cell, mother cell and the forespore, in response to nutrient limitation [160]. The forespore undergoes dehydration, while the cortex is produced between the two membranes that separate the mother cell and the forespore. Eventually the mature spore is released when the mother cell lyses. This mature spore has the ability to remain dormant for long periods of time [160]. The most important sporulation related transcriptional regulator is Spo0A which is phosphorylated via a complex network of interactions in response to nutrient limitation [140,161]. Furthermore, genes in the Res system are induced under anaerobic growth conditions which contribute to the sporulation cascade during oxygen insufficiency [162]. Low phosphate concentration results in the earlier onset of sporulation due to the response of the Pho system to phosphate starvation [162]. Magnesium sulphate, calcium carbonate and phosphate all stimulate sporulation, whereas divalent cations (particularly Ca^{2+}) assist in dehydration and mineralization of the spore [154,161]. According to Monteiro *et al.* [134], the sporulation efficiency for *B. subtilis* was found to be independent of the pH values within the range of 6.9-9.0. For several *Bacillus spp.* sporulation is highly related to O_2 supply and it has been reported that non-limited oxygen conditions during the growth phase are important to realise high spore yields [163-164].

8. Application of biological agents

A key challenge for usefulness of biological agents is the survival of the micro-organisms in the environment to which they are applied. Biological agents must thus be tolerant to the prevailing environmental conditions in which they are expected to perform, often dictated by the species being cultured for a specific aquaculture application [49]. Several methods of addition of biological agents to the host or its ambient environment exist, with each application method presenting unique challenges to the survival and efficacy of the biological agent [26,71,118,165]. A biological agent must provide actual benefit to the host, be able to survive in the environment of the intended application and should be stable and viable for prolonged storage and in the field [77]. Other factors such as natural deterioration and washout of the biological agent may necessitate the on-going addition of the treatments to maintain their positive effect [26]. Information on the robustness and functionality of biological agents in response to environmental conditions such as salinity, pH and temperature are however limited. Lalloo *et al.* [92] demonstrated that temperature had a significant influence on germination, specific growth rate and increase in cell number of *B. cereus* in shake-flask cultures, whilst salinity and pH did not have a measurable effect on growth. Changes in the above conditions influence spore germination, cell growth, survival and the functionality of *Bacillus spp.* as aquaculture biological agents [166]. A key consideration for the application of *Bacillus* based biological agents is that the spores need to germinate and grow, such that the characteristics of a metabolically active cell can be recovered [110]. The replication of vegetative cells can further enhance the bioactivity in the intended application. Spores lose their dormant properties when conditions are favourable in the presence of specific germinants such as nutrients [167]. However, the germinant has to penetrate the outer coat and cortex layers of the spore before coming into contact with specific germinant receptors [110,168]. The germination of spores is a sensitive transition state involving the initiation of metabolism [169].

For the application of spores as aquaculture biological agents, determination of their functionality as antagonists to disease or for improvement in water quality under the physiological ranges to be encountered in the aquaculture system is thus an important requisite [88]. Changes in growth conditions such as temperature constitute a key factor that influences cell growth and survival of *Bacillus spp.* in their habitats. *B. subtilis* has the ability to sustain growth across a wide temperature range from approximately 11ºC to 52ºC [139,170]. When the growth temperature for *B. subtilis* is increased rapidly, changes in gene expression occur, which is known as a heat shock response. A cold shock response is elicited when the temperature is dropped down to 15ºC from 37ºC [166]. *B. cereus* is apparently not well adapted to cold temperatures and the metabolic rate decreases drastically below 13ºC [171]. A useful method for the elucidation of temperature domains for prediction of functionality of a biological agent is by examining conformance of efficacy measures to Arrhenius and Ratkowsky functions [171-172]. The vegetative cells of *B. cereus* are more sensitive to acidic conditions than spores. However, like many other cells, vegetative cells of *B. cereus* have the ability to induce an acid tolerance response [173]. The mechanisms of resistance to acidic conditions involve three factors i.e. (i) F_0F_1 ATPase and glutamate decarboxylase (ii) metabolic modifications and (iii) protein synthesis to protect and repair macromolecules [173]. *B. cereus* spores were shown to be tolerant to the salinity and pH extremes typically encountered in the culture of ornamental *C. carpio* [174-175]. It is noteworthy that the efficacy of biological agents to environmental conditions must be assessed in line with the dynamics of the target species and the aquaculture system in response to these physiological ranges. As an example, reduced activity of a biological agent at lower temperature does not necessarily indicate a failure of the biological agent to perform, as lower temperature could translate to a lower intake of feed, waste metabolite generation and pathogen propensity in the aquaculture system.

9. Future prospects of the technology

The traditional practise of extensive land based aquaculture is under pressure, due to a limitation in available space, which has led to the increased use of more intensive reticulated systems which also offer the benefit of greater control of physiological culture conditions. While intensive systems offer the advantages of increased stocking densities and higher production throughput, challenges include water quality and increased disease prevalence among others. These are driving the adoption of environmentally friendly solutions that meet consumer expectations and comply with regulatory requirements. Biological solutions provide an attractive option. Issues that require attention to accelerate the adoption of biological solutions include the elucidation of the mode of action of commercial biological products and demonstration of clear cost-benefit advantages for commercial products.

Author details

Mulalo Edna Nemutanzhela, Yrielle Roets, Neil Gardiner and Rajesh Lalloo*

*Address all correspondence to: RLalloo@csir.co.za

CSIR Biosciences, Biomanufacturing Industry Development Centre (BIDC), Pretoria, South Africa

References

[1] Kesarcodi-Watson A, Kaspar H, Lategan MJ, Gibson L. Probiotics in Aquaculture: The Need, Principles and Mechanisms of Action and Screening Processes. Aquaculture 2008; 274: 1-14.

[2] Chinabut S, Somsiri T, Limsuwan C, Lewis S. Problems Associated with Shellfish Farming. Reveu Scientifique et Technique 2006;25(2) 627-635.

[3] Robertson LA, Kuenen JG. Combined Heterotrophic Nitrification and Aerobic Denitrification in *Thiospaera pantotropha* and other Bacteria. Antonie van Leeuwenhoek 1990;57: 139-152.

[4] Balcazar JL, de Blas I, Ruiz-Zarzuela I, Cunningham D, Vendrell D, Mu′zquiz JL. The Role of Probiotics in Aquaculture. Vet Microbiology 2006;114: 173-186.

[5] Moriarity DJW. Disease Control in Shrimp Aquaculture with Probiotic Bacteria. In: Bell CR, Brylinsky M, Johnson-Green P. (eds.) Proceedings of the Eighth International Symposium on Microbial Ecology. Atlantic Canada Society for Microbial Ecology 1999. p237–244. Halifax, NS, Canada.

[6] Chavez-Crooker P, Obreque-Contrera J. Bioremediation of Aquaculture Wastes. Current Opinion in Biotechnology 2010;21: 313-317.

[7] Crab R, Avnimelech Y, Defoirdt T, Bossier P, Verstraete W. Nitrogen Removal Techniques in Aquaculture for a Sustainable Production. Aquaculture 2007;270: 1-14.

[8] Jeney Z, Jeney G. Recent Achievements in Studies of Common Carp (*Cyprinus carpio* L.). Aquaculture 1995;129: 397-420.

[9] Irie T, Watarai S, Iwasaki T, Kodama H. Protection Against Experimental *Aeromonas salmonicida* Infection in Carp by Oral Immunisation with Bacterial Antigen Entrapped Liposomes. Fish Shellfish Immunology 2005;18: 235-242.

[10] Soevenyi JF, Bercsenyi M, Bakos J. Comparative Examination of Susceptibility of Two Genotypes of Carp (*Cyprinus carpio* L.) to Infection with *Aeromonas salmonicida*. Aquaculture 1988;70: 301-308.

[11] Fijan N. Infectious Dropsy in Carp-a Disease Complex. In: Mawdesley-Thomas LE (ed.) Diseases of Fish. Symposia of the Zoological Society of London. London. 1972. p39-51.

[12] Austin B, Austin DA. Bacterial Fish Pathogens. 3rd edition. Chichester, UK: Springer Praxis; 1999.

[13] Liao PB, Mayo RD. Intensified Fish Culture Combining Water Re-Conditioning with Pollution Abatement. Aquaculture 1974;3: 61-85.

[14] Boyd CE. Chemical Budgets for Channel Catfish Ponds. Trans Am Fish Society 1985;114: 291-298.

[15] Shimeno S, Shikata T, Hosokawa H, Masumoto T, Kheyyali D. Metabolic Response to Feeding Rates in Common Carp, *Cyprinus carpio*. Aquaculture 1997;151: 371-377.

[16] Jeney G, Nemcsok J, Jeney Z, Ola′h J. Acute Effect of Sublethal Ammonia Concentrations on Common Carp (*Cyprinus carpio* L.). II Effect of Ammonia on Blood Plasma Transaminases (GOT, GPT), G1DH Enzyme Activity, and ATP Value. Aquaculture 1992;104: 149-156.

[17] Jana BB, Jana S. The Potential and Sustainability of Aquaculture in India. Journal of Applied Aquaculture 2003;13: 283-316.

[18] de Kinkelin P, Michel C. The Use of Drugs in Aquaculture. Infofish International 1992;11: 45-49.

[19] Gatesoupe FJ. The Use of Probiotics in Aquaculture. Aquaculture 1999;180: 147-165.

[20] Skjermo J, Vadstein O. Techniques for Microbial Control in the Intensive Rearing of Marine Larvae. Aquaculture 1999;177: 333-343.

[21] Sze CP. Antibiotics Use in Aquaculture. Infofish International 2000;19: 24-28.

[22] Moriarity DJW. The Role of Microorganisms in Aquaculture Ponds. Aquaculture 1997;151: 333-349.

[23] Sanders ME, Morelli L, Tompkins TA. Sporeformers as Human Probiotics: *Bacillus, Sporolactobacillus* and *Brevibacillus*. Comprehensive Reviews Food Science and Food Safety 2003;2: 101-110.

[24] Hong HA, Duc LH, Cutting SM. The Use of Bacterial Spore Formers as Probiotics. FEMS Microbiology Reviews 2005;29: 813-835.

[25] Rico-Mora R, Voltolina D, Villaescusa-Celaya JA. Biological Control of *Vibrio alginolyticus* in *Skeletonema castatum* (*Bacillariophyceae*) Cultures. Aquaculture Engineering 1998;19: 1-6.

[26] Verschuere L, Rombaut G, Sorgeloos P, Verstraete W. Probiotic Bacteria as Biological Control Agents in Aquaculture. Microbiology and Molecular Biology Reviews 2000;64: 655-671.

[27] Lalloo R, Maharajh D, Görgens J, Gardiner N, Görgens JF. High-Density Spore Production of *B. cereus* Aquaculture Biological Agent by Nutrient Supplementation. Applied Microbiology and Biotechnology 2009;83: 59-66.

[28] The Food and Agriculture Organization of the United Nations. The State of World Fisheries and Aquaculture 2012. Rome, Italy. ISBN 978-92-5-107225-7.

[29] Food and Agriculture Organization of the United Nations. The State of World Fisheries and Aquaculture 2010. Rome, Italy. ISBN 978-92-5-106675-1.

[30] Lalloo R. Development of a bioprocess for the production of an aquaculture biological agent. PhD thesis. University of Stellenbosch; 2010.

[31] Le Morvan C, Deschaux P, Troutland D. Effects and Mechanisms of Environmental Temperature on Carp (*Cyprinus carpio*) Anti-DNP Antibody Response and Non-Specific 122 Cytotoxic Cell Activity: A kinetic Study. Developmental and Comparative Immunology 1996;20(5) 331-340.

[32] Engelsma MY, Hougee S, Nap D, Hofenk M, Rombout JHWM, van Muiswinkel WB, Verburgvan Kemenade BML. Multiple Acute Temperature Stress Affects Leucocyte

Populations and Antibody Responses in Common Carp, *Cyprinus carpio* L.. Fish Shellfish Immunology 2003;15: 397-410.

[33] Chalmers GA. Infectious Diseases of Silver Carp, *Hypophthalmichthys molitrix* (Valenciennes, 1844). Lethbridge, Alberta. 2003.

[34] Paperna I. Disease Caused by Parasites in the Aquaculture of Warm Water Fish. Annual Review of Fish Disease 1991;1: 155-194.

[35] Barwick M, Maher W. Biotransference and Biomagnification of Selenium, Copper, Cadmium, Zinc, Arsenic and Lead in a Temperate Seagrass Ecosystem from Lake Macquire Estuary, NSW, Australia. Marine Environmental Research 2003;56: 471–502.

[36] Sakai T, Murata H, Endo M, Yamauchi K, Tabata N, Fukudome M. 2-Thiobarbituric Acid Values and Contents of α-Tocopherol and Bile Pigments in the Liver and Muscle of Jaundiced Yellowtail, *Seriola quinqueradiata*. Agricultural and Biological Chemistry 1989;53(6) 1739-1740.

[37] Lansdell W, Dixon B, Smith N, Benjamin L. COMMUNICATIONS: Isolation of Several Mycobacterium Species from Fish. Journal of Aquatic Animal Health 1993;5: 73-76.

[38] McCarthy DH. Fish Furunculosis. Aquaculture 1975;6(1) 13-18.

[39] Bejerano Y, Sarig S, Horne MT, Roberts RJ. Mass Mortalities in Silver Carp, *Hypophthalmichthys molitrix* (Valenciennes) Associated with Bacterial Infection Following Handling. Journal of Fish Diseases 1979;2: 49-56.

[40] Morrison C, Cornick J, Shum G, Zwicker B. Microbiology and Histopathology of 'Saddleback' Disease of Under Yearling Atlantic Salmon. Journal of Fish Diseases 1981;4: 243-258.

[41] Matoyama H, Hoshino M, Hosoya M. Pathogenecity of Atypical *Aeromonas salmonicida* Isolated from Coloured Carp *Cyprinus carpio* Suffering from a New Ulcerative Disease. Fish Pathology 1999;34: 189-193.

[42] Shewan JM, Hobbs G, Hodgkiss W. The *Pseudomonas* and *Achromobacter* Groups of Bacteria in the Spoilage of Marine White Fish. Journal of Applied Bacteriology 1961;23: 463-468.

[43] Otte E. Die Heutigen Ansichten Uber die Atiologie der Infektiosen Bachwassersucht der Karpfen. Wiener Tierarztliche Monatsschrift 1963;50: 995-1005.

[44] Csaba G, Prigli M, Bekesi L, Kovacs-Gayer E, Bajmocy E, Fazekas B. Septicaemia in Silver Carp (*Hypophthalmichthys molitrix*, Val.) and Bighead (*Aristichthys nobilis* Rich.) Caused by *Pseudomonas fluorescens*. In: Olah J, Molnar K, Jeney S. (eds.) Fish, Pathogens and Environment in European Polycultur. SF Muller (Fisheries Research Institute); 1981. p111-123.

[45] Allen DA, Austin B, Colwell RR. Numerical Taxonomy of Bacterial Isolates Associated with a Freshwater Fishery. Journal of General Microbiology 1983;129: 2043-2062.

[46] Farzanfar A. The Use of Probiotics in Shrimp Aquaculture. FEMS Immunology and Medical Microbiology 2006;35: 43-66.

[47] Ng WJ, Kho K, Ho LM, Ong SL, Sim TS, Tay SH, Goh CC, Cheong L. Water Quality Within a Recirculating System for Tropical Ornamental Fish Culture. Aquaculture 1992;103: 123-134.

[48] Grommen R, Van Hauteghem I, Van Wambeke M, Verstraete W. An Improved Nitrifying Enrichment to Remove Ammonium and Nitrite from Freshwater Aquaria Systems. Aquaculture 2002;211: 115-124.

[49] Gross S, Nemirovsky A, Zilberg D, Khaimov A, Brenner A, Snir E, Ronen Z, Nejidat A. Soil Nitrifying Enrichments as Biofilter Starters in Intensive Re-Circulating Saline Water Aquaculture. Aquaculture 2003;223: 51-62.

[50] Jimenez-Montealegre R, Verdegem MCJ, van Dam AA, Verreth JA. Effect of Organic Nitrogen and Carbon Mineralization on Sediment Organic Matter Accumulation in Fish Ponds. Aquaculture Research 2005;36: 983-995.

[51] Larmoyeux JD, Piper RG. Effects of Water Reuse on Rainbow Trout in Hatchery. Progressive Fish-Culturist 1973;35: 2-8.

[52] Boyd CE, Tucker CS. Pond Aquaculture Water Quality Management. Boston, MA: Kluwer Academic Publishing; 1998.

[53] Jeney G, Nemcsok J, Jeney Z, Ola'h J. Acute Effect of Sublethal Ammonia Concentrations on Common Carp (*Cyprinus carpio* L.). II Effect of Ammonia on Blood Plasma Transaminases (GOT, GPT), G1DH Enzyme Activity, and ATP Value. Aquaculture 1992;104: 149-156.

[54] Frances J, Nowak BF, Allan GL. Effects of Ammonia on Juvenile Silver Perch (*Bidyanus bidyanus*). Aquaculture 2000;183: 95-103.

[55] Sakai K, Nakamura K, Wakayama M, Moriguchi M. Change in Nitrite Conversion Direction from Oxidation to Reduction in Heterotrophic Bacteria Depending on the Aeration Conditions. Journal of Fermentation and Bioengineering 1997;86: 47-52.

[56] Ferguson SJ. Denitrification and Its Control. Antonie van Leeuwenhoek 1994;66: 89-110.

[57] Frances J, Allan GL, Nowak BF. The Effects of Nitrite on the Short-Term Growth of Silver Perch (*Bidyanus bidyanus*). Aquaculture 1998;163: 63-72.

[58] Kim JD, Kim KS, Song JS, Lee JY, Jeong KS. Optimum Level of Dietary Monocalcium Phosphate Based on Growth and Phosphorous Excretion of Mirror Carp, *Cyprinus carpio*. Aquaculture 1998;161: 337-344.

[59] Auer MT, Kieser MS, Canale RP. Identification of Critical Nutrient Levels through Field Verification of Models for Phosphorus and Phytoplankton Growth. Canadian Journal of Fisheries and Aquatic Sciences 1986;43: 379-388.

[60] Kaushik SJ. Nutrient Requirements, Supply and Utilization in the Context of Carp Culture. Aquaculture 1995;129: 225-241.

[61] Moriarty DWJ. Control of Luminous *Vibrio* Species in Penaeid Aquaculture Ponds. Aquaculture 1998;164: 351-358.

[62] Barker G. Novel Methods to Reduce Disease in Aquaculture. Fish Veterinary Journal 2000;5: 66-71.

[63] Watson AK, Heinrich K, Lategan MJ, Gibson L. Probiotics in Aquaculture. The Need, Principles and Mechanisms of Action and Screening Processes. Aquaculture 2008;274: 1-14.

[64] Tarkhani R, Imanpoor MR. Mortality Response of *Xiphophorus maculatus* (*Cyprinodontisformes: Poeciliidae*) to Some Agricultural Pesticides. World Journal of Fish and Marine Sciences 2012;4(5) 512-516.

[65] Schwarz S, Kehrenberg C, Walsh TR. Use of Antimicrobial Agents in Veterinary Medicine and Food Animal Production. International Journal of Antimicrobial Agents 2001;17: 431–437.

[66] Akinbowale, OL, Peng H, Barton MD. Antimicrobial Resistance in Bacteria Isolated from Aquaculture Sources in Australia. Journal of Applied Microbiology 2006;100: 1103-1113.

[67] Turnidge J. Antibiotic Use in Animals-Prejudices, Perceptions and Realities. Journal of Antimicrobial Chemotherapy 2004;53: 26-27.

[68] Delsol AA, Randall L, Cooles S, Woodward MJ, Sunderland J, Roe MJ. Effect of the Growth Promoter Avilamycin on Emergence and Persistence of Antimicrobial Resistance in Enteric Bacteria in the Pig. Journal of Applied Microbiology 2005;98: 564–571.

[69] Focht DD, Verstraete W. Biochemical Ecology of Nitrification and Denitrification.In: Alexander M. (ed.) Advances in Microbial Ecology. Plenum Press: New York; 1977. Volume 1. p135-214.

[70] Fast AW, Menasveta P. Some Recent Issues and Innovations in Marine Shrimp Pond Culture. Reviews in Fisheries Science 2000;8: 151-233.

[71] Gomez-Gil B, Roque A, Turnbull JF. The Use and Selection of Probiotic Bacteria for Use in the Culture of Larval Aquatic Organisms. Aquaculture 2000;191: 259-270.

[72] Lalloo R, Ramchuran S, Ramduth D, Görgens J, Gardiner N. Isolation and Selection of *Bacillus spp.* as Potential Biological Agents for Enhancement of Water Quality in Culture of Ornamental Fish. Journal of Applied Microbiology 2007;103: 1471–1479.

[73] Maeda M, Nogami K, Kanematsu M, Hirayama K. The Concept of Biological Control Methods in Aquaculture. Hydrobiologia 1997;358: 285-290.

[74] Fuller R. A Review, Probiotics in Man and Animals. Journal of Applied Bacteriology 1987;66: 365-378.

[75] Gram L, Melchiorsen J, Spanggaard B, Huber I, Nielsen T. Inhibition of *Vibrio anguillarum* by *Pseudomonas fluorescens* Strain AH2-a Possible Probiotic Treatment of Fish. Applied and Environmental Microbiology 1999;65: 969-973.

[76] Salminen S, Ouwehand AC, Benno Y, Lee YK. Probiotics: How should they be Defined? Trends in Food Science and Technology 1999;10: 107-110.

[77] Irianto A, Austin B. Probiotics in Aquaculture. Journal of Fish Diseases 2002;25: 633-642.

[78] Vanbelle M, Teller E, Focant M. Probiotics in Animal Nutrition. Archives of Animal Nutrition 1990;40: 543-567.

[79] Westerdahl A, Olsson JC, Kjelleberg S, Conway P. Isolation and Characterization of Turbot (*Scophthalmus maimus*)-Associated Bacteria with Inhibitory Effects against *Vibrio anguillarum*. Applied and Environmental Microbiology1991;57: 2223-2228.

[80] Olsson JC, Westerdahl A, Conway PL and Kjelleberg S. Intestinal Colonization Potential of Turbot (*Scophthalmus maximus*) and Dab (*Limanda limanda*) Associated Bacteria with Inhibitory Effects against *Vibrio anguillarum*. Applied and Environmental Microbiology 1992;58: 551-556.

[81] Krovacek K, Faris A, Anhe W, Månsson I. Adhesion of *Aeromonas hydrophila* and *Vibrio anguillarum* to Fish Cells and to Mucus-Coated Glass Slides. FEMS Microbiology Letters 1987;42: 85-89.

[82] Garcia T, Otto K, Kjelleberg S, Nelson DR. Growth of *Vibrio anguillarum* in Salmon Intestinal Mucus. Applied and Environmental Microbiology 1997;63(3) 1034-1039.

[83] Montes AJ and Pugh DG. The Use of Probiotics in Food-Animal Practice. Veterinary Medicine 1993; 282-288.

[84] Raa J. The Use of Immunostimulatory Substances in Fish and Shellfish Farming. Reviews in Fisheries Science 1996;4: 229-288.

[85] Söderhall K, Cerenium L. Role of Prophenoloxidase-Activating System in Invertebrate Immunity. Current Opinion in Immunology 1998;10: 23-28.

[86] Sakai K, Ikehata Y, Ikenaga Y, Wakayama M. Nitrite Oxidation by Heterotrophic Bacteria under Various Nutritional and Aerobic Conditions. Journal of Fermentation and Bioengineering 1996;82: 613–617.

[87] Ringo E, Gatesoupe FJ. Lactic Acid Bacteria in Fish: A Review. Aquaculture 1998;160: 177-203.

[88] Lalloo R, Maharajh D, Görgens J. A Downstream Process for Production of a Viable and Stable *Bacillus cereus* Aquaculture Biological Agent. Applied Microbiology and Biotechnology 2010;86: 499–508.

[89] Schwyn B, Neilands JB. Universal Chemical Assay for the Detection and Determination of Siderophores. Analytical Biochemistry 1987;160: 47-56.

[90] Wooldridge KG, Williams PH. Siderophores: Structure and Function of Microbial Iron Transport Compounds. FEMS Microbiology Reviews 1993;12: 325-348.

[91] Gatesoupe FJ. Siderophore Production and Probiotic Effect of *Vibrio* sp. Associated with Turbot Larvae, *Scophthalmus maximus*. International Aquatic Research 1997;10: 239-246.

[92] Lalloo R, Maharajh D, Görgens J, Gardiner N. Functionality of a *Bacillus cereus* Biological Agent in Response to Physiological Variables Encountered in Aquaculture. Applied Microbiology and Biotechnology 2008;79: 111-118.

[93] Martienssen M, Schöps R. Population Dynamics of Denitrifying Bacteria in a Model Biocommunity. Water Research 1999;33: 639-646.

[94] Munro PD, Mclean HA, Barbour A, Birbeck TH. Stimulation or Inhibition of Growth of the Unicellular Alga *Pavlova lutheri* by Bacteria Isolated from Larval Turbot Culture Systems. Journal of Applied Bacteriology 1995;79: 519-524.

[95] Fukami K, Kawai A, Asada M, Okabe M, Hotta T, Moriyama T, Doi S, Nishijima T, Yamaguchi M and Taniguchi M. Continuous and Simultaneous Cultivation of Benthic Food Diatom *Nitzschia* sp. and Abalone *Haliotis sieboldii* by Using Deep Seawater. Journal of Marine Biotechnology 1998;6: 237-240.

[96] Queiroz JF, Boyd CE. Effects of Bacterial Inoculum in Channel Catfish Ponds. Journal of the World Aquaculture Society 1998;29: 67-73.

[97] Rengpipat S, Rukpratanporn S, Piyatiratitivorakul S, Menasveta P. Immunity Enhancement in Black Tiger Shrimp (*Penaeus monodon*) by a Probiont Bacterium (*Bacillus* S11). Aquaculture 2000;191: 271-288.

[98] Abou Seada MNI, Ottow JCG. Effect of Increasing Oxygen Concentration on Total Denitrification and Nitrous Oxide Release from Soil by Different Bacteria. Biology and Fertility of Soils 1985;1:31-38.

[99] Robertson LA, Kuenen JG. Physiological and Ecological Aspects of Aerobic Denitrification, a Link with Heterotrophic Nitrification?. In: Revsbech NP, Serensen J. (eds.) Denitrification in Soils and Sediments. Plenum Press: New York; 1990. p91–104.

[100] Su JJ, Liu BY, Lin J, Yang CP. Isolation of an Aerobic Denitrifying Bacteria Strain NS-2 from the Activated Sludge of Piggery Wastewater Treatment Systems in Taiwan Processing Denitrification Under 92% Oxygen Atmosphere. Journal of Applied Microbiology 2001;91: 853-860.

[101] Kim JK, Park KJ, Cho KS, Nam S, Park T, Bajpai R. Aerobic Nitrification-Denitrification by Heterotrophic *Bacillus* Strains. Bioresource Technology 2005;96: 1897–1906.

[102] Lin Y, Tanaka S, Kong H. Characterization of a Newly Isolated Heterotrophic Nitrifying Bacterium. Water Pr Tech 2006;1, DOI: 10.2.2166 WPT.2006052.

[103] Nakano MM, Zuber P. Anaerobic Growth of a "Strict Aerobe" (*Bacillus subtilis*). Annual Review of Microbiology 1998;52: 165-190.

[104] Illmer P, Schinner F. Solubilization of Inorganic Calcium Phosphates-Solubilization Mechanisms. Soil Biology and Biochemistry 1995;27: 257-263.

[105] Rengpipat S, Phianphak W, Piyatiratitivorakul S, Menasveta P. Effects of a Probiotic Bacterium on Black Tiger Shrimp *Penaeus monodon* Survival and Growth. Aquaculture 1998;167: 301-313.

[106] Kennedy SB, Tucker JW, Neidig CL, Vermeer GK, Cooper VR, Jarrell JL, Sennet DG. Bacterial Management Strategies for Stock Enhancement of Warm Water Marine Fish: A Case Study with Common Snook (*Centropomus undecimals*). Bulletin of Marine Science 1998;62: 573-588.

[107] Hirata H, Murata O, Yamada S, Ishitani H, Wachi M. Probiotic Culture of the Rotifer *Brachionus plicatilis*. Hydrobiologica 1998;387/388: 495-498.

[108] Wolken WAM, Tramper J, van der Werf MJ. What Can Spores Do for Us? Trends Biotechnology 2003;21: 338-345.

[109] Ugoji EO, Laing MD, Hunter CH. An Investigation of the Shelf-Life (Storage) of *Bacillus* Isolates on Seeds. Journal of Botany 2006;72: 28-33.

[110] Moir A. How Do Spores Germinate? Journal of Applied Microbiology 2006;101: 526-530.

[111] Foldes T, Banhegyi Z, Varga L, Szigeti J. Isolation of *Bacillus* Strains from the Rhizosphere of Cereals and *In Vitro* Screening for Antagonism Against Phytopathogenic, Food Borne Pathogenic and Spoilage Micro-Organisms. Journal of Applied Microbiology 2000;89(5) 840-846.

[112] Huis in't Veld JHJ, Havenaar R, Marteau PH. Establishing a Scientific Basis for Probiotic R&D. Tibtech 1994;12: 6-8.

[113] Nogami K, Maeda M. Bacteria as Biocontrol Agents for Rearing Larvae of the Crab *Portunus tribeculatus*. Canadian Journal of Fisheries and Aquatic Sciences 1992;49: 2373–2376.

[114] Sugita H, Shibuya K, Shimooka H, Deguchi Y. Antibacterial Abilities of Intestinal Bacteria in Freshwater Cultured Fish. Aquaculture 1996;145: 195–203.

[115] Riquelme C, Araya R, Vergara N, Rojas A, Guaita M, Candia M. Potential Probiotic Strains in the Culture of the Chilean Scallop *Argopecten purpuratus* (Lamarck, 1819). Aquaculture 1997;154: 17-26.

[116] Gibson LF, Woodworth J, George AM. Probiotic Activity of *Aeromonas* Media on the Pacific Oyster, Crassostrea Gigas, When Challenged with *Vibrio tubiashii*. Aquaculture 1998;69: 111-120.

[117] Dopazo CP, Lemos ML, Lodeiros C, Bolinches J, Barja JL and Toranzo AE. Inhibitory Activity of Antibiotic-Producing Marine Bacteria against Fish Pathogens. Journal of Applied Bacteriology 1988;65: 97-101.

[118] Smith P, Davey S. Evidence for the Competitive Exclusion of *Aeromonas salmonicida* from Fish with Stress-Inducible Furunculosis by a Fluorescent Pseudomonad. Journal of Fish Diseases 1993;16: 521-524.

[119] Jöborn A, Olsson JC, Westerdahl A, Conway PL, Kjelleberg S. Colonization in the Fish Intestinal Tract and Production of Inhibitory Substances in Intestinal Mucus and Faecal Extracts by *Carnobacterium* sp. Strain K1. Journal of Fish Diseases 1997;20(5) 383-392.

[120] Mayer-Harting A, Hedges AJ, and Berkeley RW. Methods for Studying Bacteriocins. Methods in Microbiology 1972;7A: 315-422.

[121] Ten Brink B, Minekus M, Bol J, Huis in 't Veld J. Production of Antibacterial Compounds by *Lactobacilli*. FEMS Microbiology Reviews 1987;46: 64.

[122] Bergh Ø. Bacteria Associated With Early Life Stages of Halibut, *Hippoglossus hippoglossus* L., Inhibit Growth of a Pathogenic *Vibrio* sp.. Journal of Fish Disease 1995;18: 31-40.

[123] Gildberg A, Mikkelsen H. Effect of Supplementing the Feed to Atlantic Cod (*Gadus morhua*) Fry With Lactic Acid Bacteria and Imuno Stimulating Peptides During a Challenge Trial With *Vibrio anguillarum*. Aquaculture 1998;167(1) 103-113.

[124] Bull AT, Slater JH. Microbial Interactions and Communities. Academic Press, Inc., New York, N.Y; 1982.

[125] Suminto K, Hirayama K. Application of a Growth-Promoting Bacterium for Stable Mass Culture of Three Marine Microalgae. Hydrobiologia 1997;358: 223-230.

[126] Preeta R, Jayaprakash NS, Philip R, Singh ISB. Optimization of Carbon and Nitrogen Sources and Growth Factors for the Production of an Aquaculture Probiotic (*Pseudomonas* MCCB 103) Using Response Surface Methodology. Journal of Applied Microbiology 2006;102: 1043-1051.

[127] Patel AK, Deshattiwar MK, Chaudari BL, Chincholkar SB. Production, Purification and Chemical Characterization of the Catecholate Siderophore from Potent Probiotic Strains of *Bacillus spp*. Bioresource Technology 2009;100: 368-373.

[128] Gouda MK, Swellam AE, Omar SH. Production of PHB by a *Bacillus megaterium* Strain Using Sugarcane Molasses and Corn Steep Liquor as Sole Carbon and Nitrogen Sources. Microbiological Research 2001;156: 201-207.

[129] Payot T, Chemaly Z, Fick M. Lactic Acid Production by *Bacillus coagulans*—Kinetic Studies and Optimization of Culture Medium for Batch and Continuous Fermentations. Enzyme and Microbial Technology 1998;24: 191-199.

[130] Rättö M, Poutanen K, Viikari L. Production of Xylanolytic Enzymes by an Alkali Tolerant *Bacillus circulans* Strain. Applied Microbiology and Biotechnology 1992;37: 470-473.

[131] Lawford HG and Rousseau JD. Corn Steep Liquor as a Cost-Effective Nutrition Adjunct in High-Performance *Zymomonas* Ethanol Fermentations. Journal of Biochemistry and Biotechnology 1997;63-35: 287-304.

[132] Kona RP, Qureshi N, Pai JS. Production of Glucose Oxidase Using *Aspergillus niger* and Corn Steep Liquor. Bioresource Technology 2001;78: 123-126.

[133] Schisler DA, Slininger PJ, Behle RW, Jackson MA. Formulation of *Bacillus spp.* for Biological Control of Plant Diseases. Phytopathology 2004;94: 1267-1271.

[134] Monteiro SM, Clemente JJ, Henriques AO, Gomes RJ, Carrondo MJ, Cunha AE. A Procedure for High-Yield Spore Production by *Bacillus subtilis*. Biotechnology Progress 2005;21: 1026-1031.

[135] Prabakaran G, Balaram K, Hoti SL, Manonmani AM. A Cost Effective Medium for the Large-Scale Production of *Bacillus sphaericus* H5a5b (VCRC B42) for Mosquito Control. Biological Control 2007;41: 379-383.

[136] Lopez JM, Dromenck A, Freese E. Response of Guanosine 5-Triphosphate Concentration to Nutritional Changes and its Significance for *Bacillus subtilis* Sporulation. Journal of Bacteriology 1981;146: 605-613.

[137] Zhang J, Greasham R. Chemically Defined Media for Commercial Fermentations. Applied Microbiology and Biotechnology 1999;51: 407-421.

[138] Liu WM, Bajpai R, Bihari V. High-Density Cultivation of Spore Formers. Annals of the New York Academy of Sciences 1994;721: 310-325.

[139] Nicholson WL, Munakata N, Horneck G, Melosh HJ, Setlow P. Resistance of *Bacillus* Endospores to Extreme Terrestrial and Extraterrestrial Environments. Microbiology and Molecular Biology Reviews 2000;64: 548-572.

[140] Sonenshein AL. Control of Sporulation Initiation in *Bacillus subtilis*. Current Opinion in Microbiology 2000;3: 561-566.

[141] Nickerson K, Bulla LA Jr. Physiology of Spore Forming Bacteria Associated with Insects: Minimal Nutritional Requirements for Growth, Sporulation and Parasporal Crystal Formation of *Bacillus thuringiensis*. Applied Microbiology 1974;28: 124-128.

[142] Gandhi PH, Ray RM, Patel RM. Exopolymer Production by *Bacillus* Species. Carbohydrate Polymers 1997;34: 323-327.

[143] Vuolanto A, von Weymarn N, Kerovuo J, Ojamo H, Leisola M. Phytase Production by High Cell Density Culture or Recombinant *Bacillus subtilis*. Biotechnology Letters 2001;23: 761-766.

[144] Calik P, Bilir E, Calik G, Özdamar TH. Bioreactor Operation Parameters as Tools for Metabolic Regulations in Fermentation Processes: Influence of pH Conditions. Chemical Engineering Science 2003;58: 759-766.

[145] Calik P, Özdamar TH. Carbon Sources Affect Metabolic Capacities of *Bacillus* Species for the Production of Industrial Enzymes: Theoretical Analyses from Serine and Neutral Proteases and α-Amylase. Biochemical Engineering Journal 2001;8: 61-81.

[146] Srivasta RAK, Baruah JN. Culture Conditions for Production of Thermostable Amylase by *Bacillus stearothermophilus*. Applied and Environmental Microbiology 1986;52: 179-184.

[147] Shinke R, Kunimi Y, Aoki K, Nishira H. Filamentation in *Bacillus cereus* During Amylase Production. Journal of Fermentation Technology 1977;55: 103-110.

[148] Sharpe ES, Bulla LA Jr. Characteristics of the Constituent Substrains of *Bacillus papillae* Growing in Batch and Continuous Cultures. Applied and Environmental Microbiology 1978;35: 601-609.

[149] Goldberg I, Sneh B, Battat E, Klein E. Optimization of a Medium for a High Yield Production of Spore-Crystal Preparation of *Bacillus thurengiensis* Effective Against the Egyptian Cotton Leaf Worms *Podopteral ittoralis boisd*. Biotechnology Letters 1980;2: 419-426.

[150] Nohata Y, Kurane R. Complete Defined Medium for Large-Scale Production of Polysaccharide Bioabsorbent from *Alcaligenes latus* B-16. Journal of Fermentation and Bioengineering 1997;83: 116-117.

[151] Miller TL, Churchill BW. Substrates for Large-Scale Fermentation. In: Demain AL, Solomon NA. (eds.) Manual of Industrial Microbiology and Biotechnology. Washington: American Society of Microbiology; 1986. p130-131.

[152] Shikata S, Saeki K, Okoshi H, Yoshimatsu T, Ozaki K, Kawai S, Ito S. Alkaline Cellulases for Laundry Detergents: Production by Alkalophilic Strains of *Bacillus* and Some Properties of the Crude Enzymes. Agricultural Biology and Chemistry 1990;54: 91-96.

[153] Kuppusamy M, Balaraman K. Effect of Corn Steep Liquor on Growth and Mosquito Larvicidal Activity of *Bacillus thuringiensis var irealensis de Barjac* and *Bacillus sphaericus Neide* 1904. Indian Journal of Experimental Biology 1997;29: 87-189.

[154] Shi F, Zhu Y. Applications of Statistically-Based Experimental Designs in Medium Optimization for Spore Production of *Bacillus subtilis* from Distillery Effluent. Biological control 2007;52: 845-853.

[155] Chandra G. Dryinid Parasitoids of Rice Leaf Hoppers and Planthoppers in the Philippines.I.Taxonomy and bionomics. Acta Oecologica, Oecologia Applicata 1980;1(2) 161-172.

[156] Mao W, Pan R, Freedman D. High Production Alkaline Protease by *Bacillus licheniformis* in a Fed-Batch Fermentation Using Synthetic Medium. Journal of Industrial Microbiology 1992;11: 1-6.

[157] Purushothaman M, Anderson RKI, Narayana S, Jayaraman VK. Industrial by Products as Cheaper Medium Components Influencing the Production of Polyhydroxyalkanoates (PHA)- Biodegradable Plastics. Bioprocess and Biosystems Engineering 2001;24: 131-136.

[158] Silveira MM, Wisbeck E, Hoch I, Jonas R. Production of Glucose-Fructose Oxidoreductase and Ethanol by *Zymomonas mobilis* ATCC 29191 in Medium Containing Corn Steep Liquor as a Source of Vitamins. Applied Microbiology and Biotechnology 2001;55: 442-445.

[159] Almedia JL, Wand L, Morrow JB, Cole KD. Requirements for the Development of *Bacillus anthracis* Spore Reference Materials Used to Test Detection Systems, J, Res. Natl. Inst. Stand. Technol. 2006:111: 205-217.

[160] Wang ST, Setlow B, Conlon TM, Lyon JL, Imamura D, Sato T, Setlow P, Losick R, Eichenberger P. The Forespore Line of Gene Expression in *Bacillus subtilis*. Journal of Molecular Biology 2006;358: 16-37.

[161] Errington J. Regulation of Endospore Formation in *Bacillus subtilis*. Nature Reviews Microbiology 2003;1: 117-126.

[162] Msadek T. When the Going Gets Tough: Survival Strategies and Environmental Signalling Networks in *Bacillus subtilis*. Trends in Microbiology 1999;7: 201–207.

[163] Dingman DW, Stahly DP. Medium Promoting Sporulation of *Bacillus* Larvae and Metabolism of Medium Components. Applied and Environmental Microbiology 1983;46: 860-869.

[164] Austin B, Stuckey LF, Robertson PAW, Effendi I, Griffith DRW. A probiotic Strain of *Vibrio alginolyticus* Effective in Reducing Diseases Caused by *Aeromonas salmonicida, Vibrio anguillarum and Vibrio ordalii*. Journal of Fish Diseases 1995;18: 93–96.

[165] Avignonne-Rossa C, Arcas J, Mignone C. *Bacillus thuringiensis* Growth, Sporulation and δ-Endotoxin Production in Oxygen Limited and Non-Limited Cultures. World Journal of Microbiology and Biotechnology 1992;8: 301-304.

[166] Budde I, Steil L, Scharf C, Völker U, Bremer E. Adaptation of *Bacillus subtilis* to Growth at Low Temperature: A Combined Transcriptomic and Proteomic Appraisal. Microbiology+ 2006;152: 831-853.

[167] Vary JC. Germination of *Bacillus* Megaterium Spores after Various Extraction Procedures. Bacteriology 1973;2: 797-802.

[168] Hornstra LM, de Vries YP, Wells-Bennik MH, de Vos WM, Abee T. Characterization of Germination Receptors of *Bacillus cereus* ATCC 14579. Applied and Environmental Microbiology 2006;72: 44-53.

[169] Stewart GSAB, Johnstone K, Hagelberg E, Ellar DJ. Commitment of Bacterial Spores to Germinate. Biochemical Journal 1981; 106-109.

[170] Holtman G, Bremer E. Thermoprotection of *Bacillus subtilis* by Exogenously Provided Glycine Betaine and Structurally Related Compatible Solutes: Involvement of Opu Transporters. Journal of Bacteriology 2004;186: 683-1693.

[171] Choma C, Clavel T, Dominguez H, Razafindramboa N, Soumille H, Nguyen-the C, Schmitt P. Effect of Temperature on Growth Characteristics of *Bacillus cereus* TZ415. International Journal of Food Microbiology 2000;55: 73-77.

[172] Ratkowsky DA, Lowry RK, McMeekin TA, Stokes AN, Chandler RE. Model for Bacterial Culture Growth Rate throughout the Entire Biokinetic Temperature Range. Journal of Bacteriology 1983;154: 1222-1226.

[173] Thomassin S, Jobin MP and Schmitt P. The Acid Tolerance of *Bacillus cereus* ATCC14579, Is Dependent on Culture pH, Growth Rate and Intracellular pH. Archives of Microbiology 2006;186: 229-239.

[174] Leguerinel I, Couvert O, Mafart P. Relationship between the Apparent Heat Resistance of *Bacillus cereus* Spores and the pH and NaCl Concentration of the Recovery Medium. International Journal of Food Microbiology 2000;55: 223-227.

[175] Jobin MP, Clavel T, Carlin F, Schmitt P. Acid Tolerance Response is Low-pH and Late Stationary Growth Phase Inducible in *Bacillus cereus* TZ415. International Journal of Food Microbiology 2002;79: 65-73.

Chapter 7

Application of Biological Agents in Abalone Aquaculture: A South African Perspective

Ghaneshree Moodley, Lethabo Mashigo,
Rajesh Lalloo and Suren Singh

Additional information is available at the end of the chapter

http://dx.doi.org/10.5772/57197

1. Introduction

1.1. The impact of abalone mariculture on developing economies

Aquatic animals are nutritionally important for human consumption, as they are an excellent source of proteins, trace elements, and polyunsaturated fatty acids [1]. There has been a significant increase in the demand for an array of both fish and shellfish products as a result of growth in the global population [2]. Fisheries itself, cannot provide sufficient amounts of aquatic products to fulfil the demands of the consumer; therefore, aquaculture provides a crucial alternative resource [1,4]. Aquaculture has become more significant and intensive over the last few decades and is presently the fastest growing food production industry. An average industrial growth of more than 6% in the period between 1985 and 2005 has been reported, with an annual increase of approximately 3.2% per annum during the period up to 2009 [1,3].

Modern aquaculture involves the intensive production of finfish, crustaceans, molluscs, and algal plants under controlled conditions [5]. Aquaculture yields far exceeds that of natural fishing, and provides an effective means for a constant, year round supply for good quality seafood and seafood products [6-8]. The practice of aquaculture not only provides local food security, but also improves the livelihoods of people in many poorly developed coastal regions [2].

Commercial abalone mariculture has become a thriving, global industry. It has a promising future due to the high prices being paid for abalone, coupled to a worldwide decline in fisheries production because of overfishing and poaching [8,9]. Abalone is one of the most valuable seafood species in the world, whereby demand far exceeds supply, especially in Asian countries such as; Hong Kong, China, Japan, Taiwan and Singapore which are major destination markets (6-8,10). Abalone is used primarily as a celebration dish, especially during weddings and other special occasions such as the Chinese New Year [11]. On account of the ever growing demand of live, dried and canned abalone, already high prices of this seafood delicacy continue to escalate.

Abalone (family *Haliotidae*) belongs to a class of marine vetigastropod molluscs, which are distributed along rocky shores and reefs of coastal temperate and tropical waters [11, 12]. The abalone family consists of about 56 species all belonging to the genus *Haliotis* [13,14]. Many members of this family have achieved commercial status as fishery and/or aquaculture species, and are of major economic importance (Table 1). In 2007, it was reported that abalone was supplied to export markets in the following product forms; dried (7%), frozen (24%), live (18%) and canned (51%). Live abalone achieves higher revenues however, it does deem problematic in terms of transportation and related logistics [15]. Due to the demand of this prestigious seafood, supply of abalone is under severe pressure; and has led to the increase in the occurrence of abalone farming facilities around the world.

SPECIES NAME	COMMON NAME	LOCATION	TYPE OF FISHERY
Haliotis rufescens	Red abalone	N. America	Farmed/ Recreational
Haliotis rufescens	Red abalone	Chile	Farmed
Haliotis cracherodii	Black abalone	N. America	Farmed
Haliotis fulgens	Green abalone	N. America	Wild/Farmed (Mexico)
Haliotis corrugata	Pink abalone	N. America	Wild/Farmed (Mexico)
Haliotis kamtschatkana	Pinto abalone	N. America	Farmed
Haliotis midae	Perlemoen	South Africa	Wild/Farmed
Haliotis laevigata	Green-lip abalone	S. Australia	Wild
Haliotis rubra	Black-lip abalone	S. Australia	Wild/Farmed
Haliotis roei	Roe's abalone	Australia	Wild
Haliotis iris	Black footed paua	New Zealand	Wild
Haliotis diversicolor supertexta	Small abalone	Taiwan	Wild/Farmed
Haliotis discus hannai	Disk abalone	Japan, China	Wild/Farmed

Table 1. Globally farmed abalone species and their location adapted from [16,17].

Cultivation of abalone is widespread in many countries, including USA, Mexico, South Africa, Australia, New Zealand, Japan, Taiwan, China, Ireland, Chile and Iceland [17-20] China is the largest producer in the world with over 300 farms and a total production of approximately 4500 metric tonnes [9]. China, Taiwan and South Africa are considered as the key production powerhouses in the abalone industry (Table 2). China is the highest contributor of live product annually, and is still the major market for abalone produced world-wide [8]. This occurrence is closely related to the economic growth and the increase in personal wealth exhibited by the

Chinese population as well as the growth of the Chinese middle class population [4,15]. It has been reported that the total abalone produce reaching markets through harvesting, illegal poaching and natural supply, does not meet demand for this seafood delicacy [9,18].

COUNTRY	CULTURE	LEGAL HARVEST	ILLEGAL HARVEST
China	4500	-	-
Taiwan	3000	-	-
South Africa	600	237	1850
Japan	200	2200	536
USA	170	0	250
Australia	290	5128	1000
Chile	200	-	-
Mexico	50	1066	550
New Zealand	3	1078	400
Other	30	442	110
Total	9 043	10 151	4696
Grand total in 2005: 24 040 tonnes live weight			

Table 2. Global production of abalone indicating world leaders in abalone production, quantity of produce legally harvested and sold, and quantity of product illegally harvested; 2004-2005 data [20].

The South African abalone industry continues to establish itself as a premium brand in Asia, and is a good example of mariculture in a developing country. Abalone farming in SA is a relatively new but dynamic industry and has demonstrated a high production capacity [15]. One of the main challenges faced by the SA industry is the loss in revenue experienced due to poaching. Reports suggest that approximately 2000 tonnes are lost to the economy [4]. The abalone mariculture industry started developing in South Africa during the 1990's and has been gaining popularity. As a result, an economic environment whereby abalone aquaculture has become increasingly attractive as a financial investment has been established [17]. Abalone rearing facilities employs an intensive system in which abalone is reared at high densities in shore-based aquaculture systems [21].

The South African abalone, *Haliotis midae,* locally known as "perlemoen" is the only one of six indigenous species that is of commercial importance [8,22]. The abalone *H. midae,* takes over 30 years to reach a maximum size of 200mm (shell length) in natural habitats [21]. Even under farmed conditions, abalone growth is slow and often varies with size and age [23]. *H. midae* takes approximately 4 to 5 years to reach a marketable size of 100 mm (shell length) before it can be sold between US$ 34 to 36 per kg on international markets [23,24]. Mariculture of abalone is thus important to ensure market supply and it is for these reasons that alternate approaches involved in the promotion of abalone growth and an increased immunity to disease of farmed abalone are required.

Figure 1. Holding tanks containing farm-produced abalone on the West coast of South Africa.

Land based aquaculture of abalone has increased over the last decade in South Africa (Figure 1), and commercially produced abalone has almost completely replaced the wild harvested product [4]. In 2010 the output of all facets of abalone harvest totalled 1015.44 metric tons.

The former status of abalone aquaculture in South Africa is outlined in Table 3. These farms produced 890 tonnes of abalone, and created direct employment to about 840 people. There was an increase in skilled individuals of approximately 7.6% over the 2 year period. Due to the high demand for this seafood delicacy, a gross turnover of approximately R200 million per annum was achieved [26]. The industry has demonstrated continued growth. In 2003/4; 19 enterprises secured permits to culture this species and by 2007, this number had increased to 24, further highlighting the growth potential of this particular sector [27]. It was estimated that by 2020 the production of abalone would amount 2895 tons with a value of R551 million, making abalone mariculture the leading subsector contributor in the aquaculture industry [4]. This growth has had a direct impact on the socio-economic growth of the country, whereby more than 1200 people with necessary skills are currently employed in the industry. Global aquaculture initiatives have shown that the success of the technology is largely dependent on government sectors for support to enable the creation of a robust and sustainable industry [15]. The mariculture of abalone and on-going growth of this industry is extremely important, as it addresses a number of challenges faced by the South African nation, which are also common to many developing countries. This practice will contribute to a number of strategic imperatives including economic and enterprise development, job creation, food security as well as the adoption of sustainable mariculture practices [15].

Year	No. of producing farms	Investment (R-million)	Tonnes produced	Annual increase in industry (%)	No. of employees
2004	13	-	576	-	556
2005	13	197	745	27	776
2006	13	182	890	21	840

Table 3. The status of abalone aquaculture and total investment in the South African abalone industry between 2004 and 2006 [26].

2. Challenges faced in abalone mariculture and conventionally used mitigation strategies

Many aquaculture farmers, including those in the abalone mariculture sub-sector are faced with a myriad of challenges [28]. The challenges are further exacerbated as abalone mariculture activities become more intensified to optimise efficiencies in land usage and productivity. Adversities faced include slow growth rate of abalone, the outbreak of diseases, waste accumulation and deterioration of environmental conditions within the culture system [29,30]. Disease occurrence is usually associated with primary invasion by pathogenic strains as well as mechanical injury coupled to stressful environmental conditions such as physiochemical changes and poor water quality [31]. These factors, in an interactive way, challenge the health and immune response of the abalone and can lead to poor growth, ill health and increased mortality. This predicament has become one of the main barriers towards the successful development in the aquaculture industry, given that it limits the production of aquaculture products in terms of quality, quantity, and regularity [23].

Disease control is an inherent part of any animal production system, however, in the aquatic environment, the intimate relationship between bacteria and their host, and the use of open production systems adds to this challenge [5]. Unpredictable mass mortalities still occur in the early life stages as a result of the proliferation of pathogens and opportunistic microorganisms, which are responsible for major economic losses [1]. Abalone like other aquatic species is susceptible to common marine pathogenic organisms such as *Vibrio parahaemolyticus, Vibrio anguillarum* and *Vibrio carchariae*, as well as prokaryotes and viruses [23,32,33]. When pathogenic bacteria or viruses are detected, farmers usually apply antimicrobial compounds to the feed and the rearing water [34]. Broad-spectrum anti-microbials have been extensively used as a means of disease control on many aquaculture facilities and unfortunately remains the method of choice for many farmers [23]. Some farmers also use antibiotics as prophylactics in large quantities, even when pathogens are not evident. This ill-advised practice has led to an increase in *Vibrios*, and other opportunistic pathogens, which possess multiple antibiotic resistance and as a result leads to the emergence of more virulent pathogens [28,35]. Plasmid-carrying resistance determinants have been transferred *in-vitro* from aquatic pathogens to human pathogens, such as from *V. cholerae* and *V. parahaemolyticus* to *Escherichia coli* by the horizontal spread of plasmids [36]. Furthermore, the presence of antibiotic residues in the tissues of animals, an imbalance of microorganisms in the gastrointestinal tract of aquatic species and the release of antibiotics into natural waters, and thereby poses further challenges. Consequently, the indiscriminate use of antibiotics confers a negative effect on the health of aquatic host species, the environment and consumers of food products [37]. Due to these concerns, more stringent regulation of antibiotic use in aquaculture has been imposed by the European Union [38]. Since the application of antibiotics is problematic, a strong demand for alternative methods of disease control is required in abalone mariculture.

Abalones are generally regarded as opportunistic herbivores that readily accept a wide range of diets. In natural ecosystems, abalone feed primarily on seaweed or kelp. This food contains a high degree of alginolytic material that is not readily digestible; as a result, enteric microflora is relied upon to effectively digest this material. If the host intestinal flora lacks the ability to produce beneficial enzymes, a very slow digestion process would result, and consequently hinder the growth of the abalone itself. The proper nutrition and resultant growth of cultured abalone are critical factors that require insight in order to successfully culture this mollusc. Appropriate mechanisms for feeding of abalone are therefore very important and it has been shown that different diets results in different growth rates [39]. Growth rates, especially at the early life stages of abalone are affected considerably by the diet and the ability of the individuals to utilize available food with a high resultant feed conversion ratio [40]. In abalone production systems abalones are fed either formulated diets or seaweed/kelp, and in some instances, a combination of both [25]. An optimum formulated diet should enable more efficient digestion consequently resulting in higher feed conversion ratios, and ultimately boost the growth of the abalone, but the reality is that diets are based on raw material availability and minimum cost formulation models. This presents a challenge in digestibility, feed conversion efficiency, animal health and waste generation into the culture environment. The development of artificial feeds and specialized feeding regimes to improve the growth of abalone has assisted in developing this practice into a more cost-effective and manageable industry [21]. It has been reported that abalone fed an artificial diet, have better canning characteristics than that of wild abalone, and canning yields have shown an increase of up to 15% [15].

Incorrectly formulated diets, may also lead to the accumulation of waste in the culture system which could cause the deterioration of water quality in the culture environment. The propensity of algal blooms and the proliferation of disease-causing parasites and pathogens increases in the event of waste accumulation due to poor husbandry and poor feed digestibility. The abalone itself then becomes highly susceptible to disease due to these negative conditions in the mariculture water and succumbs to such challenging conditions. Additionally, the digestive systems of these aquatic hosts are in constant contact with the rearing water, making the host more prone to infection.

In conventional mariculture operations, due to the high stocking densities, the generation of elevated stressful conditions in the culture environment is a frequent occurrence [41]. During the sorting process, abalones are presented with further stresses due to excessive handling and may sustain mechanical damage. Both disease and the deterioration of the environmental conditions are the most significant contributors to mass mortalities in mariculture operations [42]. Most operations employ land-based cultivation systems and use pump ashore technology which is energy intensive and costly [15]. The dilution of culture water, to reduce waste concentrations, by increasing flow rates is therefore not a feasible option. Regulatory authorities are also becoming more stringent on the poor quality of farm effluent that is returned to the sea, as a result, preservation of the surrounding environment also becomes a serious challenge to abalone farmers. Bearing in mind that these factors are interactive and ultimately; either as singular occurrences or in combination, may result in decreased production and potential negative impact on the entire aquatic system. Improving digestion, reducing the concentration of waste and disease causing agents in the surrounding water and a heightened immune response are logical mitigation considerations to address the challenges of abalone mariculture. However, classical interventions are costly and mass mortalities continue to occur, resulting in severe setbacks on both economic and social fronts. In more serious instances, some farms have had no other option but to cease operations. The abalone mariculture industry is therefore in dire need of suitable interventions that can address these challenges in an affordable and sustainable manner.

3. Biological agents as an option to address the challenges in abalone aquaculture

During the past two decades, the use of biological agents, particularly in feed and as water additives, as an alternative to the use of antibiotics and chemicals has shown to be promising in aquaculture, particularly in fish and shellfish larviculture [43]. The concept of biological agents has been traditionally associated with the use of beneficial microorganisms to restore the microbial balance in the gastro-intestinal tract of the host and to treat or prevent diseases and/or disorders [44]. Biological agents are emerging as a significant microbial supplements in the field of prophylaxis [36]. Many studies to date have revealed the potential of these beneficial organisms to combat disease in an aquaculture environment [5, 45-51].

In aquatic ecosystems there is an intimate relationship between microorganisms and other biota in the environment [47]. Apart from the aquatic animal being surrounded by water, there is also a constant flow of water through the digestive tract of the aquatic animal. This consequently affects the synergistic balance of indigenous microflora associated with the cultured animal. The classical definition of a probiotic being that of microbes added to food, has become modified with respect to aquaculture, whereby a biological agent is used as a wider term and is defined as "a live microbial adjunct which has a beneficial effect on the host by modifying the host-association or ambient microbial community, by ensuring improved use of the feed or enhancing its nutritional value, by enhancing the host response towards disease, or by improving the quality of its ambient environment" [47]. Some studies have shown that as a result of intensification of aquaculture farms a negative impact has been conferred on the composition of the different protective microbial flora interacting with the host [5]. This occurrence results in an increase in susceptibility of the host to diseases. It has become evident that augmentation of aquaculture systems with biological agents can lead to growth of beneficial bacteria thus improving overall health of the culture system and the host [5].

The use of biological agents in disease control and improvement of aquaculture is important as demand for environmentally friendly aquaculture practices is on the rise. Biological agents that may be applied in aquaculture comprise of isolates belonging to a wide range of yeast, bacteria and even phytoplankton species [52]. In abalone aquaculture, potential probionts listed to date include, *Vibrio* spp., [23,53-55] *Debaryomyces sp., Cryptococcus sp.*, and *Pseudoalteromonas sp.*, [23,24,30], *Lactobacillus* and *Enterococcus sp.*, [56]; *Pediococcus* sp. Strain Ab1 [57], *Agarivorans albus* F1-UMA [58] and *Shewanella sp* [51].

Biological agents have been found to confer beneficial effects on the host by various modes of action. These may occur as a singular or combined effect, and thus far the following have been reported; (1) the production of antimicrobial products; (2) competitive exclusion; (3) colonisation of the gut and improving microbial balance; (4) enhancement of the host immune response; (5) detoxification of harmful compounds; (6) improved growth rate of the host;(7) antiviral effects, (8) provision of nutrients and enzymatic functions; and (9) improved water quality. Further reports by [24] stated that the addition of probiotics to the diet of farmed abalone, could possibly lead to a boost in abalone growth by a number of potential strategies. Some of which include (1) increasing the nutrients accessible to the abalone for absorption in the gut, (2) increasing the pool of secreted digestive enzymes in the gut of abalone, and (3) use of bacterial supplements as an additional nutrient source.

In many instances, pathogen inhibition and/or disease control has been observed as a consequence of the release of chemical substances with bactericidal effects by probiotic bacteria [47]. The production of antibiotics, bacteriocins, enzymes, hydrogen peroxide, siderophores and the altering of the pH levels due to the generation of organic acids are all traits displayed by biological agents [47,59-61]. In addition, these biological agents compete with pathogens based on intrinsic growth rate and spacial attachment. Microbial colonisation is characterised by the attachment of the biological agent to the mucosal surface and epithelial cells of the host. This prevents the proliferation of opportunistic pathogens thereby preventing infection [62]. It is common knowledge that for a pathogen to be active and replicate in a host system, it requires attachment to these surfaces [62]. When probiotics are administered over a long period, they successfully colonize the gastrointestinal tract, even after cessation of feed supplemented with probiotics. This occurs since the multiplication rate of these probiotics is higher than the rate at which they are removed, thus a build-up in the intestinal mucosa of the host is observed [62].

Host nutrition is improved, as the applied probionts secrete high levels of hydrolytic enzymes such as amylases, proteases and lipases; as well as the provision of growth factors such as fatty acids, amino acids and vitamins [63,64]. Some isolates also have the ability to break down potentially indigestible components of the feed thus reducing toxicity and improving feed conversion efficiency [23,63]. Abalones are in most instances, fed a diet consisting mainly of kelp, which is a complex macroalgal polysaccharide deficient in many essential nutrients [64]. It is therefore imperative that enteric bacteria in the abalone gut are present in sufficient amounts which will adequately facilitate digestion by supplying highly effective polysaccharolytic enzymes [23]. Many bacteria displaying these properties have been found to exist throughout the digestive tract of *H. midae* [23,40]. Some findings indicated that enteric bacteria isolated from the gastrointestinal tract of abalone were capable of degrading agar, carrageenan, laminarin, and alginate. It was also shown that 70 - 90% of the enzyme activity was extracellular suggesting that bacterial enzymes were secreted into the lumen of the gut where they were able to hydrolyse complex algal polysaccharides [40].

Related studies have indicated that *Debaryomyces hansenii* HF1; isolated from larvae of European bass (*Dicentrarchus labrax*) demonstrated high levels of amylase and trypsin; which aided in the digestion of feed [65]. Similar studies on a combination of 3 potential probiotic strains (*Agarivorans albus* F1-UMA, *Vibrio sp.* C21-UMA and *Vibrio sp.* F15-UMA) showed significant increases in growth of abalone over a 210 day period [58]. An average monthly improvement in growth of 9.58% of length and 15.94% in weight was observed in relevant test systems. Probiotic organisms persisted in the gut up to a concentration of 10^6 CFU.g^{-1} and also remained present for 16 to 19 days in juvenile and adult abalone after cessation of feeding with a probiotic supplemented diet. Authors, [40] and [66] also reported that when probiotics were applied to a host, a higher growth rate was observed, as isolated gut bacteria produced enzymes that were able to aid in digestion thus improving the health of abalone.

An inaugural application of probiotics in abalone aquaculture was demonstrated by [23]. They reported that microbes isolated from the gastrointestinal tract of *H. midae* demonstrated an ability to improve digestion, growth and immunity of abalone. From their study it was discovered that *D. hansenii, Cryptococcus sp., V. midae, and Pseudoalteromonas sp.* reside in the intestinal tract of *H. midae* and have the ability to improve the nutritional status of the abalone feed. Further research demonstrated that these probionts were able to breakdown complex proteins and starches, hence making the subsequent assimilation by abalone easier. Studies conducted by [23] indicated that abalones that had been supplemented with probiotics had a survival rate of 62% compared to 25% of untreated abalones; in challenge trials against bacterium *V. anguillarum*. They later formulated a mixture of probiotics using two yeasts and one bacterial strain (SS1, AY1 and SY9) respectively for abalone. The probiont cocktail was added to dry feed to a final concentration of $1{\times}10^7$ cells.g^{-1}. The growth rate of small abalone (20 mm) improved by 8% and large abalone (60 mm) increased by 34%. In addition, increases in intestinal proteolytic and amylolytic activity were observed, in probiotic fed abalone when compared to abalone fed the standard feed devoid of probiotics [30].

Lactic acid bacteria (LAB) from different sources and evaluated potential probiotic effects in abalones *in-vitro, Lactobacillus sp.* strain a3 and *Enterococcus sp.* strain s6, was isolated by [56], and were shown to inhibit the growth of three abalone pathogens *viz., Listonella anguillarum, V. carchariae* and *V. harveyi.* Furthermore these organisms were able to colonize the gut of

Haliotis gigantea thus enhancing the production of volatile short chain fatty acids (VSCFA) such as acetic acid. They later showed that by supplementing commercially available abalone feed with a potential probiotic organism, *Pediococcus sp.* Ab 1, a change in host intestinal flora was observed. In addition, higher levels of alginate lyase activity and VSCFAs were recorded. All of these factors led to a combined impact by enhancing the growth of the abalone, *H. gigantea* [57].

Studies conducted by [51] revealed that within a week of supplementing the feed of *Haliotis discus hannai Ino* with two probiotic organisms, *Shewanella colwelliana* WA64 and *Shewanella olleyana* WA65, increases in cellular and humoral immune response, higher haemocytes, respiratory burst activity, serum lysozyme activity and total levels of protein were observed. It was therefore concluded that both strains may be used as a dietary probiotic supplement to improve innate immunity and disease resistance in abalone.

Studies on feed probiotics for abalone aquaculture show much promise, however the use of water bioremediation bacteria has been neglected. With intensification of abalone culture activities, increased energy costs of pumping sea water and stricter regulation on environmental pollution, the need for such biological agents will become obvious in the near future. A study based on carp was done by [67] where a consortium of three *Bacillus* isolates demonstrated the ability to reduce diseases and improve water quality. Additionally, studies revealed that when a three organism consortium was added to a culture environment, a decrease in the prevalence of pathogenic bacteria was observed. Moreover it was found that nitrate, nitrite and ammonium concentrations were significantly lower as compared to the control treatments and that the applied treatment did not alter the health, growth and oxygen sufficiency of the test systems negatively. The attractive nature of *Bacillus* spp. as biological agents should be considered for application in abalone mariculture.

4. Rationale used for the production of probiotics and biological agents

The use of biological agents in aquaculture has over the years gained momentum. It is thus, imperative that these micro-organisms be commercially produced in order to meet market needs. A comprehensive production process needs to be developed and optimised for each biological agent. This will facilitate the commercial roll-out of probiotic products of this nature, but will be largely dependent on (1) the efficiency of the production process and (2) the ability to produce large quantities of the probiotic in a suitable form with practical shelf stability [69]. Important criteria influencing the commercial use of biological products are cost, efficacy, shelf life and convenience to the end user [70,71]. The cultivation of microorganisms at a large scale is influenced by various factors such as the composition of the media, physical and chemical variables, substrate feed, oxygen availability and many others [72]; each of which have to be optimized to ensure a cost effective production process.

The growth medium that is used to support high productivities in commercial bioprocesses is predominantly formulated with inexpensive nutrient sources [73]. The choice of medium to be used in production is an essential aspect of process development as it influences the

http://dx.doi.org/10.5772/57197

economic competitiveness of the bioprocess technology [74]. Nutrient sources generally play a dominant role in the productivity of the production process since they supply nutrient and growth factors that are directly linked with the formation of biomass and metabolites [75]. It has been suggested that economical and commercially available medium options be investigated in order to reduce production costs [76,77]. The growth medium used can be either a defined or undefined medium. A defined medium has known quantities of all the ingredients that constitute the formulation. An undefined medium contains complex ingredients such corn steep liquor (CSL), which consist of a mixture of chemicals in unknown quantities that vary according to supplier and production batches. The undefined medium option is usually applied in industrial processes based on its low cost [74].

Yeast extract is a commonly used growth medium component, and has been used extensively in many production processes. It is an important nitrogen and nutrient source as it contains an array of amino acids, vitamins and other growth factors required for microbial growth [78-80]. Several studies have indicated that high cell yields and productivities have been obtained with the use of yeast extract in various production processes [81-83]. However, the disadvantage of using this nutrient source is that it results in high-priced production processes due to its associated cost [78]. The use of yeast extract in a production process is therefore regarded as a major technical hurdle that should be overcome in order to successfully minimize production costs [79]. Other nutrient sources that have been used include casamino acids and peptone, which are produced via the enzymatic digestion of meat. The use of these products results in expensive production processes; even though these have been shown to be highly effective nutrient sources. Furthermore, these nutrients sources have negative market acceptance as they are animal by-products [80]. Regulations have also exerted significant pressure on the use of these animal by-products, which have limited their availability. It is therefore imperative that cheaper, safer and readily accessible nutrient sources, capable of supporting production of biological agents, be used in order to ensure that a production process is economically attractive.

CSL has been identified as a lower cost, more effective growth medium that can be used in production, in comparison to conventional nutrient substrates such as yeast extract, peptone and casamino acids [50]. It is produced by immersing corn into dilute sulphur oxide during the starch-manufacturing processes and is a major by-product of the corn-starch processing industry [84]. It has also been shown to be a supplementary source of vitamins and nitrogen to the culture medium [85,86]. The use of CSL has had numerous successes in diverse industrial fermentation processes [76] with high cell yields and productivities being major benefits [87]. Other than the assessment of a suitable nitrogen source, alternative carbohydrate sources also need to be reviewed as they play a dominant role in the productivity of a production process. These nutrient sources are directly linked with energy provision for the formation of biomass and metabolites [75]. Different microbes utilise carbohydrate sources in varying ways. Glucose is a relatively expensive carbohydrate source, and its use in large scale process is limited as a result of high subsequent production costs [88]. When developing efficient bioprocesses, attempts are made to obtain economical and commercially available carbohydrate sources such that the production costs are minimised [74,76,77,89,90]. High test molasses (HTM) is a valuable carbohydrate used commercially due to its local availability and cost competitiveness. It has been applied extensively as an alternative carbohydrate source in various production processes [74,76,77,80,90,91]. HTM, unlike conventional molasses, is a purer product form that enhances mass transfer and reduces impurities in a production process. HTM has been used as a carbohydrate source because it consists of glucose, fructose and sucrose. Inverted HTM is also readily accessible, which contains mainly glucose and fructose in equal proportions with a small amount of residual sucrose. Other than being a carbohydrate source, HTM also provides abundant vitamins and other growth factors required for microbial growth [80,92].

In some instances, microorganisms may require vitamins to be present in the cultivation medium which can be found in the supplemented complex nutrient sources, whereas others can be cultivated in a medium devoid of vitamins [93]. Vitamins are growth factors required by most microorganisms for the production of enzyme cofactors [74,94]. Other than vitamins, microorganisms also require trace elements for their growth. Trace elements form part of enzymes and co-factors and they aid in the catalysis of reactions and maintenance of protein structures [74,95]. Supplementation of exotic trace elements and vitamins can be costly and are therefore avoided if cheaper nutrient sources can satisfy the essential requirements for growth of biological agents in production processes.

Other than an influence on growth, the type of growth medium used in a production process also influences physical parameters such as mass transfer and the formation of foam. Growth media rich in nitrogen sources usually result in increased foam formation [96]. In addition the sparging of gas through the growth medium and agitation at high speeds results in excess foam formation, in oxygen intensive processes. This reduces the efficiency of gaseous exchange at the surface of the culture, as a barrier is formed between the culture and the gases present in the headspace of the vessel [97,98]. Additionally, cells and the culture medium can be lost in the foam phase in the event of vessel overflow. The sensitivity of microorganisms to antifoam toxicity is an important factor that must be considered during the development of production processes; as it can result in a significant decrease of the process performance [98].

Once a suitable fermentation medium has been developed, optimization of physiological growth conditions such as temperature, pH and oxygen sufficiency are imperative, in order to successfully produce biological agents on a large scale. Temperature and pH have been reported to be amongst the most important environmental parameters which control the activities and growth rates of many microorganisms as it governs all the physiological processes. The impact of temperature has been observed at the cellular level, and can either increase or decrease the catalytic activity of pertinent metabolic and digestive enzymes [99-102]. It has been reported that the alteration of growth conditions to an unsuitable range can significantly increase the lag phase of a wide range of micro-organisms, which is highly undesirable when designing an efficient bioprocess strategy [104]. Since temperature affects microbial growth rate, it also affects the growth yield of a culture because the relative energy requirements for cell maintenance increases; when growth rates are reduced [105]. pH homeostasis is another important factor that needs to be considered during the growth of microorganisms [106]. For most microorganisms, there is an increase in growth rate between the minimum and the optimum pH levels and a corresponding decrease in growth rate between the optimum and the maximum pH value [107]. It is well known that pH is important in controlling initiation of growth by microorganisms [108]. The effect of pH on growth include: (1) affecting the production and activity of enzyme systems controlling growth and division, (2) altering the solubility of essential nutrients, (3) modifying the permeability of cells to substances essential for growth, (4) changing the nature of cell surfaces of envelope materials and cell morphology, and (5) modifying the composition of the cultivation medium [108-111]. Oxygen sufficiency is an additional factor to be considered in the design of an optimum bioprocess strategy. In high-cell density cultivations, oxygen limitation can be very challenging, and prevents attainment of high cell titres [112]. The method of oxygenation must be given a high degree of consideration, as excessive rates of agitation and sparging will encourage foam formation and initiate cell sheer. On the contrary, inadequate aeration causes oxygen limitation, and has been reported to be highly detrimental to process productivity, in terms of growth rates and product formation as well as cell viability [96,112]

These factors have an impact on the improved yield and productivity of a process and as a result the overall cost of the production process. In addition they also confer information of the functionality of the probiotic once it enters the host environment [50,90]. Therefore, bioprocesses are designed such that the overall process has increased cell yields, productivities and a lowered cost, which ultimately results in a feasible and economically attractive production process. It is essential that these requirements are met to ensure that biological agents can be affordably adopted for use in abalone aquaculture.

5. Processing of probiotics into market acceptable products

Once the relevant biological agents have been successfully cultivated at a large scale, the resultant fermentation broth needs to be recovered efficiently to be utilized in subsequent processing and formulation steps [113,114]. The downstream process has a major influence on product commercialization as major constraints in most processes are embedded in harvesting and formulation costs [115,116]. This includes key aspects such as maximising recovery and preservation of viability, which are essential, in terms of applying an effective biological agent, especially in aquatic systems [117]. In addition, it is also vital to ensure that the final product to be administered to the host aligns with end user requirements such as stability, consistency, easy application, efficacy and affordability [115,118,119]. As a consequence; robust cost-effective choices of process steps and ingredients, dictated by the end product characteristics, are necessary to improve the commercial success of newly developed biological agents [115].

The main objective for downstream processing is to minimise the number of unit operations involved in the process, thus reducing overall process and validation costs, while also simplifying ease and economy of process automation [115]. An additional consideration is the final anticipated form of the end product which has implications on the choice of process options while meeting customer expectations (Figure 2) [120]. The downstream process unit operations, completes the process chain; from the upstream fermentation to the end product. It is therefore considered to be an extremely important prerequisite for commercialization of probiotic technologies. Regrettably, published literature regarding downstream processing and formulation for commercially available products is very limited [115,119].

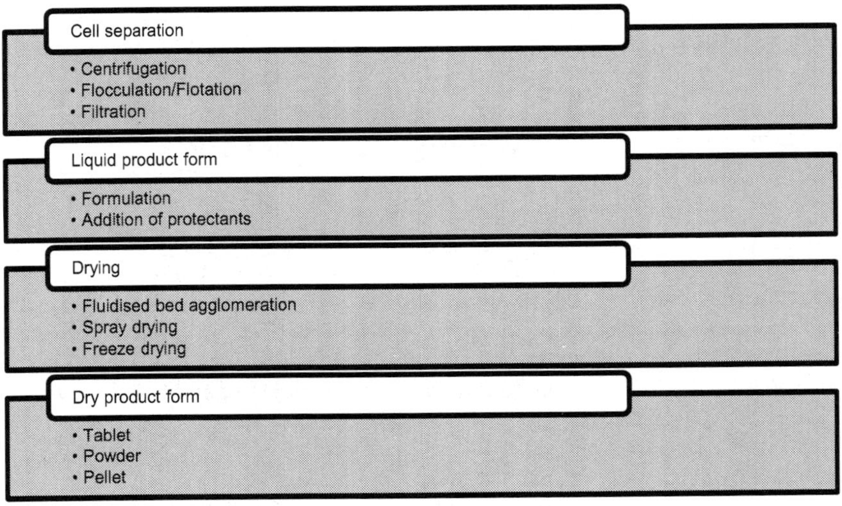

Figure 2. Schematic illustration of potential downstream process unit options [121].

The harvesting efficacy of a unit operation governs the marketability of a product, as it affects the potency and aids in any potential further processing during formulation and product development. The goal of the recovery process is to produce a product of acceptable quality, in compliance with any regulatory and safety requirements, at an acceptable cost [120]. Process options for cell harvesting from fermentation broth include microfiltration, sedimentation, flocculation and ultrafiltration (Figure 2) [116,122]. Flocculation and flotation using surface action or electrical charge have been reported to be inefficient in the separation of bacterial cells [120]. Although there have been some positive reports for harvesting using ultra-filtration [116], the most widely used process remains centrifugation [123-125]. Among all options, centrifugation appears to be the most viable alternative for cell harvesting resulting in recoveries of ~ 99% [115,126]. Usually, product intermediates are anticipated to be a high cell concentration paste containing the biological agent. Tube centrifugation has been considered to be a useful process that can yield a lower moisture paste thus minimizing the energy required in later stage drying steps if necessary [127,128].

Subsequent to cell separation, formulation generates a crucial link between production and application of probiotics (Figure 2). This crucial step dictates process-ability, economy, shelf life, and efficacy as well as ease of application and provision of a product form that commands customer appeal. Intelligent formulations will allow innovation in application techniques using unique combinations of active ingredients, adjuvants or inerts and the end product should seamlessly integrate into standard food production and farming activities [115,129]. Formulations of biological agents can be broadly classified into either dry solids in various forms, or liquid suspensions or emulsions [130]. The inclusion of additives that enhance process-ability eco-friendliness and customer acceptance of the final product are also important considerations [115]. Any potential impact to the host, environment and even end product consumer, must be thoroughly investigated prior to application [131].

In the case of a probiotic product, the formulation needs to encompass ingredients that aid viability and growth of the cells in its intended application. Sugars and proteins are normally the key nutrients that support the stability of cell preparation. It also further provides a protective layer for the cells, preventing death and assists in the recovery of injured cells during processing [115,132,133]. The addition of nutrients was also shown to improve storage of a *Pseudomonas fluorescens* F113 strain [134] and a *Bacillus megaterium* [135] for use in biocontrol applications. Appropriate formulations can facilitate easier processing and influence the stability and appeal of an end product in large scale production [70].

Processing options for abalone biological agents, will include, both dry and liquid product forms (Figure 2). Due to the intended use of the selected isolates as a living cell preparation, product options with a high stability are considered to be most appealing. The application of fresh cells that need to be routinely produced is not attractive as there is significant risk in ensuring consistent inclusion into the abalone feed [135]. This complicates the processing segment of the technology to a large extent, as innovative ways of ensuring and maintaining cell viabilities are required. Potential options for commercial processes to stabilize these probiotic products include refrigeration, freezing, freeze drying, spray drying and low temperature fluidised bed agglomeration. Refrigerated and frozen cultures occupy large storage volumes and demand higher storage and shipment costs in contrast to dry cultures which are an economic and practical alternative; however, some microorganisms are highly vulnerable to death when any form of drying is carried out [117,136,137]. An alternate approach is to concentrate the product into a convenient dosage quantity and form that reduces the bulk logistics burden for products that are not amenable to drying. Low temperature drying processes such as freeze drying are suitable for higher value, heat labile bio-products, but is costly, time consuming and discontinuous for bulk production compared with moderate temperature drying processes [131,138]. Spray drying processes are widely used for large scale drying of products; however, higher drying temperatures decrease the viability of microbes faster than lower drying temperatures [131,139]. Spray drying requires high temperatures to facilitate water evaporation, which can cause irreversible changes to structural and functional integrity of the intended biological product and reduce viability and activity of the organisms itself [140,141]. Spray drying also has a high energy demand requiring 2500 to 10 000 $J.g^{-1}$ of evaporated water and is therefore not the most attractive process option for drying of abalone probiotics [122].

There are several reports on the use of agglomeration as a commercially viable process option for moderate temperature drying of biological material, mainly due to excellent mass and heat transfer characteristics [133,139,142]. During agglomeration, a mixture is atomised to form droplets at lower temperature (typically 30-40°C) which results in coating of the probiotic cells on the surface of suitable carrier particles. Probiotic cell cultures are subjected to evaporative cooling during the warming up and constant-rate drying periods and therefore have a substantially lower temperature than the drying air, resulting in increased viability [139.]. Advantages of fluidised bed drying over freeze and spray drying include lower investment and maintenance costs, ease of large scale continuous production, rapid exchange of heat, minimising heat damage, rapid mixing providing near isothermal conditions and uniform end product [122,139,143]. Due to these reasons fluidized bed drying has become an accepted method for large scale production of heat labile biological materials, however, viability losses have still been reported [142].

In addition to production and formulation of user-friendly product through a downstream process, the stability and consistency of product intermediates and the end product itself are crucial requirements for successful commercialization [144]. A loss of bioactivity in a product, that is intended to be applied in a viable state, will definitely incur a great deal of process complications and as a result impart a direct increase in production cost [141]. In a typical production process, the lag time between process operations can vary due to process integration and scheduling during manufacture. Thus storage conditions and the addition of specific stabilizers may be required to prevent vegetative growth or the appearance of contamination in the probiotic product or its relevant intermediates [145].

The problems of stability during processing, storage and after application have stalled development of biologically based products [146]. Accelerated aging studies based on the methodology of death rate plots at different temperatures to generate thermal resistance curves have been shown to be a useful technique for predicting stability [147]. Temperature dependant half-life plots can be generated to predict stability of the probiotic product intermediates as well as the end product. This approach has however only been used to a limited extent [121,123,135].

After addressing the considerations of the actual production process, success of the technology is still not a certainty. It is imperative, that in order to realise the success of using this new technology, the probiotic product must be supplied as a live cell preparation, and must be able to survive not only the feed production process, but also maintain viability in the digestive tract of the host [44,47]. Many probiotics have been successfully applied to land-based animal production practices, however, the aquaculture industry are faced with further limitations as a result of continuous water exposure [148]. The method of incorporation selected must overcome challenges faced in feed production and the mariculture system itself; such as major losses of viability, in order to achieve the desired effect of the probiotic technology.

There have been various methods applied to successfully administer viable probiotics to a host in aquaculture environments. These include mixing, soaking, spraying, vacuum infusion, extrusion and bathing [148]. Incorporation of the probiotic into the feed is almost always the method of choice, except when bioremediation agents are added directly to the water. Mixing is the most commonly used method and involves the incorporation of the probiotic into the dry ingredients of the feed during the feed production process. Many researchers [56,150,151] have successfully used this method; however, probiotics that are susceptible to excessive heating and drying during the feed production process do not show high rates of survival

[117]. The soaking method uses preformed feed pellets, which are soaked in a saline broth containing the probiotic organism at a desired concentration [57,149]. Soaked pellets are then dried and stored appropriately for further use. A modified method of soaking, whereby actual fronds of macroalgal species, *Macrocystis integrifolia* were soaked in preconditioned tanks containing bacterial cells were used by [58]. Upon aeration of these tanks, the bacteria were allowed to colonize the surface of the fronds, and were thereafter fed to test abalone. In other studies, the spraying of feeds with probiotic cells was also carried out [153]. In addition, [154] described methods of spraying dried feeds with cells that were placed onto plastic trays. Lastly, the bathing option of probiotics involving the application of living cells directly to the rearing water has been used [155]. All potential mechanisms of probiotic inclusion into the feed must be suitably ratified in order to maximize the potential of the technology. The method selected, should have the ability to integrate easily into existing feed production process, and should in no way negatively impact on the host or the rearing environment. The journey taken to produce a commercially viable probiotic product is by no means forthright. It encompasses innovative process design, effective cell production and formulation technologies, as well as successful maintenance of cell viability and stability. Once all the identified challenges have been effectively overcome, the uptake of this technology and the associated boom in abalone export by means of aquaculture will be inevitable.

6. Considerations for application of biological agents in abalone aquaculture

Over the past two decades, the applicability of probiotics as solutions to various aquaculture related challenges have been widely reported. However, it is still imperative to consider the safety issues associated with the use of these probiotic products [29,52,156]. Safety is the state of being certain that a biological agent used will not have undesirable effects under defined conditions. The production system in which the cell cultivations are conducted must also maintain high levels of sterility to easily facilitate a monoseptic culture, as well as reduce any potential contamination by common food pathogenic bacteria [68].

Once the selected culture has been accurately identified and deposited into a culture repository, extensive literature searches and relevant scientific experimentation must be carried out in order to obtain information on the biological agent of interest. As the number of isolated probiotic species increase, it is important not to assume biosafety levels and characteristics of each probiotic strain. Furthermore, it is recommended that the exact mode of action of the probiotic organisms be elucidated, in order to achieve the desirable effect when applied to the host system. It has been suggested that prior to incorporation of these organisms in abalone aquaculture, it is important to carefully assess the probiotics for pathogenicity, infectivity, toxicity and their resultant metabolite production for quality assurance [157]. These critical factors have sometimes been overlooked, and have consequently led to the failure of probiotic technologies in some instances [41]. In many case studies, the use of LAB as probiotics have been rendered safe, however, in recent times there have been reports of disease-causing members belonging to *Lactococcus*, *Vagococcus* or *Carnobacterium* families [158]. Additionally, strain testing of potential probiotics should encompass the robustness of the product against process fluctuations under farm conditions and confirmation of non-transmission of drug resistance genes or virulence plasmids [159]. Another barrier preventing the worldwide adoption of this technology, relates to the absence of efficacy data, which as a result casts a shadow of doubt over the technology, thereby hindering its uptake by the aqua-culturists.

7. The impact and benefits of the application of biological agents in abalone aquaculture

Most aquaculture industries are leaning towards the use of probiotic technology as a solution to many of the challenges faced by the industry. The basis for the inclusion of probiotics into the farming environments include higher survival rate of juvenile and adult abalone, improved feed uptake and conversion ratios resulting in faster growth rates, improved resistance to disease and reduced contribution to water pollution [47]. Using probiotics is more environmentally friendly because the effluent water is cleaner and there are significant improvements in the gut flora thus enhancing the overall immune response of the host and an increase in food assimilation [160]. However other factors such as temperature, enzyme levels, water quality and genetic resistance may have an effect on the success of the technology in the farming facility [29].

Thus far, the uptake of the technology is slow-moving. This is due to the fact that farmers expect the probiotic technology to operate using the same basis as antibiotic treatment technology, in that they require and anticipate fast rapid results [160]. However changes in the microbial ecosystems present in the environment is a gradual one; and requires the continuous addition of beneficial microorganisms to compound the desired effect [158]. In addition, ineffective and costly probiotic products previously offered in aquaculture has negatively tainted the impact of this technology. Some products include *Clostridium spp., Pseudomonas putida* and other potential human pathogens, and others consist of cell densities that are too low to deliver any sort of benefit to the host [160].

The commercial aquaculture sector will make a notable difference in terms creating jobs and economic development in most developing countries by embracing this activity. To date, South Africa has validated itself to be a key player in the abalone mariculture arena. With support from government, this industry could experience a further boom, and as a result, assist in reducing the high levels of unemployment that exists [27], particularly in coastal areas that can effectively participate in aquaculture practices. Abalone industries not only include direct employment at the farm level, it also indirectly supports interlinked businesses such as the seaweed and abalone processing industries [9]. The challenge is to ensure long term sustainable growth of the abalone mariculture industry. The use of appropriate and safe biological agents in abalone mariculture has excellent potential to meet the new challenges of this important industry.

http://dx.doi.org/10.5772/57197

Author details

Ghaneshree Moodley[1,2*], Lethabo Mashigo[1], Rajesh Lalloo[1] and Suren Singh[2]

*Address all correspondence to: GMoodley@csir.co.za

1 CSIR Biosciences, Pretoria, South Africa

2 Department of Biotechnology & Food, Faculty of Applied Sciences, Durban University of Technology, Durban, South Africa

References

[1] Marques A., Dinh T., Ioakeimidis C., Huys G., Swings J., Verstraete W., Dhont P., Sorgeloos, P. and Bossier, P. Effects Of Bacteria On *Artemia* Cultured In Different Gnotobiotic Environments. Journal of Applied and Environmental Microbiology 2005; 71(8) 4307-4317.

[2] Subasinghe RP. Epidemiological Approach to Aquatic Animal Health Management: Opportunities and Challenges for Developing Countries to Increase Aquatic Production through Aquaculture. Preventative Veterinary Medicine 2005; 67 117-124.

[3] FAO. The State Of World Fisheries and Aquaculture. Rome. 2012; 209pp.

[4] Britz PJ. A study on the Status of Aquaculture Production and Trade in South Africa. Volume 2: Growth Potential of the South African Aquaculture Industry and Recommendations for Sector Development. A report for the Department of trade and Industry produced by Enviro-fish Africa (Pty.) Ltd. 2007; 24p.

[5] Olafsen JA. Interactions between Fish Larvae and Bacteria in Marine Aquaculture. Aquaculture 2001; 200(1-2) 223-247.

[6] Gordon HR and Cook PA. World Abalone Fisheries and Aquaculture Update: Supply and Market Dynamics. Journal of Shellfish Research 2004; 23 935-939.

[7] Stanford J. Aqua Culture Ambition: South African Aquaculture Industry Is Heading For Export Boom. Engineering News 16-17 August. 2004; 6-12.

[8] Reddy-Lopata K., Auerswald L. and Cook P. Ammonia Toxicity and Its Effect on the Growth of the South African Abalone *Haliotis Midae Linnaeus*. Aquaculture 2006; 261(2) 678-687.

[9] Troell M., Robertson-Andersson D., Anderson R.J., Bolton J.J., Maneveldt G., Halling C and Probyn T. Abalone Faming In South Africa: An Overview With Perspectives On Kelp Resources, Abalone Feed, Potential For On-Farm Seaweed Production And Socio- Economic Importance. Aquaculture 2006; 257 266-281.

[10] Aquaculture Association of South Africa. AASA. www.aasa-aqua.co.za (accessed on 10 May 2013).

[11] Raemaekers S and Britz PJ. Profile of the Illegal Abalone Fishery (*Haliotis Midae*) In the Eastern Cape Province, South Africa: Organised Pillage and Management Failure. Fisheries Research 2009; 97(3) 183-195.

[12] Degnan SM, Imron Geiger DL and Degnan BM. Evolution in Temperate and Tropical Seas: Disparate Patterns in Southern Hemisphere Abalone (Mollusca: Vetigastropoda: Haliotidae). Molecular Phylogenetics and Evolution 2006; 41(1) 249-256.

[13] Geiger DL. A Total Evidence Cladistic Analysis Of The Family Haliotidae (Gastropoda: Vetigastropoda). Ph.D. Thesis, University Of Southern California, Los Angeles. 1999; 423.

[14] Geiger DL and Poppe GT and Groh K. Haliotidae. In: (Eds), *Conchological Iconography*, Conchbooks, Hackenheim 2000; 135 83.

[15] Western Cape aquaculture development initiative; WCADI (2012) Draft Western Cape aquaculture market analysis and development programme/strategy. Draft Document. 1-167.

[16] Fishtech Inc: www.fishtech.com (accessed on 10 May 2013).

[17] Oakes FR and Ponte RD. The Abalone Market: Opportunities for Cultured Abalone. Aquaculture 1996; 140(1-2) 187-195.

[18] Gordon HR and Cook PA. World Abalone Supply, Markets and Pricing: Historical, Current and Future. Journal of Shellfish Research 2001; 29(3) 569-571.

[19] Flores-Aguilar RA., Gutierrez A., Ellwanger AS and Searcy-Bernal R. Development and Current Status of Abalone Aquaculture in Chile. Journal of Shellfish Research 2007 26(3) 705-711.

[20] PAUA Industry Council Limited. www.paua.org.nz (Accessed 23 November 2009.).

[21] Sales J and Britz PJ. Research on Abalone Cultivation in South Africa. Aquaculture Research 2001; 32(11) 863-874.

[22] Evans BS., Sweijd NA., Bowie RCK., Cook PA and Elliott NG. Population Genetic Structure of the Perlemoen *Haliotis Midae* in South Africa: Evidence of Range Expansion and Founder Events. Marine Ecology Progress Series 2004; 270 163-172.

[23] Macey BM and Coyne VE. Improved Growth Rate and Disease Resistance in Farmed *Haliotis midae* through Probiotics Treatment. Marine Biotechnology 2005; 245(1-4) 249-261.

[24] ten Doeschate K and Coyne VE. Improved Growth Rate in Farmed *Haliotis midae* through Probiotic Treatment. Aquaculture 2008; 284(1-4) 174-179.

[25] Aquaculture Annual Report: South Africa (2011) ISBN: 973-1-868-71-355-4. 1-40

[26] Andersson D, Troell M, Halling C., Anderson R., Maneveldt G. and Bolton JJ. Abalone (*Haliotis midae*) Farming and Seaweed Harvesting in South Africa: Industry Interdependencies and Socio-Economic Importance. Slideshow Presented at the Phycological Society of Southern Africa Meeting in Maputo, Mozambique, 2006.

[27] Shipton T and Britz PJ. A Study on the Status of Aquaculture Production and Trade in South Africa. Volume 1: Industry Status and Diagnostic Report. A Report for the Department of Trade and Industry produced by Enviro-Fish Africa (Pty.) Ltd. 2007 90p.

[28] Moriarty DJW. 1999. Disease Control In Shrimp Aquaculture with Probiotic Bacteria, Microbial Interactions in Aquaculture, In: Bell CR, Brylinsky M (Eds) Proceedings of the 8th International Symposium on Microbial Ecology, Canada.

[29] Balcázar JL., Blas ID., Ruiz ZI., Cunningham D., Vendrell D., & Múzquiz JL. The Role of Probiotics in Aquaculture. Veterinary Microbiology 2006; 114(3-4) 173-186.

[30] Macey MB and Coyne VE. Colonization of the Gastrointestinal Tract of the Farmed South African Abalone *Haliotis midae* by the Probionts *Vibrio midae* SY9, *Cryptococcus Sp.* SS1 and *Debaryomyces hansenii* AY1. Marine Biotechnology 2006; 8 246-259.

[31] Jeney Z and Jeney G. Recent Achievements in Studies of Common Carp (*Cyprinus carpio* L.). Aquaculture 1995; 129(1-4) 397-420.

[32] Nicolas JL., Basuyaux O and Mazurie J. The bault, a *Vibrio carchariae,* a Pathogen of the Abalone, *Haliotis tuberculata*. Diseases of Aquatic Organisms. 2002; 50 35-43.

[33] Bower SM. Update on Emerging Abalone Diseases and Techniques for Health Assessment. Journal of Shellfish Research. 2003; 22 805-810.

[34] Gram L, Løvald T., Nielsen J., Melchiorsen J and Spanggaard B. In-vitro Antagonism of Probiont *Pseudomonas Fluorescens* against *Aeromonas Salmonicida* Does Not Confer Protection of Salmon Against Furunculosis. Aquaculture 2001; 199(1-2) 1-11.

[35] Nomoto K. Prevention of Infections by Probiotics. Journal of Bioscience and Bioengineering 2005; 100(6) 583-592.

[36] Gomez GRD, Balcazar JL and Shen MA. Probiotics as Control Agents in Aquaculture. Journal of Ocean University of China 2007; 6(1) 76-79.

[37] Moriarty, DJW. Control of Luminous *Vibrio* Species in Aquaculture Ponds. Aquaculture 1998; 164(1-4) 351-358.

[38] Ronson P.J and Medina R. *Probioticos en al Acuicultura, Ciencia y Mar Notas.* 2002.

[39] Naidoo K, Maneveldt G, Ruck K and Bolton JJ. A Comparison of Various Seaweed-Based Diets and Formulated Feed on Growth Rate of Abalone in a Land-Based Aquaculture System. Journal of Applied Phycology 2006; 18(3-5) 437-443.

[40] Erasmus JH, Cook PA and Coyne VE. The Role of Bacteria in the Digestion of Seaweed by the Abalone *Haliotis midae*. Aquaculture 1997; 155 377-386.

[41] Wang Y, Li J, and Lin J. Probiotics in Aquaculture: Challenges and Outlook. Aquaculture. 2008; 281(1-2) 1-4.

[42] Bondad-Reantuso MG, Subasinghe RP and Arthur JR. Disease and Health Management in Asian Aquaculture. Veterinary Parasitology 200; 132(3-4) 249-272.

[43] Tinh NTN, Dierckens K, Sorgeloos P. and Bossier P. A Review of the Functionality of Probiotics in the Larviculture Food Chain. Marine Biotechnology. 2007; 10 1-12.

[44] Gatesoupe FJ. The Use of Probiotics in Aquaculture. Aquaculture 1999; 180(1-2) 147-165.

[45] Rengpipat S, Rukpratanporn S, Piyatiratitivorakul S and Menasaveta P. Immunity Enhancement in Black Tiger Shrimp (*Penaeus monodon*) by a Probiont Bacterium (*Bacillus* S11). Aquaculture 2000; 191(4) 271-288.

[46] Robertson PAW, O'Dowd C, Burrells C, Williams P and Austin B. Use of *Carnobacterium* sp. As a Probiotic for Atlantic Salmon *Salmo salar* L. and rainbow Trout *Oncorhynchus mykiss,* Walbaum. Aquaculture 2000; 185(3-4) 235-243.

[47] Vershuere L, Rombant G, Sorgeloos P. and Verstraete W. Probiotic Bacteria as Biological Agents in Aquaculture. Microbiology and Molecular Biology Reviews 2000; 64(4) 655-671.

[48] Chythanya R, Karunasagar I and Karunasagar I. Inhibition Of Shrimp Pathogenic *Vibrios* By a Marine *Pseudomonas* I-2 Strain. Aquaculture 2002; 208(1-2) 1-10.

[49] Bautista-Teruel MN, Fermin AC and Koshio SS. Diet Development and Evaluation for Juvenile Abalone, *Haliotis asinina*: Animal and Plant Protein Sources. Aquaculture 2003; 219(1-4) 645-653.

[50] Lalloo R, Maharajh D, Gorgens J and Gardiner N. Functionality of a *Bacillus Cereus* Biological Agent in Response to Physiological Variables Encountered in Aquaculture. Applied Microbiology and Biotechnology 2008; 79 112- 113.

[51] Jiang HF, Lui XL. Chang YQ, Lui MT and Wang GX. Effects of Dietary Supplementation of Probiotic *Shewanella colwelliana* WA64, *Shewenella olleyana* WA65 on the Innate Immunity and Disease Resistance of Abalone, *Haliotis discus hannai Ino.* Fish and Shellfish Immunology 2013; 1-4.

[52] Kesarcodi-Watson A, Kaspar H, Lategan MJ and Gibson L. Probiotics in Aquaculture: The Need, Principles and Mechanisms of Action and Screening Processes. Aquaculture 2008; 274(1) 1-14.

[53] Sawabe T, Sugimura I, Ohtsuka M, Nakano K, Tajima K, Ezura Y and Christen R. *Vibrio halioticoli* sp. nov., A Non-Motile Alginolytic Marine Bacterium Isolated From

The Gut of the Abalone *Haliotis discus hannai*. International Journal of Systematic Bacteriology 1998; 48(2) 573-580.

[54] Tanaka R, Sugimura I, Sawabe T, Yoshimizu M and Ezura Y. Gut Microflora of Abalone *Haliotis discus hannai* in Culture Changes Coincident with a Change in Diet. Fish Science 2003; 69(3) 951-958.

[55] Sawabe T, Inoue S, Fukui Y, Yoshie K, Nishihara Y and Miura H. Mass Mortality of Japanese Abalone *Haliotis discus hannai* Caused by *Vibrio harveyi* Infection. Microbes Environment 2007; 22(3) 300-308.

[56] Iehata S, Inagaki T, Okunishi S, Nakano M, Tanaka R and Maeda H. Colonization and Probiotic Effects of Lactic Acid Bacteria in the Gut of the Abalone *Halotis gigantea*. Fish Science 2009; 75 1285-1293.

[57] Iehata S, Ingaka T, Okunishi S, Nakano M, Tanaka R and Maeda H. Improved Gut Environment ff Abalone *Haliotis gigantea* Through *Pediococcus* sp. Ab1 treatment. Aquaculture 2010; 245 249-261.

[58] Silva-Aicares FR, Carvajal PO, Mejías CA and Riquelme CE. Use of Macroalgae Supplemented with Probiotics in the *Haliotis rufescens* Culture in Northern Chile. Aquaculture Research 2011; 42 953-961.

[59] Hong HA, Duc LH and Cutting SM. The use of bacterial spore formers as probiotics. Federation of European Microbiological Societies Microbiology Review 2005; 29(4) 813-835.

[60] Lalloo R, Moonsamy G, Ramchuran S, Görgens and Gardiner N. Competitive Exclusion as a Mode of Action of a Novel *Bacillus cereus* Aquaculture Biological Agent. Letters in Applied Microbiology 2010; 50(6) 563-567.

[61] Pandiyan, P., Deivasigamani, B., Thirunavukkarasu, R., George, E.G.J., Subaramaniyan, K., Manikkam, S and Sadayappan, B. Probiotics in aquaculture. Drug Invention Today 2003; 5(1) 55-59.

[62] Ouwehand AC, Kirjavainen PV, Grönlund MM, Isolauri E and Salminen S. Adhesion of Probiotic Microorganisms to Intestinal Mucus. International Dairy Journal 1999; 9(9) 623-630.

[63] Irianto A and Austin B. 2002. Probionts in Aquaculture. Review Journal of Fish Diseases 2002; 25(11) 633-642.

[64] Ghosh S, Sinha A and Sahu C. Effect of Probiotic on Reproductive Performance in Female Livebearing Ornamental fish. Aquaculture Research 2007; 38 518-526.

[65] Simpson BJA. An Investigation of the Diet Management Strategies for the Culture of the South African Abalone, *Haliotis midae*. MSc Thesis. University of Cape Town; 1994.

[66] Tovar D, Zambonino J, Cahu, Gatesoupe FJ, Vázquez-Juárez R and Lèsel R. Effect of Live Yeast Incorporation in Compound Diet on Digestive Enzyme Activity in Sea Bass (*Dicentrachus labrax*) Larvae. Aquaculture 2002; 204(1-2) 113-123.

[67] El-Shanshoury AR, Mona MH, Shoukr FA and El-Bossery, AM. The Enumeration and Characterization of Bacteria and Fungi Associated with Marine Wood-Boring Isopods, and the Ability of these Microorganisms to Digest Cellulose and Wood. Marine Biology 1994; 119 321-326.

[68] Lalloo R, Ramchuran S, Ramduth D, Görgens J and Gardiner N. Isolation and Selection of *Bacillus* Spp. as Potential Biological Agents for Enhancement of Water Quality in Culture of Ornamental Fish. Journal of Applied Microbiology 2007; 103(5) 1471-1479.

[69] Patel VJ, Tendulkar SR and Chattoo BB. Bioprocess Development for the Production of an Antifungal Molecule by *Bacillus licheniformis* BC98. Journal of Biosciences and Bioengineering 2008; 98(4) 231-235.

[70] Amer GA and Utkhede RS. Development of Formulations of Biological Agent for Management of Root Rot of Lettuce and Cucumber. Canadian Journal of Microbiology 2000; 46(9) 809-816.

[71] Keller K, Friedmann T and Boxman A. The Bioseparation Needs for Tomorrow. Trends in Biotechnology 2001; 19(11) 438-441.

[72] Maharajh D, Roth R, Lalloo R, Simpson C, Mitra R, Gorgens J and Ramchuran S. Multi-Copy Expression And Fed Batch Production Of *Rhodotorula araucariae* Epoxide Hydrolase in *Yarrowia lipolytica*. Applied Microbiology and Biotechnology 2008; 79 235-244.

[73] Zhang J and Greasham R. Chemically Defined Media for Commercial Fermentations. Applied Microbial Biotechnology 1999; 51 407-421.

[74] Prescott LM, Harley JP and Klein DA. Microbiology. New York: Mc Graw Hill 2005.

[75] Lopez JLC, Perez JAS, Sevilla JMF, Fernandez FGA, Grima EM and Chisti Y. Production of Lovastatin by *Aspergillus terreus*: Effects of the C:N Ratio and the Principal Nutrients on Growth and Metabolite Production. Enzyme and Microbial Technology 2003; 33(2-3) 270-277.

[76] Lawford HG and Rousseau JD. Corn Steep Liquor as a Cost Effective Nutrition Adjunct in High-Performance *Zymomonas* Ethanol Fermentations. Applied Biochemistry and Biotechnology 1997; 63-65 287-304.

[77] Kona RP, Qureshi N and Pai JS. Production of glucose oxidase using Aspergillus niger and corn steep liquor. Bioresource technology 2001; 78(2) 123-126.

[78] Payot T, Chemaly Z and Fick M. Lactic Acid Production by *Bacillus coagulans* – Kinetic Studies and Optimization of Culture Medium for Batch and Continuous Fermentations. Enzyme and Microbial Technology 1998; 24(3-4) 191-199.

[79] Salgado JM, Rodriguez N, Cortes S and Dominguez JM. Development of Cost-Effective Media to Increase the Economic Potential for Larger-Scale Bioproduction of Natural Food Additives by *Lactobacillus rhamnosus, Debaryomyces hansenii* and *Aspergillus niger*. Journal of Agriculture and Food Chemistry 2009; 57(21) 10414-10428.

[80] Lalloo R, Maharajh D, Görgens J and Gardiner N. High-Density Spore Production of a *Bacillus cereus* Aquaculture Biological Agent by Nutrient Supplementation. Applied Microbiology and Biotechnology 2009; 83 59-66.

[81] Nohata Y and Kurane R. Complete Defined Medium for Large-Scale Production of Polysaccharide Bioabsorbent from *Alcaligenes latus* B-16. Journal of Fermentation and Bioengineering 1997; 83(1) 116-117.

[82] Vuolanto A, von Weymarn N, Kerovuo J, Ojamo H and Leisola M. Phytase Production by High Cell Density Culture or Recombinant *Bacillus subtillus*. Biotechnology Letters 2001; 23 761-766.

[83] Prabakaran G, Balaraman K, Hoti SL and Manonmani AM. A Cost-Effective Medium for the Large-Scale Production of *Bacillus sphaericus* H5a5b (VCRC B42) for Mosquito Control. Biological Control 2007; 41(3) 379-383.

[84] Niwa T, Doi U, Kato Y and Osawa T. Antioxidative Properties of Phenolic Antioxidants Isolated From Corn Steep Liquor. Journal of Agricultural and Food Chemistry 2001; 49(1) 177-182.

[85] Amartey S. and Jeffries TW. Comparison of Corn Steep Liquor with Other Nutrients in the Fermentation of D-Xylose by *Pichia stipitis* CBS 6054. Biotechnology Letters 1994; 16(2) 211-214.

[86] Silveira MM, Wisbeck E, Hoch I and Jonas R. Production Of Glucose-Fructose Oxidoreductase and Ethanol by *Zymomonas mobilis* ATCC 29191 in Medium Containing Corn Steep Liquor as a Source of Vitamins. Applied Microbial Biotechnology 2001; 55 442-445.

[87] Srivastava RAK, Baruah JN. Culture conditions for production of thermostable amylase by *Bacillus stearothermophilus*. Applied Environmental Microbiology 1986; 52(1) 179-184.

[88] Yahiro K, Shibata S, Jia SR, Park YS and Okabe M. Efficient Itaconic Acid Production from Raw Corn Starch. Journal of Fermentation Bioengineering 1997; 84(4) 375-377.

[89] Lee SY and Chang HN. Production of Poly(hydroxyalkanoic acid). Advances in Biochemical Engineering 1995; 52 27-58.

[90] Burkert JFM, Maugeri F and Rodrigues MI. Optimization of Extracellular Lipase Production by *Geotrichum sp*. using Factorial Design. Bioresource Technology 2004; 91(1) 77-84.

[91] Gouda MK, Swellam AE, Omar SH. Production of PHB by a *Bacillus megaterium* Strain Using Sugarcane Molasses and Corn Steep Liquor as Sole Carbon and Nitrogen Sources. Microbiology Research 2001; 156(3) 201–207.

[92] Malathi S. and Chakraborty R. Production of Alkaline Protease by a New *Aspergillus flavus* Isolate Under Solid-Substrate Fermentation Conditions for Use as a Depilation Agent. Applied Environmental Microbiology 1991; 57(3) 712-716.

[93] Kask S, Laht TM, Pall T. and Paalme T. A Study on Growth Characteristics and Nutrient Consumption of *Lactobacillus plantarum* In A-Stat Culture. Antonie van Leeuwenhoek 1999; 75 309-320.

[94] Chan ECS. Microbial Nutrition and Basic Metabolism: in the Handbook of Water and Wastewater Microbiology. London UK: Academic Press. 2003.

[95] Burrows W. The Nutritional Requirements of Bacteria, Chicago: The University of Chicago Press. 1936.

[96] Morao A, Maia CI, Fonseca MMR, Vasconcelos JMT and Alves SS. Effect of Antifoam Addition on Gas-Liquid Mass Transfer in Stirred Fermenters. Bioprocess Engineering 1999; 20(2) 165-172.

[97] Huang TK, Wang PM. and Wu WT. Cultivation of *Bacillus thuringiensis* in an Airlift Reactor with Wire Mesh Draft Tubes. Biochemical Engineering Journal 2001; 7(1) 35-39.

[98] Holmes W, Smith R. and Bill R. Microbial Cell Factories, Aston Academy of Life Sciences, Birmingham, UK, Aston University, Aston Triangle. 2006.

[99] Ratkowsky DA, Olley J, McMeekin TA and Ball A. Relationship between Temperature and Growth Rate of Bacterial Cultures. Biotechnology 1982; 149(1) 1-5.

[100] Mayo AW and Noike T. Effects of Temperature and pH on the Growth of Heterotrophic Bacteria in Waste Stabilisation Ponds. Water Research 1995; 30(2) 447-455.

[101] Betts GD, Linton P. and Betteridge RJ. Synergistic Effect of Sodium Chloride, Temperature and pH on Growth of a Cocktail of Spoilage Yeast: A Research Note. Food Microbiology 2000; 17(1) 47-52.

[102] Kurita O and Yamazaki E. Growth under Alkaline Conditions of Salt Tolerant Yeast *Debaryomyces hansenii* IFO 10939. Current Microbiology 2002; 45 277-280.

[103] Bonaiti C, Leclercq-Perlat MN, Latrille E and Corrieu G. Deacidification by *Debaryomyces hansenii* of Smear Soft Cheeses Ripened under Uncontrolled Conditions: Relative Humidity and Temperature Influences. Dairy Science 2004; 87(11) 3976-3988.

[104] Xiang G, Li J, Duan J, Shao F, Xu J, Fu, S, and Gong H. Acceleration Effect of Amino Acid Supplementation on Glycerol Assimilation by *Escherichia Coli* in Minimal Medium. Biotechnology Letters 2013; DOI 10.1007/s10529-013-1232-4.

[105] Knoblauch C. and Jorgensen BB. Effect of Temperature on Sulphate Reduction, Growth Rate and Growth Yield in Five Psychrophilic Sulphate Reducing Bacteria from Arctic Sediments. *Environmental Microbiology* 1999; 1 457-567.

[106] Hutkins RW. and Nannen NL. pH Homeostasis in Lactic Acid Bacteria. Journal Series Nebraska Agricultural Experiment Station 1992; 76 2354-2365.

[107] Todar K. Online Textbook of Bacteriology. University of Wisconsin-Madison Department of Bacteriology. 2007.

[108] Lichstein HC. Symposium on Initiation of Bacterial Growth III. Physiological Aspects of Growth Initiation. Bacteriology Reviews 1959; 23 261-266.

[109] Cochrane VW. Physiology of Fungi. John Wiley and Sons, Inc. New York. 1958.

[110] Mitchell P. Physical Factors Affecting the Growth and Death in Bacterial Physiology Werkman CH and Wilson PW. New York: Academic Press, Inc. 1951.

[111] Pirt SJ. Principles of Microbe and Cell Cultivation. Oxford, London: Blackwell Scientific Publications, Inc. 1975.

[112] Fike R. Nutrient Supplementation Strategies for Biopharmaceutical Production, Part 3. Scaling Strategies for Rapid Nutrient Supplement Prototyping. 2010. Bioprocess Technical, Bioprocess International, January 2010; 24-31.

[113] Tsun HY, Liu CM and Tzeng YM. Recovery and Purification of Thuringiensin from the Fermentation Broth of *Bacillus thuringiensis*. Bioseparation 1999; 7 309-316.

[114] Rowe GE. and Margaritis A. Bioprocess Design and Economic Analysis for the Commercial Production of Environmentally Friendly Bio-Insecticides from *Bacillus thuringiensis* HD-1 Biotechnology and Bioengineering 2004; 86(4) 377-388.

[115] Brar SK, Verma M, Tyagi RD and Valéro JR. Recent Advances in Downstream Processing and Formulations of *Bacillus thuringiensis* based on Biopesticides. Process Biochemistry 2006; 41(2) 323-342.

[116] Prabakaran G, Balaraman K, Hoti SL. and Manonmani AM. A Cost-Effective Medium for the Large-Scale Production of *Bacillus sphaericus* H5a5b (VCRC B42) For Mosquito Control. Biological Control 2007; 41(3) 379-383.

[117] Lacroix C. and Yildirim S. Fermentation Technologies for the Production of Probiotics with High Viability and Functionality. Current Opinion in Biotechnology 2007; 18 176-183.

[118] Burges HD. and Jones KA. Formulation of Bacteria, Viruses and Protozoa to Control Insects H.D. Burges (Ed.), Formulation of Microbial Biopesticides: Beneficial Microorganisms, Nematodes and Seed Treatments, Kluwer, Dortrecht 1998. pp. 33-127.

[119] Schisler DA, Slininger PJ, Behle RW, Jackson MA. Formulation of *Bacillus spp.* for Biological Control of Plant Diseases. Phytopathology 2004; 94 1267-1271.

[120] Lydersen BK, D'Elia NA, Nelson KL. Bioprocess Engineering: Systems, Eequipment and Facilities. Chemical Engineering Science. 1994; 50(6) 1069 1070

[121] Lalloo R, Maharajh D, Gorgens J. and Gardiner N. A Downstream Process for Production of A Viable and Stable *Bacillus Cereus* Aquaculture Biological Agent. Applied Microbiology and Biotechnology 2010; DOI 10.1007/s00253-009-2294-2.

[122] Luna-Solano G, Salgado-Cervantes MA, Rodrígu

[123] ez-Jimenes GC and García-Alvarado MA. Optimization of Brewer's Yeast Spray Drying Process. Journal of Food Engineering 2005; 68(1) 9-18.

[124] Puziss M, Manning LC, Lynch JW, Barclay E, Abelow I, Wright GG. Large-Scale Production of Protective Antigen of *Bacillus anthracis* in Anaerobic Cultures. Applied Microbiology 1963; 11(4) 330-334.

[125] Zamola B, Valles P, Meli G, Miccoli P, Kajfez F. Use Of The Centrifugal Separation Technique in Manufacturing a Bioinsecticide Based on *Bacillus thuringiensis.* Biotechnology Bioengineering 1981; 23(5) 1079-1086.

[126] Rojas JV, Gutierrez E, de la Torre M. Primary Separation of the Entomopathogenic Products Of *Bacillus thuringiensis*. Biotechnology Progress 1996; 12(4) 564-566.

[127] Torres-Anjel MJ and Hedrick TI. Spore Removal by Centrifugation and Its Effect on Ultra-High Temperature Commercial Sterilization of Milk. Journal of Dairy Research 1970; 54(3) 326-330.

[128] Rivière J.Industrial applications of microbiology. Wiley, London 1977

[129] Berovic M. New Products and New Areas of Bioprocess Engineering. Chemical Engineering Science 1998; 50(6)1069-1070

[130] de Medeiros FPM, de Melo Santos MAV, Regis L, Rios EMM, Neto PJM. Development of a *Bacillus sphaericus* Tablet Formulation and its Evaluation as a Larvicide in the Biological Control of *Culex Quinquefasciatus*. Memórias do Instituto Oswaldo Cruz 2005; 100(4) 431-434.

[131] Rhodes DJ. Exploitation of Microorganisms, London: Chapman & Hall; 1993.

[132] Werner L, Latzko F, Hampel W. Spray Drying of Yeast-Lytic Enzymes From *Arthrobacter* sp. Biotechnology Techniques 1993; 7(9) 663-666.

[133] Costa E, Teixidó N, Usall J, Fons E, Gimeno V, Delgado J, Viñas I. Survival of *Pantoea agglomerans* Strain CPA-2 in Spray-drying Process. Journal of Food Protection 2001; 65(1) 185-191.

[134] Larena I, de Cal A, Liñán M, Melgarejo P. Drying of *Epicoccum nigrum conidia* for Obtaining a Shelf-Stable Biological Product against Brown Rot Disease. Journal of Applied Microbiology 2003; 94(3) 508-514.

[135] Moene-Loccoz Y, Tichy HV, O'Donnell A, Simon R and O'Gara F. Impact of 2,4-Diacetylphloroglucinol-Producing Biocontrol Strain *Pseudomonas fluorescens* F113 on Intraspecific Diversity of Resident Culturable Fluorescent Pseudomonads Associated with the Roots of Field-Grown Sugar Beet Seedlings. Applied Environmental Microbiology. 1999; 67(8) 3418-3425.

[136] Wiwattanapatapee R, Pengnoo A, Kanjanamaneesathian M, Matchavanich W, Nilratana L, Jantharangsri A. Floating Pellets Containing Bacterial Antagonists for Control Sheath Blight of Rice: Formulations, Viability and Bacterial Release Studies. Journal Controlled Release 2004; 95(3) 455-462.

[137] Klein N. and Lortal S. Attenuated Starters: An Efficient Means to Influence Cheese Ripening-A Review. International Dairy Journal 1999; 9(11) 751-762.

[138] Meng XC, Stanton C, Fitzgerald GF, Daly C, Ross RP. Anhydrobiotics: The Challenges of Drying Probiotic Cultures. Food Chemistry 2008; 106 1406-1416.

[139] Knorr D. Technology Aspects Related to Microorganisms in Functional Foods. Trends in Food Science and Technology 1998; 9(8-9) 295-306.

[140] Bayrock D. and Ingledew WM. Mechanism of Viability Loss during Fluidized Bed Drying of Baker's Yeast. Food Research International 1997; 30(6) 417-425.

[141] Tamez-Guerra P, McGuire MR, Medrano-Roldan H. and Galan-Wong LJ. Sprayable Granule Formulations of *Bacillus thuringiensis*. Journal of Economic Entymology 2000; 93(2) 219-225.

[142] Chen XD, and Patel KC. Microorganism Inactivation During Drying of Small Droplets or Thin-Layer Slabs-A Critical Review of Existing Kinetics Models and an Appraisal of the Drying Rate Dependent Model. Journal of Food Engineering 2007; 82(1) 1–10.

[143] Grabowski JA, Truong VD. and Daubert CR. Spray-Drying of Amylase Hydrolyzed Sweet Potato Puree and Physicochemical Properties of Powder. Journal of Food Science 2009; 71 (5) E209–217.

[144] Mille Y, Obert J, Beney L. and Gervais P. New Drying Process for Lactic Bacteria Based on Their Dehydration Behaviour in Liquid Medium. Biotechnology and Bioengineering 2004; 88(1) 71-76.

[145] Lewis JA. and Papavizas GC. Biocontrol of Plant Diseases: The Approach for Tomorrow. Crop Protection 1991; 10(2) 95-105.

[146] Soper RS. and Ward MG. Beltsville Symposia In Agricultural Research. Biological Control in Crop Production 1981; 5 161-180.

[147] Tsuji K, Watanuki T, Kondo F, Watanabe MF, Nakazawa H, Suzuki M, Uchida H and Harada K. Stability of Microcystins from Cyanobacteria-iv. Effect of Chlorination on Decomposition. Toxicon 1997; 35(7) 1033-1041.

[148] Ramaswamy H. and Marcotte M (Eds.), Food Processing: Principles and Applications, CRC Press, Taylor and Francis Group, Boca Raton, FL, USA, 2006 (xvi C420 ISBN 1-58716-008-0)

[149] Venkat HK, Sahu NP. and Jain KK. Effect of Feeding *Lactobacillus*-Based Probiotics on the Gut Microflora, Growth And Survival Of Postlarva of *Macrobrachium rosenbergii* (de Man). Aquaculture Research 2004; 35(5) 501-507.

[150] Keysami MA, Mohammadpor M. and Saad CR. Probiotic Activity of *Bacillus subtilis* In Juvenile Freshwater Prawn, *Macrobrachium rosenbergii* (de Man) at different Methods of Administration to the Feed. Aquaculture International. 2012; 20 499-511.

[151] Nikoskelainen S, Ouwenhand A, Salminen S. and Bylund G. Protection of Rainbow Trout *Oncorhynchus Mykiss* from Furunculosis by *Lactobacillus rhamnosus*. Aquaculture 2001; 198(3-4) 229-236.

[152] Meunpol O, Lopinyosiri K. and Menasveta P. The Effects of Ozone and Probiotics on The Survival of Black Tiger Shrimp (*Penaeus monodon*). Aquaculture 2003; 220(1-4) 437-448.

[153] Robertson, P.A.W., O'Dowd, C., Burrells, C., Williams, P and Austin, B. Use of Carnobacterium sp. as a probiotic for Atlantic salmon *Salmo salar* L. and rainbow trout *Oncorhynchus mykiss*, Walbaum. Aquaculture 2000; 185(3-4) 235–243.

[154] García de La Banda, I., Loboa, C., León-Rubiob, J.M., Tapia-Paniagua, S., Balebona, M.C., Moriñigo, M.A., Moreno-Ventas, X., Lucas L.M., Linares,F., Arce, F., and Arijo S. Influence Of Two Closely Related Probiotics On Juvenile Senegalese Sole (*Solea Senegalensis*, Kaup 1858) Performance And Protection Against *Photobacterium Damselae Subsp. Piscicida*. Aquaculture 2012; 306 281-288.

[155] Gildberg A. and Mikkelsen H. Effects of Supplementing the Feed of Atlantic Cod (*Gadus morhua*) Fry with Lactic Acid Bacteria and Immune-Stimulating Peptides During A Challenge Trial With *Vibrio anguillarum*. Aquaculture 1998; 167(1-2) 103-113.

[156] Gullian M, Thompson F. and Rodriguez J. Selection of Probiotic and Study of the Immunostimulatory Effect on *Penaeus vannamei*. Aquaculture 2004; 233(1-4) 1-14.

[157] Courvalin P. Antibiotic Resistance: The Pros and Cons of Probiotics. Digestive and Liver Disease 2006; 38(2) S261-S265.

[158] Ishibashi N. and Yamazaki S. Probiotics and Safety. American Journal of Clinical Nutrition 2001; 73(2) S465-S470.

[159] Ringo E. and Gatesoupe FJ. Lactic Acid Bacteria in Fish: A Review. Aquaculture 1998; 160(3-4) 177-203.

[160] Moubareck C, Gavini F, Vaugien L, Butel MJ. and Doucer-Popularie F. Antimicrobial susceptibility of *Bifidobacteria.* Journal of Antimicrobial Chemotherapy 2005; 55(1) 38-44.

[161] Moriarty DJW, Decamp O. and Lavens P. Probiotics In Aquaculture. Shrimp Culture management in AQUA Culture Asia Pacific Magazine. September/October 2005. 14-16

Chapter 8

Use of Probiotic Bacteria against Bacterial and Viral Infections in Shellfish and Fish Aquaculture

Héctor Cordero, María Ángeles Esteban and Alberto Cuesta

Additional information is available at the end of the chapter

http://dx.doi.org/10.5772/57198

1. Introduction

The term "probiotic" was firstly used to denominate microorganisms that have effects on other microorganisms [1]. Etymologically, the term "probiotic" was originated from the Latin word "pro" which means "for" and the Greek word "bios" which means "life". The best known definition for probiotics was developed by the Food and Agriculture Organization (FAO), that defined them as live microorganisms which when administered in adequate amounts confer a health benefit on the host [2]. According to this description, the potential benefits are varied, and if probiotics were administered to shellfish or fish under intensive culture they could improve their production. It is known that virus and bacterial diseases/infections are one of the most important problems in aquaculture production at present. Probiotics can provide some solutions to this problem through different mechanisms or properties such as the production of inhibitory compounds such as bacteriocins, competition for adhesion sites with opportunistic or pathogen microorganisms, competition for nutrients with other bacteria or an improvement of the immune status (e.g. increase of production of immunoglobulins, acid phosphatase, antimicrobial peptides, improvement of cellular activities, etc.) [3-10]. Several reviews have already documented the benefits of probiotics in shellfish and fish but they mainly focused on their effects in the immune response. Thus, hypothetical and desired results of administering probiotics to shellfish or fish in culture will be improving their antiviral and antibacterial defences, which is the focus of the present review. Firstly, a brief description of probiotics is included, and then a review of the main used probiotics against pathogenic virus and bacteria for shellfish and finally, the same for fish. The novelty of this review is based on the shared ability of probiotics to control both viral and bacterial diseases in shellfish and fish often share, which could be the basis for sustainable aquaculture.

2. Probiotic bacteria

There is a great diversity of tested probiotic bacteria, but only few of them have become in commercial probiotics (Table 1). Thus, further studies are mandatory to expand the use of laboratory described microorganisms with probiotic effects to the commercial level and then be used in the aquaculture industry. The procedure to test and market a probiotic is resumed in Figure 1.

Commercial name	Animal/Human	Reference/Comments
AlCare™	Mammalian	Contains *Bacillus licheniformis*
Alibio®	Fish	[30]
Bactisubtil®	Human	Contains *Bacillus cereus*
Bactocell® PA 10	Fish	[42]
BaoZyme-Aqua	Fish	Contains *Bacillus subtilis*
BGY-35	Fish	[51]
Biogrow®	Mammalian	Contains *Bacillus subtilis* and *B. licheniformis*
Bio-Kult®	Human	Contains *B. subtilis*
BioPlus® 2B	Fish	[73]
Biosporin®	Human	Contains *B. subtilis* and *B. licheniformis*
Biostart®	Fish	Contains a mix of *Bacillus* spp. and *Paenobacillus* sp.
Biovicerin®	Human	Contains *B. cereus*
Bispan®	Human	Contains *Bacillus polyfermenticus*
Cernivet®	Fish	[85]
Domuvar	Human	Contains *Bacillus* spp.
Ecomarine®	Shellfish	
Esporafeed Plus®	Swine	Contains *B. cereus*
Lactobacil	Fish	[45]
Lactopure	Mammalian	Contains *Lactobacillus sporogenes*
Liqualife®	Fish	Contains *Bacillus* spp.
Neoferm BS 10	Mammalian	Contains *Bacillus clausii*
Neolactoflorene	Human	Contains *Lactobacillus* spp. and *Bacillus* spp.
Promarine®	Shellfish	
SanoCare®	Fish	Contains *Bacillus* spp.
SanoGuard®	Fish	Contains *Bacillus* spp.
SanoLife®	Fish	Contains *Bacillus* spp.
Sporolac	Fish	[45]
Sustenex®	Human	Contains *Bacillus coagulans*
Toyocerin®	Fish	[85]

Table 1. List of commercial probiotics, including those for shellfish and fish.

http://dx.doi.org/10.5772/57198

Probiotics are usually consisting on bacteria but some other microorganisms such as yeast, microalgae or even some fungi. They are mainly used as living cells but some studies have also shown their benefits when supplied as heat-inactivated cells (also known as heat-killed cells), formalin-killed (FKC), freeze-dried, dead cells or cell-free supernatant (CFS). Among the vast number of probiotic species used most information relies on the use of *Bacillus* sp. and *Lactobacillus* sp. Different administration modes have been checked, as bath, intraperitoneal or intramuscular injection and in diet being the bath and diet those preferred for the use in the aquaculture. Moreover, more recently, for oral dietary administration the probiotics can be encapsulated in different ways. Besides that, *Artemia* and rotifers (two main diets larvae in marine larviculture) are usually enriched with probiotics in order to produce benefits in the fish/shellfish larvae.

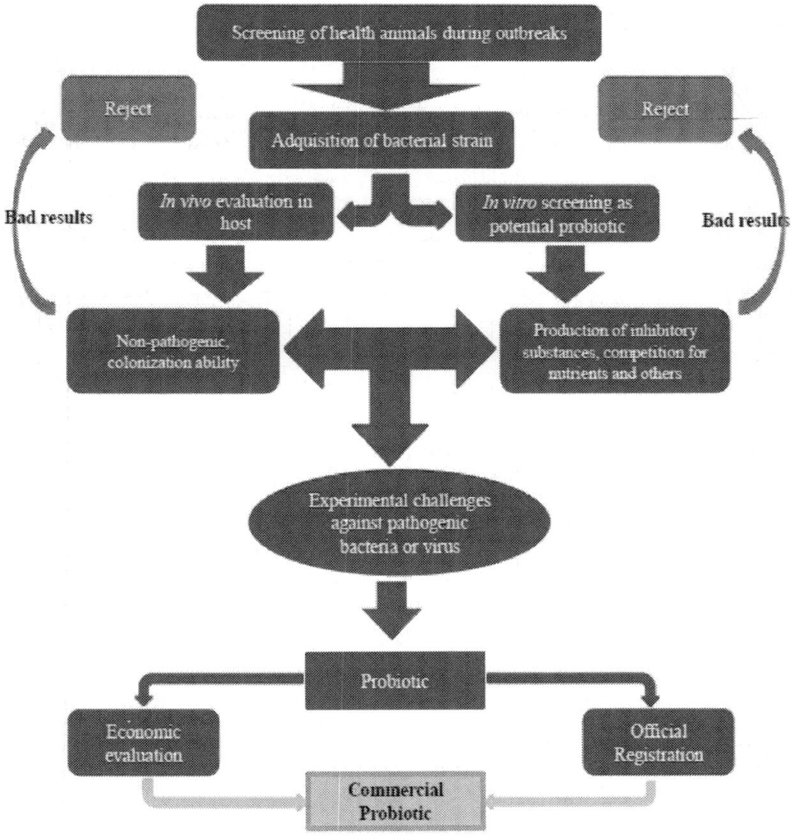

Figure 1. Process for making commercial probiotics.

3. Probiotics against virus in shellfish

Viral infections are one of the most important problems in aquaculture production. In the case of shellfish, probiotics might provide a good preventive solution to this problem since they promote the innate immune response, which is the only one attributed to be responsible for the resistance in these animals.

Mainly seven viral diseases are known in shellfish which are: white spot syndrome virus (WSSV), lymphocystis disease virus (LCDV), infectious hypodermal and hematopoietic necrosis virus (IHHNV), taura syndrome virus (TSV), yellow head disease virus (YHV), infectious myonecrosis virus (IMNV) and *Macrobrachium rosenbergii* nodavirus (MrNV). Unfortunately, all the studies have focused on the potential preventive effects of few probiotics on the pacific white shrimp (*Litopenaus vannamei*) resistance against WSSV. In a single study it was demonstrated that bath treatment of *L. vannamei* specimens with the probiotic *Vibrio alginolyticus* at a dose of 10^5 cfu ml^{-1} showed a higher rate survival against WSSV compared to those non exposed to the probiotic [11]. Interestingly, most of the information comes from studies using dietary administration of the probiotics which results the most desired for aquaculture of shellfish. It has been reported that survival of *L. vannamei* specimens fed supplemented diets containing 10^5 cfu g^{-1} of a mixture formed by lactic acid bacteria (BAL3, BAL7, BC1 and CIB1) failed to protect against WSSV infections [12]. By contrast, dietary administration of 10^{10} cfu g^{-1} of *Bacillus* OJ in *L. vannamei* specimens produced significantly higher survival after challenge by WSSV [13]. It has also been reported that dietary administration of *Pediococcus pentosaceus* and *Staphylococcus hemolyticus* to *L. vannamei* specimens showed a decrease in the prevalence of WSSV, but not IHHNV [14]. Further studies including more shellfish species and virus are necessary in order to find potential solutions for the viral diseases found under their intensive culture.

4. Probiotics against bacteria in shellfish

In the case of bacterial diseases much more studies have focused on the benefits of the use of probiotics for shellfish species. Moreover, and in contrast to the viral pathogens described above, more shellfish species have focused the studies about the use of probiotics. Herein we will summarize the main findings about the potential use of probiotics against bacterial diseases grouped by shellfish species.

A first attempt to describe the probiotic potential of a microorganism comes from *in vitro* studies. Thus, it has been demonstrated that *Pseudoalternomonas* sp. strains DIT09, DIT44 and DIT46 isolated from *Peromytilus purpuratus* showed bacteriostatic anti-*Vibrio parahaemolyticus* activity [15] but their *in vivo* effects have not been tested yet. In a similar way, *Roseobacter* sp. strain BS107 isolated form the scallop (*Pecten maximus*) showed antibacterial activity against several pathogenic *Vibrio* sp. [16] as well as the probiotic *Alteromonas haloplanktis* obtained from *Argopecten purpuratus* larvae specimens [17]. Further preliminary studies of this kind are worthy to be taken in the future and prior to those conducted *in vivo*.

Several studies have been conducted in bivalves. In the case of Pacific oyster larvae (*Crassostrea gigas*) exposed to 10^5 cfu ml^{-1} of the pathogenic *Vibrio tubiashii* reached a total mortality in just 2 days, whilst in combination with 10^4 cfu ml^{-1} of the probiotic *Aeromonas media* A199 strain the larvae prolonged their viability up to 144 hours indicating its benefits when used by bath [18]. By contrast, *C. virginica* specimens fed supplemented diets containing 10^4 cfu ml^{-1} of *Vibrio* sp. OY15 for three weeks showed no effect in survival ratio after challenge with *Vibrio* sp. M183 [19] It has been reported [20] that abalone (*Haliotis discus hannai*) specimens fed supplemented diet with 10^9 cfu g^{-1} of *Shewanella colwelliana* WA64 and *Shewanella oyellana* WA65 for four weeks showed a better survival rate (with mortalities of 27%-50% in WA64, and 30%-43% in WA65 compared with 77%-80% in the control group) when infected with *Vibrio harveyi*. In other research with other abalone specie, *Haliotis midae* specimens fed supplemented diet with a mix of three unknown probiotic strains (SY9, SS1 and AY1) at doses of 10^7 cfu ml^{-1} for two weeks showed a better survival ratio (62%) than control group specimens after intra-mantle injection of *Vibrio anguillarum* [21]. Further studies are still needed to broad the use of probiotics in bivalves against bacterial diseases.

Among the shellfish, most of the studies have at this respect focused on shrimps. Thus, western king prawn (*Penaeus latisulcatus*) specimens fed 20×10^5 cfu kg^{-1} diet of *Pseudomonas aeruginosa* and *Pseudomonas synxantha* for eighty-four days and afterwards challenged with *V. harveyi*. *P. aeruginosa*-supplemented diet improved the survival rate of the western king prawns more effectively than *P. synxantha*-supplemented diet, and furthermore, administration of both probiotics in combination resulted in better results than when administering separately [22-23].

Most of the studies administering probiotics have been developed in white shrimp (*Litopenaeus vannamei*) at different development stages. For example, *Bacillus subtilis* E20 administered in the diet at 10^6, 10^7 and 10^8 cfu kg^{-1} increased the survival rates at 13.3%, 16.7% and 20% respectively, after the injection of pathogenic *V. alginolyticus* [24]. In juvenile specimens, commercial white shrimp fed supplemented diet with 10^5 cfu g^{-1} diet of *Bacillus subtilis* UTM126 achieved a mortality of 18% against pathogenic infection of vibrios (including *V. harveyi, V. alginolyticus* and *V. parahaemolyticus*) while the control group mortality exceeded of 50% [25]. In other research, juvenile specimens fed supplemented diets containing *V. alginolyticus* UTM 102, *B. subtilis* UTM 126, *Roseobacter gallaeciensis* SLV03 or *Pseudomonas aestumarina* SLV22, separately, at doses of 10^5 cfu g^{-1} diet for four weeks showed low mortality (between 17%-22%) after immersion with *Vibrio parahaemolyticus* PS-017 compared with the control group (33%) [26]. In adult specimens of *L. vannamei* fed supplemented diet with 3×10^5 cfu of the probiotic *Vibrio gazogenes* per shrimp showed a decrease of mortality after infection with *Vibrio* spp. (including *V. harveyi, V. anguillarum* and *V. alginolyticus*) [27]. In addition, the inhibitory effect was also demonstrated in a *in vitro* assay [27]. Other recent work [28] has been carried out with white shrimp fed a supplemented diet containing 10^5 cfu g^{-1} (BM5) and 10^8 cfu g^{-1} (BM8) (two *Bacillus subtilis* strains) for 2 months, and afterwards each shrimp was injected with 10^7 cfu of *Vibrio harveyi*. Results indicate that cumulative mortality of the control group was 63.3%, whereas in the groups fed probiotics were of 20% and 33.3%, for the group fed BM8 or BM5 strains, respectively. Cumulative mortality also decreased in white shrimp fed a supplemented diet with 10^{10} cfu kg^{-1} of *Lactobacillus plantarum* after injection with *V. alginolyticus* [29].

Moreover, the administration of a mixture of *Bacillus* (*B. endophyticus* YC3-b, *B. endophyticus* C2-2 and *B. tequilensis* YC5-2) to the water at doses of 0.1×10^6 cfu ml^{-1} to juvenile specimens resulted in a high survival ratio (33%) compared with the control group (9.5%) after challenge with *V. parahaemolyticus*. However, a commercial probiotic (Alibio) at the same dose that the *Bacillus* mix had no effect in survival ratio compared with the control group in *Litopenaeus vannamei* specimens [30]. *L. vannamei* specimens fed diet supplemented with two potential probiotics (strains C2 and B6) achieved a better survival ratio (44% and 50%) than control group (21%) after infection with *Vibrio harveyi* in stages from Myosis 3 to postlarvae 1 [31]. Strikingly, other microorganisms such as yeast have been also assayed as potential probiotics. Unfortunately, *L. vannamei* specimens fed *Saccharomyces cerevisiae, Phaffia rhodozyma* and *Saccharomyces exiguus* showed no significant different in survival ratio after infection with *V. harveyi* compared with control group specimens [32].

Black tiger shrimp (*Penaeus monodon*) has also received much attention. Thus, *P. monodon* specimens exposed to 10^6 cfu ml^{-1} of *B. subtilis* BT23 for 5 days (long-term treatment) or for 1 hour (short-term treatment), and thereafter challenged with *V. harveyi*, showed a decrease in their cumulative mortality in both groups (32% and 60%, respectively) [33]. In other research, *P. monodon* juvenile specimens fed *Bacillus* sp. S11 at 10^{10} cfu g^{-1} diet for one month and infected with *V. harveyi*, combined with ozone addition, showed a significant increase in the survival ratio (75%) compared with the control group and not fed with probiotics [34]. Also in juvenile specimens fed supplemented diet containing *Lactobacillus acidophilus* 04 at dose of 10^5 cfu g^{-1} for one month showed a higher survival ratio (80%) than the control group (13.3%) after challenged with *Vibrio alginolyticus* [35]. In postlarvae specimens, dietary administration of *Paenibacillus* sp. EF012164 and *Bacillus cereus* DQ915582 at doses of 10^4 and 10^5 cfu ml^{-1} caused lower mortality after infection with *Vibrio harveyi* and *Vibrio* spp. (without statistical analysis) [36]. In other work, *Penaeus monodon* postlarvae specimens fed supplemented diet with 10^9 cfu g^{-1} diet of two strains of *Synechocystis* sp. (C51 and C54) separately for twenty days showed significantly better survival after infection with *Vibrio harveyi* MCCB 111 than those fed without probiotics [37]. Also in postlarvae specimens, dietary administration of *Bacillus* sp. P11 at 10^9 cfu g^{-1} caused a high survival ratio (66%) compared with the control group (0%) after 9 days of infection with *Vibrio harveyi* and *Vibrio* spp. [38]. Dietary administration of *Artemia*-encapsulated *Bacillus* sp. S11 showed an increased survival of *Penaeus monodon* when infected with *Vibrio harveyi* D331 [39]. Finally, dietary administration to *P. monodon* with 10^3 cfu ml^{-1} of *Pseudomonas* sp. PM11 and *Vibrio fluvialis* PM17 for 45 days did not alter the mortality after challenge with *Vibrio anguillarum* [40]. As it has been widely shown in shellfish and fish the use of low or suboptimal dosages of probiotics have no biological role, and in this case protective effect against pathogens.

Other shrimp species have received little attention. In the Indian white shrimp (*Penaeus indicus*) juvenile specimens fed diets supplemented with *Lactobacillus acidophilus, Streptococcus cremoris, Lactobacillus bulgaricus* 56 or *L. bulgaricus* 57 at doses of 5×10^6 cfu g^{-1} for 4 weeks and infected with *Vibrio alginolyticus* showed a higher survival rate (56% - 72%) compared with that observed in specimens of the control group (20%) [41]. Similarly, in blue shrimp (*Litopenaeus stylirostris*) specimens fed supplemented diet of 10^7 cfu g^{-1} of *Pediococcus acidilactici* for 4 weeks and infected with *Vibrio nigripulchritudo* SFn1 showed a mortality level of 25% in the probiotic-treated group while in non-treated group the mortality was of 41.7% [42]. It was also reported that *Penaeus chinensis* postlarvae specimens exposed to *Arthrobacter* XE-7 at dose of 10^6 cfu ml^{-1} and pathogenic *Vibrios* sp. (*Vibrio parahaemolyticus, Vibrio anguillarum* and *Vibrio nereis*) showed a significant higher survival ratio than specimens exposed to pathogenic *Vibrios* spp. alone [43].

Marron (*Cherax tenuimanus*) specimens fed five probiotics (*Bacillus* sp. AQ2, *Bacillus mycoides* A10, *Shewanella* sp. A12, *Bacillus subtilis* PM3 and *Bacillus* sp. PM4) separately showed no significant differences in survival rate. However, the total haemocyte count was significantly higher in all probiotic-treated groups compared with the control group after injection with 2×10^8 cfu ml^{-1} of *Vibrio mimicus* [44].

Overall, studies have shown that probiotics are good alternative to protect shellfish against pathogenic bacteria, namely against *Vibrio* sp. pathogens, the most important in the culture of shellfish. However, further studies are necessary to broad the probiotic candidates and the shellfish species prior they are applied to aquaculture from a practical point of view. Moreover, the mechanisms behind this protection are generally ignored and deserve deeper evaluation.

5. Probiotics against virus in fish

Viral diseases are major problems in fish farming since there is a lack of suitable antiviral agents and a very limited number of effective vaccines. Moreover, there are few studies about the effects of probiotics against viral infections in fish. Olive flounder (*Paralychthys olivaceus*) and grouper (*Epinephelus coioides*) are the two main species which have been studied. Olive flounder specimens fed 2.4×10^8 cfu g^{-1} of Lactobacil and/or Sporolac (commercial acid lactic bacteria) were infected with lymphocystis disease virus (LCDV) [45]. Lowest mortality rate was seen in groups fed Lactobacil (30%) or Lactobacil and Sporolac (25%) supplemented diets followed by groups receiving Sporolac alone (45%) compared to those groups fed without probiotics that showed a mortality of 80%. Evaluating the disease resistance of grouper through probiotics against virus infection, a recent study has demonstrated that specimens fed a supplemented diet with 10^8 cfu g^{-1} of *B. subtilis* E20 for 28 days showed a survival rate of 50% higher than the control group for seven days post-infection with iridovirus [46]. In another study, grouper specimens fed a diet containing *L. plantarum* at 10^8 cfu kg^{-1} and challenged with an iridovirus showed an increase in the survival of 36.7% compared to the survival rate in control group [47]. Similar results were obtained when grouper specimens were fed *S. cerevisiae* supplemented diet (5.3×10^7 cfu kg^{-1} for four weeks) and afterwards infected with a grouper iridovirus (GIV). Specimens of treated group showed a higher survival ratio (43.3%) than specimens in the control group (16.7%) [48]. Viral pathogens diversity and impact in the actual aquaculture deserves further characterization of the potential benefits of probiotics for economically important cultured fish world-wide.

6. Probiotics against bacteria in fish

By far, the effects of probiotics on fish have received most of the investigations. Among the fish studied, the rainbow trout (*Oncorhynchus mykiss*) has been the most evaluated. Many different probiotic bacteria have been tested and two of the best studied are *Bacillus subtilis* and *Lactobacillus acidophilus,* two lactic acid bacteria which showed *in vitro* inhibition against *Aeromonas hydrophila* [49]. Furthermore, *B. subtilis* avoids the development of *Pseudomonas fluorescens* while *L. acidophilus* had also antimicrobial activity against *Streptococcus iniae*. The information relative to the use of probiotics as a beneficial treatment of fish against bacterial pathogens is described below and summarized (Table 2).

Fish tested	Probiotic	Pathogen	Survival	Cites
Anguilla anguilla	*Enterococcus faecium* SF68 *Bacillus toyoi*	*Edwardsiella tarda 981210L1*	Significant increase for SF68 and no difference for *B. toyoi*	[85]
Anguilla japonica	*Lactobacillus pentosus* PL11	*Edwarsiella tarda*	Significant increase	[87]
Carassius auratus	*Aeromonas hydrophila* A3-51 formalin-inactivated	*Aeromonas salmonicida*	Significant increase	[90]
Carassius auratus *Xiphophorus helleri*	*Bacillus* sp., *Lactobacillus* sp., *Streptococcus faecium*, and *Saccharomyces cerevisiae*	*Pseudomonas fluorescens* 58C	No differences	[89]
Clarias gariepinus	*Lactobacillus acidophilus*	*Staphylococcus xylosus* *Aeromonas hydrophila* gr2 *Streptococcus agalactiae*	Significant increase	[91]
Dicentrarchus labrax	*Vagococcus fluvialis*	*Vibrio anguillarum*	Significant increase	[107]
Epinephelus coioides	*Lactobacillus plantarum*	*Streptococcus* sp.	Significant increase	[47]
	Saccharomyces cerevisiae	*Streptococcus* sp.	Significant increase	[48]
	Bacillus subtilis E20	*Streptococcus* sp.	Significant increase	[46]
Gadus morhua	*Carnobacterium divergens*	*Vibrio anguillarum* *Aeromonas salmonicida*	Significant increase	[57]
Labeo rohita	*Bacillus subtilis*	*Aeromonas hydrophila*	No difference	[96]
	Pseudomonas aeruginosa VSG-2	Aeromonas hydrophila MTC1739	Significant increase	[98]
	Lactobacillus plantarum VSG-3	*Aeromonas hydrophila*	Significant increase	[97]
Miichthys miiuy	*Clostridium butyricum* CB2 as alive and dead cells	*Vibrio anguillarum* *Aeromonas hydrophila*	Significant increase	[94]
Mycteroperca rosacea	*Debariomyces hansenii* CBS-8000339	*Aeromonas hydrophila* AH-315	No difference	[50]

http://dx.doi.org/10.5772/57198

Fish tested	Probiotic	Pathogen	Survival	Cites
Oncorhynchus mykiss	*Clostridium botyricum*	*Vibrio anguillarum*	Significant increase	[95]
	Streptococcus iniae Dan-1 formalin inactivated	*Streptococcus iniae* virulent	Significant increase	[80]
	Pseudomonas fluorescens AH2	*Vibrio anguillarum*	Significant increase	[72]
	Lactobacillus rhamnosus ATCC 53103	*Aeromonas salmonicida* ssp. *salmonicida*	Significant increase	[67]
	Aeromonas hydrophila A3-51 *Vibrio fluvialis* A3-47S *Carnocterium* sp. BA211 Unidentified coccus A1-6	*Aeromonas salmonicida*	Significant increase	[60]
	Aeromonas hydrophila A3-51 *Vibrio fluvialis* A3-47S *Carnocterium* sp. BA211 Unidentified coccus A1-6 formalin-inactivated	*Aeromonas salmonicida*	Significant increase	[62]
	Bacillus subtilis Bacillus licheniformis	*Yersinia ruckeri*	Significant increase	[73]
	Carnobacterium maltaromaticum B26 *Carnobacterium divergens* B33	Yersinia ruckeri Aeromonas salmonicida	Significant increase	[75]
	Lactococcus lactis ssp. *lactis* CFLP100 *Leuconostoc mesenteroides* CLFP196 *Lactobacillus sakei* CLFP201	*Aeromonas salmonicida* ssp. *salmonicida* CLFP501	Significan increase	[63]
	Bacillus sp. JB-1 *Aeromonas sobria* GC2	*Streptoccocus iniae* *Lactococcus garvieae* *Vibrio anguillarum* *Vibrio ordalii* *Aeromonas salmonicida* *Yersinia ruckeri*	Significant increase	[64]
	Bacillus subtilis AB1 as live, sonicated and formalized cells and cell-free supernatant	*Aeromonas* sp.	Significant increase	[82]
	Brochothrix thermophasta BA211 *Aeromonas sobria* GC2	*Aeromonas bestiarum* ORN2	Significant increase	[65]
	Brochothrix thermophasta BA211 *Aeromonas sobria* GC2	*Ichthyophthrius multifiliis*	Significant increase for GC2 and no difference for BA211	[65]
	Leuconostoc mesenteroides CLFP196 Lactobacillus plantarum CLFP238	*Lactococcus garvieae*	Significant increase	[68]

Fish tested	Probiotic	Pathogen	Survival	Cites
	Enterobacter cloacae *Bacillus mojavensis*	*Yersinia ruckeri*	Significant increase	[74]
	Kocuria SM1	*Vibrio anguillarum*	Significant increase	[69-71]
	Lactobacillus plantarum CLFP238 *Lactococcus lactis* CFLP100 *Leuconostoc mesenteroides* CLFP196	*Lactococcus garvieae* CLFP LG1	Significant increase	[66]
	Pseudomonas sp. M174 and M162	*Flavobacterium psychrophilum*	Significant increase	[79]
	Enterococcus faecalis inactivated	*Aeromonas salmonicida*	Significant increase	[81]
Oplegnathus fasciatus	*Lactobacillus sakei* BK19	*Edwarsiella tarda*	No difference	[88]
Oreochromis niloticus	*Lactobacillus acidophilus, Bacillus subtilis, Clostridium butyricum* and *Saccharomyces cerevisiae*	*Edwardsiella tarda*	Significant increase	[86]
	Bacillus subtilis *Lactobacillus acidophilus*	*Aeromonas hydrophila,* *Pseudomonas fluorescens* *Streptococcus iniae*	Significant increase	[49]
Oreochromis	*Saccharomyces cerevisiae*	*Aeromonas hydrophila* *Pseudomonas fluorescens* *Flavobacterium columnare*	Significant increase	[51]
Paralichthys olivaceus	*Zooshikella* sp. JE-34	*Stretoccocus iniae*	Significant increase	[93]
	Bacillus subtilis *Bacillus pumilus* *Bacillus licheniformis*	*Streptococcus iniae*	Significant increase (except for B. licheniformis)	[92]
Salmo salar	*Vibrio alginolyticus*	*Aeromonas salmonicida* 256/81 *Vibrio anguillarum* VIB256 *Vibrio ordalii* 17K	Significant increase	[52]
	Vibrio alginolyticus	*Yersinia ruckeri* Ex5	No difference	[52]
	Pseudomonas fluorescens AH2	*Aeromonas salmonicida*	No difference	[55]
Salmo trutta	*Lactococcus lactis* ssp. *lactis* CLFP100 *Leuconostoc mesenteroides* CLFP196	*Aeromonas salmonicida*	Significant increase	[83]
Salvelinus fontinalis	S1, S5, S9 and S10	*Flavobacterium columnare*	Significant increase	[84]

http://dx.doi.org/10.5772/57198

Fish tested	Probiotic	Pathogen	Survival	Cites
Scophthalmus maximus	*Roseobacter* sp. strain 27-4	*Vibrio anguillarum*	Significant increase	[108]
	Phaeobacter sp. *Ruegeria* sp.	*Vibrio anguillarum*	Unmeasured	[102]
	Lactobacillus plantarum *Carnobacterium* sp. *Roseobacter* sp.	*Vibrio* sp.	Significant increase	[99]
Solea senegalensis	*Shewanella putrefaciens* Pdp11 *Shewanella baltica* Pdp13	*Photobacterium damselae* ssp. *piscicida*	Significant increase	[104-105]
Sparus aurata	*Shewanella putrefaciens* Pdp11	*Vibrio anguillarum* DC11R2	Significant increase	[103]
	Bacillus subtilis	*Photobacterium damselae* ssp. *piscicida*	No effect	[109]

Table 2. Overview of the effects of probiotics against bacteria in fish.

Few works have evaluated the disease resistance of grouper (*Epinephelus coioides*) through probiotics against the pathogenic *Streptococcus* sp. Thus, dietary treatment of grouper specimens fed *Lactobacillus plantarum* at 10^6 to 10^8 cfu kg^{-1} [47] or 10^8 cfu g^{-1} of *Bacillus subtilis* E20 [46] showed a better survival rate than the control. Moreover, the yeast *Saccharomyces cerevisiae* has shown probiotic effects in the grouper. Feeding with 5.3×10^7 cfu kg^{-1} yeasts four weeks showed a higher survival ratio (56.6%) than the control group (20%) after infection with *Streptococcus* sp. [48].

Leopard grouper (*Mycteroperca rosacea*) specimens fed supplemented diet with 10^6 cfu g^{-1} of *Debaryomyces hansenii* CBS 8339 for five weeks showed an increase in immunoglobulin M (IgM), catalase (CAT) and superoxide dismutase (SOD) after infection with *Aeromonas hydrophila* AH-315 and there was no mortality in any group [50].

Nile tilapia (*Oreochromis niloticus*) fed supplemented diet containing 0.5×10^7 cfu g^{-1} of a mixture of *B. subtilis* and *L. acidophilus*, or 10^7 cfu g^{-1} of each bacteria alone, for two months showed a higher relative level of protection against *Aeromonas hydrophila*, *Pseudomonas fluorescens* and *Streptococcus iniae* compared to the control group [49]. The results were even better when fish were fed a commercial probiotic supplemented diet containing *S. cerevisae*. Similar results were also obtained in another two experiments using as a challenge an injection of 2×10^7 cfu ml^{-1} of *P. fluorescens* and fish immersion with 2×10^9 cfu ml^{-1} of *Flavobacterium columnare* [51].

Probiotic bacteria identified as *Vibrio alginolyticus* was inoculated intramuscular or intraperitoneally in atlantic salmon (*Salmo salar*) at doses of 4×10^6 cfu ml^{-1} followed by a bath for ten minutes in a suspension of the same probiotic with 10^8 cfu/ml and seven days later fish were challenged with *Aeromonas salmonicida* 256/81, *Vibrio anguillarum* VIB256, *Vibrio ordalii* 17K or *Yersinia ruckeri* Ex5 [52]. So, this work indicated that application of the probiotic to salmon specimens induced a decrease in mortalities after challenge with *Aeromonas salmonicida* 256/81, and to a lesser extent with *Vibrio anguillarum* VIB256 and *Vibrio ordalii* 17K and does not reduce mortality with *Yersinia ruckeri* Ex5. In this sense, competition *in vitro* studies will help to elucidate these *in vivo* results. In other work [53] atlantic salmon specimens were fed a supplemented diet with 5×10^8 cells ml^{-1} of the microalgae *Tetraselmis suecica* for 14 days were challenged with fish pathogens. Results showed that use of *T. suecica* as a probiotic supplement was successful in preventing mortalities caused by *Aeromonas hydrophila, Aeromonas salmonicida* (strains LL and NG), *Serratia liquefaciens, Vibrio anguillarum, Vibrio salmonicida* and *Yersinia ruckeri* type I. *Salmo salar* fry specimens which were fed *Lactobacillus plantarum* at dose of 2.5×10^9 cfu g^{-1} and infected with *Aeromonas salmonicida* AL2020 showed a cumulative mortality lower than infected control group [54]. *Pseudomonas fluorescens* AH2 at doses of 10^3-10^5 cfu ml^{-1} in water did not confer protection against *Aeromonas salmonicida* in *Salmo salar* specimens [55]. It has been also reported *in vitro* that the pathogen *Vibrio anguillarum* LFI1243 showed a complete inhibition of growth in presence of *Carnobacterium divergens* strains [56]. This is in accordance with another study showing that *Carnobacterium* sp. isolated from salmon inhibited the growth of both *Vibrio anguillarum* and *Aeromonas salmonicida* in intestinal fish mucus [57]. Interestingly, *Carnobacterium divergens* isolated from *Salmo salar* specimens were also tested as fed probiotics in atlantic cod (*Gadus morhua*) specimens which showed lower mortalities.

The most studied fish specie regarding the potential benefits of probiotics is the rainbow trout (*Oncorhynchus mykiss*). *In vitro* studies have demonstrated the competitive adhesion and production of antagonistic compounds by some lactic acid bacteria (*Lactococcus lactis* ssp. *lactis* CLFP100, *Lactococcus lactis* ssp. *cremoris* CLFP102 and *Lactobacillus curvatus* CLFP150) against fish pathogens, including *Aeromonas salmonicida* ssp. *salmonicida* CLFP 501, *Carnobacterium piscicola* CLFP 601, *Lactococcus garvieae* CLFP LG1, *Vagococcus salmoninarum* CLFP 602, *Yersinia ruckeri* ATCC 29473 and *Vibrio anguillarum* La192 [58]. In another *in vitro* assay authors checked the inhibitory effect of *Carnobacterium* sp. and *Pseudomonas* sp. isolated from gut of rainbow trout against *Vibrio anguillarum,* although there was no correlation with the *in vivo* study since the same probiotic failed to protect them against *Vibrio anguillarum* infection [59]. In rainbow trout specimens fed 10^7 cfu g^{-1} of four putative probiotics (*Aeromonas hydrophila, Vibrio fluvialis, Carnobacterium* sp. and an unidentified coccus) showed a better survival after intraperitoneal injection of *Aeromonas salmonicida* [60]. However, the same dietary doses of *Carnobacterium inhibens* and *Vibrio alginolyticus* conferred a lower protection against *Aeromonas salmonicida.* These results were correlated with other two studies [52, 61]. In rainbow trout fingerlings, the same four putative probiotics seen previously [60] but administered as formaline-inactivated bacteria showed a lower mortality (4%, 4%, 8% and 0%, respectively) after challenge with *Aeromonas salmonicida* [62] suggesting that the use of dead probiotics has also many benefits for fish. Dietary administration of lactic acid bacteria (*Lactococcus lactis* ssp. *lactis* CLFP 100, *Leuconostoc mesenteroides* CLFP 196, and *Lactobacillus sakei* CLFP 202) at doses of 10^6 cfu g^{-1} for 2 weeks showed a survival rate of 97.8%-100% (versus 65.6% in the control group) when trout specimens were challenged with *Aeromonas salmonicida* ssp. *salmonicida* CLFP 501 [63]. It has been reported that dietary supplementation with *Bacillus* sp. JB-1 and *Aeromonas sobria* GC2 at doses of 2×10^8 and 10^7 cfu g^{-1}, respectively for two weeks led to a higher survival rates in trout after challenge with *Streptococcus iniae* and *Lactococcus garvieae* at doses of 2×10^7 cfu ml^{-1}, and *Vibrio anguillarum, Vibrio ordalii, Aeromonas salmonicida* and *Yersinia ruckeri* at doses of 3×10^8 cfu ml^{-1} [64]. Thus, survival rates in specimens fed control diets were

0%-20% whereas in specimens fed probiotic-diets survival rate was 100% in all treatments (with JB-1 and GC2) with all pathogens bacteria except for *Vibrio anguillarum* (87% and 94% respectively) and *Yersinia ruckeri* (94% in GC2 diet). In other study it has been found that dietary administration of *Aeromonas sobria* GC2 at dose of 10^8 cfu g^{-1} and *Brochothrix thermosphasta* BA211 at dose of 10^{10} cfu g^{-1} for two weeks showed a higher survival rate (76% and 88%) than in control group (22%) after intramuscular injection with *Aeromonas bestiarum* ORN2 [65]. In the same experiment, it was demonstrated that GC2 probiotic exerts resistance also against ichthyophthiriasis (caused by the parasite *Ichthyophthirius multifiliis*) however BA211 strain had no effect against this pathogen. An *in vitro* assay tested the inhibitory ability of *Lactobacillus plantarum* strains, *Lactococcus lactis* strains and *Leuconostoc mesenteroides* strains against *Lactococcus garvieae* CLFP LG1 [66]. Other research [67] reported that rainbow trout specimens fed *Lactobacillus rhamnosus* ATCC 53103 at doses of 10^9 and 10^{12} cfu g^{-1} for fifty-one days obtained a reduced mortality (18.9% and 46.3%, respectively) compared with the control group (52.6%) when were infected with *Aeromonas salmonicida* ssp. *salmonicida.* An *in vivo* assay against lactococcosis, dietary administration with lactic acid bacteria (*Leuconostoc mesenteroides* CLFP 196, and *Lactobacillus plantarum* CLFP 238) at doses of 10^6 cfu g^{-1} for four weeks showed a decrease in cumulative mortality (46% and 54%) compared with the control group (78%) in trout specimens after injection with *Lactococcus garvieae* [68]. Following with the development of protection in rainbow trout, specimens were fed a supplemented diet with 10^3 cfu g^{-1} of *Kokuria* SM1 for four weeks and after replacement for control diet they were infected with *Vibrio anguillarum* every week [69]. Interestingly, this relative protection was maximum (87%) just after the end of the probiotic-supplemented diet that was disappearing with the time and was of 71%, 68%, 62% and 36% after two, three, four and five weeks after cessation of probiotic, respectively, representing a sign of gradual loss of effect [70, 71]. In other research, *O. mykiss* specimens exposed to *Pseudomonas fluorescens* AH2 at 10^5 cfu ml^{-1} for 5 days or added *in situ* when challenged with *Vibrio anguillarum* showed a higher survival ratio (56% and 65%, respectively) than specimens exposed to *Vibrio anguillarum* without probiotic (50%) [72]. Dietary administration of BioPlus2B, wich contains two probiotic bacteria (*Bacillus subtilis* and *Bacillus licheniformis*) for four weeks resulted in a better survival ratio (41.7%) compared with Ergosan-diet (8.9%) and control diet (9%) in trout specimens after intraperitoneal injection of *Yersinia ruckeri* [73]. Following with the protection against yersiniosis, dietary administration of 10^8 cfu g^{-1} of *Enterobacter cloacae* and *Bacillus mojavensis* separately for two months achieved a high survival ratio (99.2%) compared with the control group (35%) when infected with *Yersinia ruckeri* [74]. In addition, in other research, dietary administration of 10^7 cfu g^{-1} of *Carnobacterium maltaromaticum* B26 and *Carnobacterium divergens* B33 separately for two weeks conferred protection against *Yersinia ruckeri* with a high survival ratio of 73% and 80% respectively, compared with the control group (13%); and the same probiotics (B26 and B33) also provided protection against *Aeromonas salmonicida* with a survival ratio of 80% in both cases compared with the control group (20%) [75]. *Flavobacterium psychrophilum* is the causative agent of coldwater disease (CWD), also known as rainbow trout fry syndrome (RTFS). Although many types of salmonids are susceptible to RTFS, rainbow trout can be especially impacted due to direct mortality or deformities in surviving specimens leading to economic losses in aquaculture [76, 77]. In order to establish strategies of resistance against

CWD with probiotics, in two studies [78, 79] it was demonstrated the ability of *Pseudomonas* sp. M174 and M162 to inhibit *Flavobacterium psychrophilum in vitro*. In addition, others *in vivo* experiments, rainbow trout specimens fed supplemented diet with *Pseudomonas* sp. M174 (at 4×10^{6}) and M162 (at doses of 5×10^{7}-2×10^{9} cfu g^{-1}) showed a decrease in cumulative mortality after infection with *Flavobacterium psychrophilum* JIP02/86. Thus, cumulative mortality was 41% in the M174-diet group, 35% in the M162-diet group, and 57% in control groups. In an interesting study, oral vaccines with formalin-killed *Streptococcus iniae* Dan-1 at doses of 3×10^{11} cfu ml^{-1} were inoculated in *Oncorhynchus mykiss* specimens provided them protection against *Streptococcus iniae* virulent at doses of 10^{5} cfu ml^{-1} until six months later. The survival ratio was 90% in the treated group and 20% in the control group [80]. As seen in the vast literature the benefits of many probiotics in the culture of rainbow trout is achieved. Furthermore, some papers also demonstrate that probiotics do not need to be alive exclusively. Thus, trout specimens fed supplemented diet with inactivated *Enterococcus faecalis* at dose of 5g kg^{-1} feed showed lower cumulative mortality (40%) than the control group (83%) after challenge with *Aeromonas salmonicida* [81]. Other probiotic forms of *Bacillus subtilis* AB1 such as live cells, sonicated cells, formaline-dead cells and cell-free supernatant were applied as supplement in diets to rainbow trout specimens which achieved a survival of 100% in all forms of probiotic-treatments whereas the survival in control groups was 10-15% after intraperitoneal injection with a pathogenic *Aeromonas* sp. [82].

Other trout species have been slightly evaluated. Thus, brown trout (*Salmo trutta*) specimens fed diets containing lactic acid bacteria (*Lactococcus lactis* ssp. *lactis* CLFP 100 or *Leuconostoc mesenteroides* CLFP 196) at doses of 10^{6} cfu g^{-1} for four weeks separately, reduced the cumulative mortality after challenge with *Aeromonas salmonicida* from 37% in the control group to 15% and 9%, respectively. [83]. In the case of brook trout (*Salvelinus fontinalis*), specimens exposed to four potential probiotics (S1, S5, S9 and S10) separately at doses of 10^{5} cfu ml^{-1} and one pathogen (*Flavobacterium columnare*) showed a higher survival ratio than specimens exposed to *Flavobacterium columnare* (without probiotics) being S9 the most successful with a cumulative mortality of only 4% [84].

Edwardsiellosis, a bacterial septicaemia caused by the Gram-negative bacterium *Edwardsiella tarda*, is one of the most serious bacterial diseases in cultured eels [85]. So, in a study with European eel (*Anguilla anguilla*), dietary administration with *Enterococcus faecium* SF68 from Cernivet® and *Bacillus toyoi* from Toyocerin® for 2 weeks was followed by challenge with *Edwardsiella tarda* 981210L1. *Bacillus toyoi* did not protected against Edwardsiellosis whilst *Enterococcus faecium* SF68 showed higher rate of survival (73%) compared with the control (45%). In the resistance of Nile tilapia (*Oreochromis niloticus*) against edwardsiellosis, dietary administration of a commercial mix of probiotics that contained *Lactobacillus acidophilus* (1.2×10^{8} cfu g^{-1}), *Bacillus subtilis* (1.6×10^{7} cfu g^{-1}), *Clostridium butyricum* (2×10^{7} cfu g^{-1}) and *Saccharomyces cerevisiae* (1.6×10^{7} cfu g^{-1}) for 30 days following infection with *Edwardiella tarda*, provided a cumulative mortality lower than positive control group [86]. Recently, it has been also reported [87] that dietary supplementation of 10^{8} cfu g^{-1} of *Lactobacillus pentosus* PL11 in Japanese eel (*Anguilla japonica*) challenged with *Edwardsella tarda* showed an increase in growth performance compared with the control group. In the case of rock bream (*Oplegnathus fasciatus*) it has been also shown that dietary supplementation with 2.2×10^7 cfu g^{-1} of *Lactobacillus sakei* BK19 and challenged with *Edwardsiella tarda* produced a non-significant decrease in the cumulative mortality [88].

Dietary supplementation of different species of *Bacillus* sp., *Lactobacillus* sp., *Streptococcus faecium* and *Saccharomyces cerevisae* had no effect in survival ratio of ornamental fishes (*Carassius auratus* and *Xiphophorus helleri*) specimens after challenge with *Pseudomonas fluorescens* 58C [89]. However, other study with *Carassius auratus* fed a supplemented diet of formalin-inactivated *Aeromonas hydrophila* A3-51 for twenty days showed a decrease in cumulative mortality compared with the control group after infection with *Aeromonas salmonicida* [90].

African catfish (*Clarias gariepinus*) juvenile specimens were fed a commercial diet supplemented with 3×10^7 cfu g^{-1} of *Lactobacillus acidophilus* for 12 weeks. Then, fish were intraperitoneally injected with 2×10^6 cfu ml^{-1} of *Staphylococcus xylosus, Aeromonas hydrophila* gr2 and *Streptococcus agalactiae* separately [91]. At one week post infection, the fish survival rate in control group and in infected groups treated with probiotic diet was 100%, whilst in the groups infected with *Staphylococcus xylosus, Aeromonas hydrophila* gr2 and *Streptococcus agalactiae* fed the non-probiotic diet, fish survival recorded was 83.3%, 76.6% and 80.0% respectively.

Olive flounder (*Paralichthys olivaceus*) specimens fed supplemented diet with *Bacillus subtilis, Bacillus pumilus* and *Bacillus licheniformis,* separately and at doses 10^{10} cfu g^{-1} for eight weeks showed a higher survival ratio in the case of *Bacillus subtilis* and *Bacillus pumilus* (97.3% and 98.7%, respectively) than specimens in the control group (77.3%) after immersion with *Streptoccocus iniae* [92]. For *Bacillus licheniformis* diet, specimens did not show statistically significant differences in survival ratio (86.7%) compared with the control group (77.3%). In another study, *Paralichthys olivaceus* specimens were fed a diet containing 3.4×10^4 (low dose), 3.5×10^6 (medium dose) and 3.4×10^8 cfu ml^{-1} (high dose) of *Zooshikella* sp. JE-34 and challenged with *Streptococcus iniae* showed their mortality reduced from 85 to htose of the controls 25-40% [93].

Chinese drum (*Miichthys miiuy*) specimens were also fed commercial diet supplemented with 10^8 cfu g^{-1} of *Clostridium botyricum* CB2 in the form of alive cells (CB) or dead cells (D-CB) for 30 days and then challenged with *Vibrio anguillarum* and *Aeromonas hydrophila,* separately. Result showed that survival in chinese drum specimens increased in both groups of probiotic diet compared with the control for both pathogen bacteria [94]. These results are according to other study [95] which demonstrated that dietary administration of *Clostridium botyricum* in rainbow trout (*Oncorhynchus mykiss*) achieved resistance against vibriosis.

Tropical freshwater fish (*Labeo rohita*) specimens were fed a supplemented diet with 0.5×10^7, 10^7 or 1.5×10^7 cfu g^{-1} of *Bacillus subtilis* for two weeks. After challenge by intraperitoneal injection of *Aeromonas hydrophila* O:18, specimens showed increased serum bactericidal activity and granulocyte numbers in probiotic-fed groups compared with the control group [96]. In other work [97] it has been reported that *L. rohita* specimens fed dietary supplementation with 10^6, 10^8 or 10^{10} cfu g^{-1} of *Lactobacillus plantarum* VSG3 for two months showed a higher survival rate (37%, 77% and 63%, respectively) than the control group (14%) after injection of *Aeromonas hydrophila*. In addition, dietary supplementation of 10^7 or 10^9 cfu g^{-1} of *Pseudomonas aerugino-*

sa VSG-2 for two months showed a higher survival rate (66% and 55%, respectively) than in the control group (11%) after injection with *Aeromonas hydrophila* MTCC1739. So, the appropriate administration dose was 10^7 cfu g^{-1} of *Pseudomonas aeruginosa* VSG-2 and 10^8 cfu g^{-1} of *Lactobacillus plantarum* VSG-3 which achieved the better survival rate (66% and 77%, respectively) after challenge with *Aeromonas hydrophila* MTCC1739 [97, 98], demonstrating that probiotics are only effective when administered in adequate doses.

Turbot (*Scophthalmus maximus*) larvae specimens fed rotifers enriched with *Lactobacillus plantarum* and *Carnobacterium* sp. at doses of 10^7-2×10^7 cfu ml^{-1} showed a higher survival ratio (53%) than specimens fed rotifers without probiotics (8%) [99]. Similarly, larvae specimens exposed to *Roseobacter* sp. strain 27-4 at dose of 10^7 cfu ml^{-1} showed a significant decrease in cumulative mortality compared with control larvae specimens. In addition, this *Roseobacter* sp. strain 27-4 was previously tested as antagonist to *Vibrio anguillarum* [100]. When specimens were fed rotifers enriched with *Roseobacter* sp. strain 27-4 and infected with *Vibrio anguillarum*, achieved a decrease in cumulative mortality compared with specimens only infected [101]. It was demonstrated in an *in vitro* assay that *Phaeobacter* sp. and *Ruegeria* sp. are also potential probiotics against *Vibrio anguillarum* in turbot [102].

Gilthead seabream (*Sparus aurata*) specimens were fed a commercial diet supplemented with 10^8 cfu g^{-1} of *Shewanella putrefaciens* (Pdp11) for 15 days and challenged with 3.7×10^7 cfu ml^{-1} of *Vibrio anguillarum* DC11R2a [103]. The mortality of the fish which receiving the diet supplemented with the potential probiotic Pdp11 was 10%, lower than the mortality of the fish that received the control diet (56%).

In other works [104, 105] it has been described the effect of the dietary administration of 10^9 cfu g^{-1} of *Shewanella putrefaciens* (Pdp11) and *Shewanella baltica* (Pdp13) to sole (*Solea senegalensis*) against *Photobacterium damselae* ssp. *piscicida*. The mortality decreased after one and two months with dietary administration of both bacteria compared with the control diet.

In european seabass (*Dicentrarchus labrax*) juvenile specimens, it has been demonstrated that dietary intake of *Artemia* with an acid lactic bacteria (*Lactobacillus delbrueckii* ssp. *delbrueckii*) improved growth of specimens [106]. Dietary administration of 10^9 cfu g^{-1} of *Vagococcus fluvialis* during 20 days in adults resulted in a mortality of 17.3% while in control group (without probiotic) was 30% after exposure to *Vibrio anguillarum* 975-1 [107].

7. Conclusions

Probiotics are usually live microorganisms that administered at adequate doses confer health benefits to the host. In this review we have focused only in those probiotics conferring protection to shellfish and fish species important for the aquaculture against viral and bacterial diseases. Some of the main conclusions are summarized below:

- The most studied probiotics are usually *Bacillus* and *Lactobacillus* species.

http://dx.doi.org/10.5772/57198

- Dietary administration of probiotics is the preferred for the researchers and farmers. However, bioencapsulation through *Artemia* might be considered a good solution, mainly at larval stages.
- Most of the studies have used live bacteria but other forms such as inactivated, killed, homogenized or even supernatants have also presented good probiotic properties.
- Bacteria are the most known probiotics but other microorganisms such as yeast or micro-algae are also suitable and good candidates.
- Although probiotics have probed protection against pathogenic bacteria further evaluation of their potential against virus and parasites is deserved.
- The concentration of the administered probiotic is essential and needs to be optimized for every situation.
- The time of administration is also a very important factor and periods of 2 to 4 weeks of dietary administration seem to be the optimal.
- Only a few potential probiotics tested *in vitro* become in effective probiotics *in vivo* and in commercial probiotics.

Further studies are still necessary to increase our knowledge about the use of probiotics to control bacterial infections in shellfish and fish but much more efforts are needed in the case of viral diseases. This is an important issue for the aquaculture industry that is continuously growing due to the fish and shellfish demand for human consume. Apart from the discovery of new or better probiotic formulations, improvement of their benefits may be helpful. Thus, better and cheaper production methods, administration ways or combination with other preventive/therapeutic measures are welcomed.

Acknowledgements

H. Cordero wishes to thank the *Ministerio de Economía y Competitividad* (MINECO) for a *F.P.I.* fellowship. This work has been funded by grants AGL2010-20801-C02-02 (MINECO and FEDER), AGL2011-30381-C03-01 (MINECO) and 04538/GERM/06 (*Fundación Séneca, Grupo de Excelencia de la Región de Murcia*).

Author details

Héctor Cordero, María Ángeles Esteban and Alberto Cuesta

Fish Innate Immune System Group, Department of Cell Biology and Histology, Faculty of Biology, Regional Campus of International Excellence "Campus Mare Nostrum". University of Murcia, Murcia, Spain

References

[1] Lilly DM, Stilwell H. Probiotics: Growth-Promoting Factors Produced by Microorganisms. Science 1965;147(3659) 747-748.

[2] Food and Agriculture Organization (FAO) and World Health Organization (WHO). Health and Nutritional Properties of Probiotics in Food Including Powder Milk with Live Lactic Acid Bacteria. Joint FAO/WHO Expert Consultation on Evaluation of Health and Nutritional Properties of Probiotics in Food Including Powder Milk with Live Lactic Acid Bacteria 2001.

[3] Gram L. Prospects of Fish Probiotics. In: Microbial Ecology in Growing Animal. 2005. p379-417

[4] Panigrahi A, Kiron V, Puangkaew J, Kobayashi T, Satoh S, Sugita H. The Viability of Probiotic Bacteria as a Factor Influencing the Immune Response in Rainbow Trout *Oncorhynchus mykiss*. Aquaculture 2005;243 241-254.

[5] Balcázar JL, De Blas I, Ruíz-Zarzuela I, Cunningham D, Vendrell D, Múzquiz JL. The Role of Probiotics in Aquaculture. Veterinary Immunology 2006;114 173-186.

[6] Vine NG, Leukes WD, Kaiser H. Probiotics in Marine Larviculture. FEMS Microbiology Reviews 2006;30 404-427.

[7] Balcázar JL, Decamp O, Vendrell D, De Blas I, Ruíz-Zarzuela I. Health and Nutritional Properties of Probiotics in Fish and Shellfish. Microbial Ecology in Health and Disease 2006;18 65-70.

[8] Nayak SK. Probiotics and Immunity: a Fish Perspective. Fish and Shellfish Immunology 2010;29 02-14.

[9] Mohapatra S, Crakraborty T, Kumar V, Deboeck G, Mohanta KN. Aquaculture and Stress Management: a Review of Probiotic Intervention. Journal of Animal Physiology and Animal Nutrition 2013;97 405-430.

[10] Pandiyan P, Balaraman D, Thirunavukkarasu R, George EGJ, Subaramaniyan K, Manikkam S, Sadayappan B. Probiotics in Aquaculture. Drug Invention Today 2013;5 55-59.

[11] Rodríguez J, Espinosa Y, Echeverria F, Cárdenas G, Román R, Stern S. Exposure to Probiotics and β-1,3/1,6-Glucans in Larviculture Modifies the Immune Response of *Penaeus vannamei* Juveniles and Both the Survival to White Spot Syndrome Virus Challenge and Pond Culture. Aquaculture 2007;273 405-415.

[12] Partida-Arangure BO, Luna-González A, Fierro-Coronado JA, Flores-Miranda C, González-Ocampo HA. Effect of Inulin and Probiotic Bacteria on Growth, Survival, Immune Response, and Prevalence of White Spot Syndrome Virus (WSSV) in *Litope-*

naeus vannamei Cultured under Laboratory Conditions. African Journal of Biotechnology 2013;12 3366-3375.

[13] Li J, Tan B, Mai K. Dietary Probiotic *Bacillus* OJ and Isomaltooligasaccharides Influence the Intestine Microbial Populations, Immune Responses and Resistance to White Spot Syndrome Virus in Shrimp (*Litopenaeus vannamei*). Aquaculture 2009;291 35-40.

[14] Leyva-Madrigal KY, Luna-González A, Escobedo-Bonilla CM, Fierro-Coronado JA, Maldonado-Mendoza IE. Screening for Potential Probiotic Bacteria to Reduce Prevalence of WSSV and IHHNV in Whiteleg Shrimp (*Litopenaeus vannamei*) under Experimental Conditions. Aquaculture 2011;222 16-22.

[15] Aranda CP, Valenzuela C, Barrientos J, Paredes J, Leal P, Maldonado M, Godoy FA, Osorio CG. Bacteriostic anti-*Vibrio parahaemolyticus* Activity of *Pseudoalteromonas* sp. Strains DIT09, DIT44, DIT46 Isolated from Southern Chilean Intertidal *Perumytilus purpuratus*. World Journal of Microbiology and Biotechnology 2012;28 2365-2374.

[16] Ruíz-Ponte C, Samain J, Sánchez J, Nicolas J. The Benefit of a *Roseobacter* Species on the Survival of Scallop Larvae. Marine Biotechnology (New York) 1999;1 52-59.

[17] Riquelme C, Araya R, Vergara N, Rojas A, Guaita M, Candia M. Potential Probiotic Strains in the Culture of the Chilean Scallop *Argopecten purpuratus* (Lamarck,1819). Aquaculture 1997;154 17-26.

[18] Gibson L, Woodworth J, George A. Probiotic Activity of *Aeromonas media* on the Pacific Oyster, *Crassostrea gigas*, when Challenged with *Vibrio tubiashii*. Aquaculture 1998;169 111-120.

[19] Kapareiko D, Lim HJ, Schott EJ, Hanif A, Wikfors GH. Isolation and Evaluation of New Probiotic Bacteria for Use in Shellfish Hatcheries: II. Effects of a *Vibrio* sp. Probiotic Candidate Upon Survival of Oyster Larvae (*Crassostrea virginica*) in Pilot-Scale Trials. Journal of Shellfish Research 2011;30 617-625.

[20] Jiang H, Liu X, Chang Y, Liu M, Wang G. Effects of Dietary Supplementation of Probiotic *Shewanella colwelliana* WA64, *Shewanella olleyana* WA65 on the Innate Immunity and Disease Resistance of Abalone, *Haliotis discus hannai* Ino. Fish and Shellfish Immunology 2013; 35 86-91.

[21] Macey BM, Coyne VE. Improved Growth Rate and Diseases Resitance in Farmed *Haliotis midae* through Probiotic Treatment. Aquaculture 2005;245 249-261.

[22] Van Hai N, Fotedar R. Comparison of the Effects of the Prebiotics (BioMos® and β-1,3-D-Glucan) and the Customised Probiotics (*Pseudomonas synxantha* and *Pseudomonas aeruginosa*) on the Culture of Juvenile Western King Prawns (*Penaeus latisulcatus*, Kishinouye 1896). Aquaculture 2009;289 310-316.

[23] Van Hai N, Buller N, Fotedar R. The Use of Customised Probiotics in the Cultivation of Western King Prawns (*Penaeus latisulcatus*, Kishinouye 1896). Fish and Shellfish Immunology 2009;27 100-104.

[24] Tseng DY, Ho P, Huang SY, Cheng SC, Shiu YL, Chiu CS, Liu CH. Enhancement of Immunity and Disease Resistance in the White Shrimp, *Litopenaeus vannamei*, by the Probiotic, *Bacillus subtilis* E20. Fish and Shellfish Immunology 2009;26 339-344.

[25] Balcázar JL, Rojas-Lunas T. Inhibitory Activity of Probiotic *Bacillus subtilis* UTM 126 against *Vibrio* Species Confers Protection against Vibriosis in Juvenile Shrimp (*Litopenaeus vannamei*). Current Microbiology 2007;55 409-412.

[26] Balcázar JL, Rojas-Luna T, Cunningham DP. Effect of the Addition of Four Potential Probiotic Strains on the Survival of Pacific White Shrimp (*Litopenaeus vannamei*) Following Immersion Challenge with *Vibrio parahaemolyticus*. Journal of Invertebrate Pathology 2007;96 147-150.

[27] Thompson J, Gregory S, Plummer S, Shields RJ, Rowley AF. An *in Vitro* and *in Vivo* Assessment of the Potential of *Vibrio* spp. as Probiotics for the Pacific White Shrimp, *Litopenaeus vannamei*. Journal of Applied Microbiology 2010;109 1177-1187.

[28] Zokaeifar H, Balcázar JL, Saad SC, Kamarudin MS, Sijam K, Arshad A, Nejat N. Effects of *Bacillus subtilis* on the Growth Performance, Digestive Enzymes, Immune Gene Expression and Disease Resistance of White Shrimp, *Litopenaeus vannamei*. Fish and Shellfish Immunology 2012;33 683-689.

[29] Chiu CH, Guu YK, Liu CH, Pan TM, Cheng W. Immune Responses and Gene Expression in White Shrimp, *Litopenaeus vannamei*, Induced by *Lactobacillus plantarum*. Fish and Shellfish Immunology 2007;23 364-377.

[30] Luis-Villaseñor IE, Castellanos-Cervantes T, Gómez-Gil B, Carrillo-García AE, Campa-Córdova AI, Ascensio F. Probiotics in the Intestinal Tract of Juvenile Whiteleg Shrimp *Litopenaeus vannamei*: Modulation of the Bacterial Community. World Journal of Microbiology and Biotechnology 2013;29 257-265.

[31] Vieira FN, Pedrotti FS, Neto CCB, Mouriño JLP, Beltrame E, Martins ML, Ramírez C, Arana LAV. Lactic-Acid Bacteria Increase the Survival of Marine Shrimp. Brazilian Journal of Oceanographic 2007;55 251-255.

[32] Scholz U, Díaz GG, Ricque D, Suárez LEC, Albores FV, Latchford J. Enhancement of Vibriosis Resistance in Juvenile *Penaeus vannamei* by Supplementation of Diets with Different Yeast Products. Aquaculture 1999;176 271-283.

[33] Vaseeharan B, Ramasamy P. Control of Pathogenic *Vibrio* spp. by *Bacillus subtilis* BT23, a Possible Probiotic Treatment for Black Tiger Shrimp *Penaeus monodon*. Letters in Applied Microbiology 2003;36 83-87.

[34] Meunpol O, Lopinyosiri K, Menasveta P. The Effects of Ozone and Probiotics on the Survival of Black Tiger Shrimp (*Penaeus monodon*). Aquaculture 2003;220 437-448.

[35] Sivakumar N, Sundararaman M, Selvakumar G. Probiotic Effect of *Lactobacillus acidophilus* against Vibriosis in Juvenile Shrimp (*Penaeus monodon*). African Journal of Biotechnology 2012;11 15811-15818.

[36] Ravi A, Musthafa KS, Jegathammbal G, Kathiresan K, Pandiyan SK. Screening and Evaluation of Probiotics as a Biocontrol Agent against Pathogenic Vibrios in Marine Larviculture. Letters in Applied Microbiology 2007;45 219-223.

[37] Preetha R, Jayaprakash NS, Singh ISB. *Synechocystis* MCCB 114 and 115 as Putative Probionts for *Penaeus monodon* Post-Larvae. Disease of Aquatic Organims 2007;74 243-247.

[38] Utiswannakul P, Sangchai S, Engipat SR. Enhanced Growth of Black Tiger Shrimp *Penaeus monodon* by Dietary Supplementation with *Bacillus* (BP11) as a Probiotic. Journal of Aquaculture Research and Development 2011;01 1-9.

[39] Rengipat S, Rukpratanporn S, Piyatiratitivorakul S, Menasveta P. Probiotics in Aquaculture: a Case Study of Probiotics for Larvae of the Black Tiger Shrimp (*Penaeus monodon*). In Flegel TW (Ed) Advances in Shrimp Biotechnology. National Center for Genetic Engineering and Biotechnology,Bangkok 1998;1 1-5.

[40] Alavandi SV, Vijayan KK, Santiago TC, Poornima M, Jithendran KP, Ali S, Rajan JJS. Evaluation of *Pseudomonas* sp. PM11 and *Vibrio fluvialis* PM17 on Immune Indices of Tiger Shrimp, *Penaeus monodon.* Fish and Shellfish Immunology 2004; 17 115-120.

[41] Ajitha S, Sridhar M, Sridhar N, Singh ISB, Varghese V. Probiotic Effects of Lactic Acid Bacteria against *Vibrio alginolyticus* in *Penaeus* (*Fenneropenaeus) indicus*. Asian Fisheries Science 2004;17 71-80.

[42] Castex M, Lemaire P, Wabete N, Chim L. Effect of Probiotic *Pediococcus acidilactici* on Antioxidant Defences and Oxidative Strees of *Litopenaeus stylostris* under *Vibrio nigripulchritudo* Challenge. Fish and Shellfish Immunology 2010;28 622-631.

[43] Li J, Tan B, Mai K, Ai Q, Zhang W, Xu W, Liufu Z, Ma H. Comparative Study between Probiotic Bacterium *Arthrobacter* XE-7 and Chloramphenicol on Protection of *Penaeus chinensis* Post-Larvae from Pathogenic Vibrios. Aquaculture 2006;253 140-147.

[44] Ambas I, Suriawan A, Fotedar R. Immunological Response of Customised Probiotics-Fed Marron, *Cherax tenuimanus,* (Smith 1912) when Challenged with *Vibrio mimicus.* Fish and Shellfish Immunology 2013;35 262-270.

[45] Harikrishnan R, Balasundaram C, Heo MS. Effect of Probiotics Enriched Diet on *Paralichthys olivaceus* Infected with Lymphocystis Disease Virus (LCDV). Fish and Shellfish Immunology 2010;29 868-874.

[46] Liu CH, Chiu CH, Wang SW, Cheng W. Dietary Administration of the Probiotic, *Bacillus subtilis* E20, Enhances the Growth, Innate Immune Responses, and Disease Re-

sistance of the Grouper, *Epinephelus coioides*. Fish and Shellfish Immunology 2012;33 699-706.

[47] Son VM, Chang CC, Wu MC, Guu YK, Chiu CH, Cheng W. Dietary Administration of the Probiotic, *Lactobacillus plantarum*, Enhanced the Growth, Innate Immune Responses, and Disease Resistance of the Grouper *Epinephelus coioides*. Fish and Shellfish Immunology 2009;26 691-698.

[48] Chiu CH, Cheng CH, Gua WR, Guu YK, Cheng W. Dietary Administration of the Probiotic *Saccharomyces cerevisiae* P13, enhanced the Growth, Innate Immune Responses, and Diseases Resistance of the Grouper (*Epinephelus coioides*). Fish and Shellfish Immunology 2010;29 1053-1059.

[49] Aly SM, Abdel-Galil AY, Abdel-Aziz GA, Mohamed MF. Studies on *Bacillus subtilis* and *Lactobacillus acidophilus*, as Potential Probiotics, on the Immune Response and Resistance of Tilapia Nilotica (*Oreochromis niloticus*) to Challenge Infections. Fish and Shellfish Immunology 2008;25 128-136.

[50] Reyes-Becerril M, Tovar-Ramírez D, Ascensio-Valle F, Civera-Cerecedo R, Gracia-López V, Barbosa-Solomieu V, Esteban MA. Effects of Dietary Supplementation with Probiotic Live Yeast *Debaromyces hansenii* on the Immune and Antioxidant System of Leopard Grouper *Mycteroperca rosacea* Infected with *Aeromonas hydrophila*. Aquaculture Research 2011;42 1676-1686.

[51] Abu-Elala N, Marzouk M, Moustafa M. Use of Different *Saccharomyces cerevisiae* Biotic Forms as Immune-Modulator and Growth Promoter for *Oreochromis niloticus* Challenged with Some Fish Pathogens. International Journal of Veterinary Science and Medicine 2013;1 21-29.

[52] Austin B, Stuckey LF, Robertson PAW, Effendi I, Griffith DRW. A Probiotic Strain of *Vibrio alginolyticus* Effective in Reducing Diseases Caused by *Aeromonas salmonicida, Vibrio anguillarum* and *Vibrio ordalii*. Journal of Fish Diseases 1995;18 93-96.

[53] Austin B, Baudet E, Stobie M. Inhibition of Bacterial Fish Pathogens by *Tetraselmis suecica*. Journal of Fish Diseases 1992;15 55-61.

[54] Gildberg A, Johansen A, Jar B. Growth and Survival of Atlantic Salmon (*Salmo salar*) Fry Given Diets Supplemented with Fish Protein Hydrolysate and Lactic Acid Bacteria during a Challenge Trial with *Aeromonas salmonicida*. Aquaculture 1995;138 23-34.

[55] Gram L, Løvold T, Nielsen J, Melchiorsen J, Spanggaard B. *In Vitro* Antagonism of the Probiont *Pseudomonas fluorescens* Strain AH2 against *Aeromonas salmonicida* Does Not Confer Protection of Salmon against Furunculosis. Aquaculture 2001;199 1-11.

[56] Gildberg A, Mikkelsen H. Effects of Supplementing the Feed of Atlantic Cod (*Gadus morhua*) Fry with Lactid Acid Bacteria and Immunostimulating Peptides during a Challenge Trial with *Vibrio anguillarum*. Aquaculture 1998;169 103-113.

[57] Jöborn A, Dorsch M, Olsson ZJC, Westerdahl A, Kjelleberg S. *Carnobacterium inhibens* sp. Nov., Isolated from the Intestine of Atlantic Salmon (*Salmo salar*). International Journal of Systematic Bacteriology 1999;49 1891-1898.

[58] Balcázar JL, De Blas I, Ruiz-Zarzuela I, Vendrell D, Gironés O, Muzquiz JL. *In Vitro* Competitive Adhesion and Production of Antagonistic Compounds by Lactic Acid Bacteria against Fish Pathogens. Veterinary Microbiology 2007b;122 373-380.

[59] Spanggaard B, Huber I, Nielsen J, Sick EB, Pipper CB, Martinussen T, Slierendrecht WJ, Gram L. The Probiotic Potential against Vibriosis of the Indigenous Microflora of Rainbow Trout. Environmental Microbiology 2001;3 755-765.

[60] Irianto A, Austin B. Use of Probiotics to Control Furunculosis in Rainbow Trout (*Oncorhynchus mykiss,* Walbaum). Journal of Fish Diseases 2002;25 333-342.

[61] Robertson PAW, O'Dowd C, Burrells C, William P, Austin B. Use of *Carnobacterium* sp. as a Probiotic for Atlantic Salmon (*Salmo salar* L.) and Rainbow Trout (*Oncorhynchus mykiss,* Walbaum). Aquaculture 2000;185 235-243.

[62] Irianto A, Austin B. Use of Dead Probiotic Cells to Control Furunculosis in Rainbow Trout, *Oncorhynchus mykiss* (Walbaum). Journal of Fish Diseases 2003;26 59-62.

[63] Balcázar JL, De Blas I, Ruíz-Zarzuela I, Cunningham D, Vendrell D, Gironés O, Múzquiz JL. Enhancement of the Immune Response and Protection Induced by Probiotics Lactic Acid Bacteria against Furunculosis in Rainbow Trout (*Oncorhynchus mykiss*). FEMS Immunology and Medical Microbiology 2007;51 185-193.

[64] Brunt J, Newaj-Fyzul A, Austin B. The Development of Probiotics for the Control of Multiple Bacterial Diseases of Rainbow Trout (*Oncorhynchus mykiss,* Walbaum). Journal of Fish Diseases 2007; 30 573-579.

[65] Pieters N, Brunt J, Austin B, Lyndon AR. Efficacy of In-Feed Probiotics against *Aeromonas bestiarum* and *Ichthyophthirius multifiliis* Skin Infections in Rainbow Trout (*Oncorhynchus mykiss,* Walbaum). Journal of Applied Microbiology 2008;105 723-732.

[66] Pérez-Sánchez T, Balcázar JL, Garcia Y, Halaihel N, Vendrell D, De Blas I, Merrifield DL, Ruíz-Zarzuela I. Identification and Characterization of Lactic Acid Bacteria Isolated from Rainbow Trout, *Oncorhynchus mykiss* (Walbaum), with Inhibitory Activity against *Lactococcus garvieae*. Journal of Applied Microbiology 2011;34 499-507.

[67] Nikoskelainen S, Ouwehand A, Salminen S. Protection of Rainbow Trout (*Oncorhynchus mykiss*) from Furunculosis by *Lactobacillus rhamnosus*. Aquaculture 2001;198 229-236.

[68] Vendrell D, Balcázar JL, de Blas I, Ruiz-Zarzuela I, Gironés O, Múzquiz JL. Protection of Rainbow Trout (*Oncorhynchus mykiss*) from Lactococcosis by Probiotic Bacteria. Comparative Immunology, Microbiology and Infectious Diseases 2008;31 337-345.

[69] Sharifuzzaman SM, Austin B. Influence of Probiotic Feeding Duration on Disease Resistance and Immune Parameters in Rainbow Trout. Fish and Shellfish Immunology 2009;27 440-445.

[70] Sharifuzzaman SM, Austin B. *Kocuria* SM1 Controls Vibriosis in Rainbow Trout (*Oncorhynchus mykiss*, Walbaum). Journal of Applied Microbiology 2010;108 2162-2170.

[71] Sharifuzzaman SM, Austin B. Development of Protection in Rainbow Trout (*Oncorhynchus mykiss*) to *Vibrio anguillarum* Following Use of the Probiotic *Kocuria* SM1. Fish and Shellfish Immunology 2010;29 212-216.

[72] Gram L, Melchiorsen J, Spangaard B, Huber I, Nielsen TF. Inhibition of *Vibrio anguillarum* by *Pseudomonas fluorescens* AH2, a Possible Probiotic Treatment of Fish. Applied and Environmental Microbiology 1999;65 969-973.

[73] Raida MK, Larsen JL, Nielsen ME, Buchmann K. Enhanced Resistance of Rainbow Trout, *Oncorhynchus mykiss* (Walbaum), against *Yersinia ruckeri* Challenge Following Oral Administration of *Bacillus subtilis* and *Bacillus licheniformis* (BioPlus2B). Journal of Fish Diseases 2003;26 495-498.

[74] Capkin E, Altinok I. Effects of Dietary Probiotic Suplementations on Prevention/ Treatment of Yersiniosis Disease. Journal of Applied Microbiology 2006;106 1147-1153.

[75] Kim DH, Austin B. Innate Immune Responses in Rainbow Trout (*Oncorhynchus mykiss*, Walbaum) Induced by Probiotics. Fish and Shellfish Immunology 2006;21 513-524.

[76] Nematollahi A, Decostere A, Pasmans F, Haesebrouck F. *Flavobacterium psycrophilum* Infections in Salmonid Fish. Jornal of Fish Diseases 2003;26 563-574.

[77] Starliper CE. Bacterial Coldwater Disease of Fishes Caused by *Flavobacterium psychrophilum*. Journal of Advanced Research 2011;2 97-108.

[78] Korkea-aho TL, Papadopoulou A, Heikkinen J, Von Wright A, Adam A, Austin B, Thompson KD. *Pseudomonas* M162 Confers Protection against Rainbow Trout Fry Syndrome by Stimulating Immunity. Journal of Applied Microbiology 2012;113 24-35.

[79] Korkea-aho TL, Heikkinen J, Thompson KD, Von Wright A, Austin B. *Pseudomonas* sp. M174 Inhibits the Fish Pathogen *Flavobacterium psycrophilum*. Journal of Applied Microbiology 2011;111 266-277.

[80] Eldar A, Horovitcz A, Bercovier H. Development and Efficacy of a Vaccine against *Streptococcus iniae* Infection in Farmed Rainbow Trout. Veterinary Immunology and Immunopathology 1997;56 175-183.

[81] Rodríguez-Estrada U, Satoh S, Haga Y, Fushimi H, Sweetman J. Effects of Inactivated *Enterococcus faecalis* and Mannan Oligosaccharide and their Combination on Growth,

Immunity, and Disease Protection in Rainbow Trout. North American Journal of Aquaculture 2013;75 416-428.

[82] Newaj-Fyzul A, Adesiyun AA, Mutani A, Ramsubhag A, Brunt J, Austin B. *Bacillus subtilis* AB1 Controls *Aeromonas* Infection in Rainbow Trout (*Oncorhynchus mykiss*, Walbaum). Journal of Applied Microbiology 2007;103 1699-1706.

[83] Balcázar JL, Vendrell D, de Blas I, Ruiz-Zarzuela I, Múzquiz JL. Effect of *Lactococcus lactis* CLFP100 and *Leuconostoc mesenteroides* CLFP 196 on *Aeromonas salmonicida* Infection in Brown Trout (*Salmo trutta*). Journal of Molecular Microbiology and Biotechnology 2009;17 153-157.

[84] Boutin S, Bernatchez L, Audet C, Derôme N. Antagonistic Effect of Indigenous Skin Bacteria of Brook Charr (*Salvelinus fontinalis*) against *Flavobacterium columnare* and *F. psychrophilum*. Veterinary Microbiology 2012;155 355-361.

[85] Chang C, Liu W. An Evaluation of Two Probiotic Bacterial Strain, *Enteroccocus faecium* SF68 and *Bacillus toyoi*, for Reducing Edwardsiellosis in Cultured European Eel, *Anguilla anguilla* L. Journal of Fish Diseases 2002;25 311-315.

[86] Taoka Y, Maeda H, Jo J, Kim SM, Park SI, Yoshikawa T, Sakata T. Use of Live and Dead Probiotic Cells in Tilapia *Oreochromis niloticus*. Fisheries Science 2006;72 755-766.

[87] Lee JS, Cheng H, Lee SJ, Kim JC, Rhee MH, Suh JW, Park SC. Effects of Dietary Supplementation of *Lactobacillus pentosus* PL11 on the Growth Performance, Immune and Antioxidant Systems of Japanese Eel *Anguilla japonica* Challenged with *Edwardsiella tarda*. Fish and Shellfish Immunology 2013;34 756-761.

[88] Harikrishnan R, Kim MC, Kim JS, Balasundaran C, Heo MS. Protective Effect of Herbal and Probiotics Enriched Diet on Haematological and Immunity Status of *Oplegnathus fasciatus* (Temminck and Schlegel) against *Edwarsiella tarda*. Fish and Shellfish Immunology 2011;30 886-893.

[89] Abraham TJ, Mondal S, Babu CS. Effect of Commercial Probiotic and Fish Gut Antogonistic Bacterial Flora on the Growth and Disease Resistance of Ornamental Fishes *Carassius auratus* and *Xiphophorus helleri*. Journal of Fisheries and Aquatic Sciences 2008;25 27-30.

[90] Irianto A, Robertson PAW, Austin B. Oral Administration of Formaline-Inactivated Cells of *Aeromonas salmonicida* A3-51 Controls Infection by Atypical *A. salmonicida* in Goldfish, *Carassius auratus* (L.). Journal of Fish Diseases 2003;26 117-120.

[91] Al-Dohail MA, Hashim R, Aliyu-Paiko M. Evaluating the Use of *Lactobacillus acidophilus* as a Biocontrol Agent against Common Pathogenic Bacteria and the Effects on the Haematology Parameters and Histopathology in African Catfish *Clarias gariepinus* Juveniles. Aquaculture Research 2011;42 196-209.

[92] Cha JH, Rahimnejad S, Yang SY, Kim KW, Lee KJ. Evalutations of *Bacillus* spp. as Dietary Additives on Growth Performance, Innate Immunity and Disease Resistance of

Olive Flounder (*Paralichthys olivaceus*) against *Streptococcus iniae* and as Water Additives. Aquaculture 2013;402 50-57.

[93] Kim JS, Harikrishnan R, Kim MC, Balasundaram C, Heo MS. Dietary Administration of *Zooshikella* sp. to Enhance the Innate Immune Response and Disease Resistance of *Paralichthys olivaceus* against *Streptococcus iniae*. Fish and Shellfish Immunology 2010;29 104-110.

[94] Pan X, Wu T, Song Z, Tang H, Zhao Z. Immune Responses and Enhanced Diseases Resistance in Chinese Drum, *Miichthys miiuy* (Basilewsky), After Oral Administration of Live or Dead Cells of *Clostridium butyricum* CB2. Journal of Fish Diseases 2008;31 679-686.

[95] Sakai M, Yoshida T, Atsuta S, Kobayashi M. Enhancement of Resistance to Vibriosis in Rainbow Trout, *Oncorhynchus mykiss* (Walbaum), by Oral Administration of *Clostridium butyricum* bacterin. Journal of Fish Diseases 1995;18 187-190.

[96] Kumar R, Mukherjee SC, Ranjan R, Nayak SK. Enhanced Innate Immune Parameters in *Labeo rohita* (Ham.) Following Oral Administration of *Bacillus subtilis*. Fish and Shellfish Immunology 2008;24 168-172.

[97] Giri SS, Sukumaran V, Oviya M. Potential Probiotic *Lactobacillus plantarum* VSG3 Improves the Growth, Immunity, and Disease Resistance of Tropical Freshwater Fish, *Labeo rohita*. Fish and Shellfish Immunology 2013;34 660-666.

[98] Giri SS, Sen SS, Sukumaran V. Effects of Dietary Supplementation of Potential Probiotic *Pseudomonas aeruginosa* VSG-2 on the Innate Immunity and Disease Resistance of Tropical Freswater Fish, *Labeo rohita*. Fish and Shellfish Immunology 2012;32 1135-1140.

[99] Gatesoupe FJ. The Use of Probiotics in Aquaculture. Aquaculture 1999;180 147-165.

[100] Hjelm M, Bergh O, Riaza A, Nielsen J, Melchiorsen J, Jensen S, Duncan H, Ahrens P, Birkberk H, Gram L. Selection and Identification of Autochthonous Potential Probiotic Bacteria from Turbot Larvae (*Scophthalmus maximus*) Rearing Units. Systematic and Applied Microbiology 2004;27 360-371.

[101] Planas M, Pérez-Lorenzo M, Hjelm M, Gram L, Fiksdal IU, Bergh O, Pintado J. Probiotic Effect *In Vivo* of *Roseobacter* Strain 27-4 against *Vibrio* (*Listonella*) *anguillarum* Infections in Turbot (*Scophthalmus maximus*) Larvae. Aquaculture 2006;255 223-233.

[102] Porsby CH, Nielsen KF, Gram L. Phaeobacter and Ruegeria Species of the Roseobacter Clade Colonize Separate Niches in a Danish Turbot (*Scophthalmus maximus*)-Rearing Farm and Antagonize *Vibrio anguillarum* under Different Growth Conditions. Applied and Environmental Microbiology 2008;74 7356-7364.

[103] Chabrillón M, Arijo S, Díaz-Rosales P, Balebona MC, Moriñigo MA. Interference of *Listonella anguillarum* with Potential Probiotic Microorganisms Isolated from Farmed Gilthead Seabream (*Sparus aurata* L.). Aquaculture Research 2006;37 78-86.

[104] García De La Banda I, Lobo C, León-Rubio JM, Tapia-Paniagua S, Balebona MC, Moriñigo MA, Moreno-Ventas X, Lucas LM, Linares F, Arce F, Arijo S. Influence of Two Closely Related Probiotics on Juvenile Senegalese Sole (*Solea senegalensis*, Kaup 1858) Performance and Protection against *Photobacterium damselae* ssp. *piscicida*. Aquaculture 2010;306 281-288.

[105] Díaz-Rosales P, Arijo, S, Chabrillón M, Alarcón, FJ,Tapia-Paniagua ST, Martínez-Manzanares E, Balebona MC, Moriñigo MA. Effects of Two Closely Related Probiotics on Respiratory Burst Activity of Senegalese Sole (*Solea senegalensis*, Kaup) Phagocytes, and Protection against *Photobacterium damselae* ssp. *piscicida*. Aquaculture 2009;293 16-21.

[106] Carnevali O, De Vivo L, Sulpizio R, Gioacchini G, Olivotto I, Silvi S, Cresci A. Growth Improvement by Probiotic in European Sea Bass Juveniles (*Dicentrarchus labrax* L.), with Particular Attention to IGF-1, Myostatin and Cortisol Gene Expression. Aquaculture 2006;258 430-438.

[107] Sorroza L, Padilla D, Acosta F, Román L, Grasso V, Vega J and Real F. Characterization of the Probiotic Strain *Vagacoccus fluavialis* in the Protection of European Sea Bass (*Dicentrarchus labrax*) against Vibriosis by *Vibrio Anguillarum*. Veterinary Microbiology 2012;155 369-373.

[108] Planas M, Pérez-Lorenzo M, Vázquez, Pintado J. A Model for Experimental Infections with *Vibrio* (*Listonella*) *anguillarum* in First Feeding Turbot (*Scophthalmus maximus* L.) Larvae under Hatchery Conditions. Aquaculture 2005;250 232-243.

[109] Cerezuela R, Guardiola FA, Meseguer J, Esteban MA. Increases in Immune Parameters by Inulin and *Bacillus subtilis* Dietary Administration to Gilthead Seabream (*Sparus aurata* L.) Did Not Correlate with Disease Resistance to *Photobacterium damselae*. Fish and Shellfish Immunology 2012;32 1032-1040.